Modern Methods in Protein Chemistry
Review Articles Volume 2

Modern Methods in Protein Chemistry

Volume 2

Review Articles

including those from an
International Conference held in
Bielefeld, F.R. of Germany, June 1-2, 1984

Editor
Harald Tschesche

Walter de Gruyter · Berlin · New York 1985

Editor

Harald Tschesche, Dr. rer. nat.
Professor für Biochemie
Lehrstuhl für Biochemie
Fakultät für Chemie
Universität Bielefeld
D-4800 Bielefeld, F.R. of Germany

CIP-Kurztitelaufnahme der Deutschen Bibliothek

Modern methods in protein chemistry: review articles.
Vol. 2 Review articles including those from an international conference
held in Bielefeld, FR of Germany, June 1–2, 1984.
– Berlin; New York: de Gruyter, 1985.
ISBN 3-11-010180-7 (Berlin . . .)
ISBN 0-89925-079-3 (New York)

ISBN 3 11 010180 7 Walter de Gruyter · Berlin · New York
ISBN 0-89925-079-3 Walter de Gruyter, Inc., New York

Printing: Gerike GmbH, Berlin. – Binding: Dieter Mikolai, Berlin. – Printed in Germany.

Preface

This book is a continuation of the series of articles published in 1983 in a previous volume by de Gruyter Publishers*. It is again the intention of the editor to attempt a survey of the present status in the different fields of analytical methods available to the protein chemist for analysing and characterizing a distinct protein. The editor hopes that the book will satisfy the current need of all those involved in the aims of protein chemistry to keep up with the rapid development and perfection of all analytical and preparative methods. The articles review recent methodological progress, present the most advanced applications and give a number of key references. The references in particular should enable the reader to orient himself among the literature and adapt the method to his own particular problem.

Most of the papers were presented at the Conference on "Modern Methods in Protein Chemistry" held at the Center of interdisciplinary Research (ZiF) in Bielefeld, June 1-2, 1984. This meeting was kindly supported by the Gesellschaft für Biologische Chemie and organized for the study group of chemical protein analysis. Further support was provided by: the Center of interdisciplinary Research, Applied Biosystems GmbH, Bayer AG, Behringwerke AG, Boehringer Mannheim GmbH, Ciba-Geigy, Degussa Pharma Gruppe, Diamalt AG, Du Pont de Nemours GmbH, Hoechst AG, LKB Instrument GmbH and E. Merck.

The increasing demand for micro and submicro scale analytical procedures already observed at the 1982 meeting in Damp/Kiel was fully confirmed. This trend is reflected again in many of the articles in this volume. It is the hope of the authors and the editor that the present articles will serve to extend the present knowledge and methodological experience of the reader, thereby stimulating scientific progress and saving unnecessary time and effort.

Bielefeld, May 1985 Harald Tschesche

*Modern Methods in Protein Chemistry. Review Articles. Editor: H. Tschesche 1983. X, 464 pages. Hardcover. ISBN 3 11 009514 9

CONTENTS

MODERN MICROMETHODS

Volker Neuhoff
Max-Planck-Institute for Experimental Medicine
D-3400 Göttingen

INTRODUCTION

Micromethods do not seem to be widely applied in neurochemical analysis, perhaps because of a widespread prejudice that micromethods are only useful for "microminded", technically perfect experimenters. In addition, there may be worries about the reproducibility of micro techniques. A microgel, however, with many clearly separated and well-defined bands, each representing some nanograms of protein (see Fig. 10), is rather impressive when seen for the first time, especially when it is realized that the time taken for separation, staining, and destaining is only about one-tenth of the time necessary for the equivalent procedure using a macroscale method. A saving of experimental time is inherent in most of the micromethods described in some detail in this chapter. Such micromethods are also reproducible if they are performed correctly. The use of micromethods is therefore to be recommended even when the amount of material to be analyzed is not limited. Furthermore, now that it is harder to get sufficient funds for relevant research, micromethods are useful because the equipment needed for the analysis is not expensive.

The correct procedure for a micromethod is no more difficult to learn than is any other method. The time needed to learn any method is strongly dependent on the handiness and experience of the experimenter, and this is the same for learning a micro- or a corresponding macromethod. Obviously,

---

Reproduced with permission from Neuhoff, V.: Selected Micromethods for use in Neurochemistry, in: Handbook of Neurochemistry, 2nd Edition, Vol. 2 (A. Lajtha Ed.). Plenum Publishing Corporation New York, 1982 pp. 349-395.

Modern Methods in Protein Chemistry, Volume 2

every experimenter will take some time to be satisfied with the application of a micromethod and to clearly under stand its critical features and limitations. But these are points that apply to every method. Generally, there are no arguments against micromethods, and it is surprising that in biochemical research, where so often only small sample volumes are available, the application of micromethods is not more widespread. On the other hand, it is not surprising that progress in microscale methods is mostly connected with neurochemical applications. For example, the first application of polyacrylamide gel electrophoresis on the microscale was carried out in 1964 when Pun and Lombrozo (1) fractionated brain proteins. Since then, microscale methods for determination of RNA (2,3) and RNA base composition (4,5), measurement of enzyme activity and substrates of single neurons (6-12), mass determination (13,14), micro-flame photometry (15,16), picoliter volumes (17), microchromatography on thin-layer plates for dansyl amino acids (18-21), phospholipids (22-24), isoenzymes (25), glycoproteins, (26) and sugars (27), as well as other auxiliary methods such as microhomogenization (28) have mostly come from neurochemical laboratories.

This chapter, because of the limited space available, will neither review all micromethods possible nor describe techniques for preparation of minute amounts of defined tissue material or single nerve cells (see ref. 29). The methods selected for description are those with which the author has personal experience and therefore the necessary critical competence to describe. Very seldom are micromethods developments of their own. Usually they are deduced from macro procedures adapted to a micro scale, thereby more or less automatically increasing the sensitivity by a factor of between 10 and 1000. Most normal methods can be rather simply converted to micro versions, although one has to accept that every method will have its special merits and disadvantages.

PROTEIN DETERMINATION WITH MICROLITER VOLUMES

The quantitative determination of protein is a common prerequisite for many biochemical analyses. The different methods currently available often have a drawback in that agents used for preparation of biological tissues, e.g., sodium dodecyl sulfate (SDS), Triton X-100, NP 40, mercaptoethanol, urea, interfere with the analysis. In such cases, a protein precipitation step is often required, which is not only time consuming but also uncertain with respect to quantification. The volume routinely used for protein determinations is often rather large, and for microdeterminations, special and often cumbersome modifications are necessary. The method described here avoids all of these disadvantages and furthermore allows quantitative determination of small amounts of protein, even if the volume used (in the range between 1 and 5 µl) is unknown.

The method is, in principle, a spot analysis, a type of analytical method first introduced in 1859 by H. Schiff. Such methods had their peak approximately 40-50 years ago but are nowadays almost forgotten. However, spot methods still have very many advantages (30). One major point is the fact that by spotting sample solutions onto a suitable layer, the contents will be concentrated over a very small area. During fixation, staining, and washing out of the excess stain, all interfering components not firmly bound to the sample being analyzed, will be almost completely washed out. With this protein determination procedure, as described recently (31), it is possible to determine the protein concentration in a microliter volume when neither the volume nor the protein content in that volume are known.

Performance of the Protein Determination

Cellulose acetate strips commonly used for electrophoresis are used in this procedure (see Appendix, Scheme 1 for details). Prior to use, the acetate strips are stored in a moist chamber to facilitate the application of the sample. If a series of determinations is to be performed, the sample number may be indicated simply by cutting the corners of the acetate layer for the first five samples or by combining cut corners with arrowlike cuts on the sides. The number of strips that are stained and destained together is not critical. Sample application, staining, and destaining are done at room temperature, and commercially available 0.5-, 1-, 2-, or 5-µl capillaries are used. Cleaning of the capillaries (32) prior to use is recommended. The capillaries are filled by capillary attraction simply by being dipped into the solution to be analyzed. Complete filling has to be confirmed, and for routine determinations, 2-µl capillaries have been found to be the best. The standard deviation for a given protein con centration depends on the capillary volume as follows: 0.5 µl, 6.5%; 1 µl, 2.5%; 2 µl, 1.5%; 5 µl, 2.4%; 10µl, 7.5%. In most instances, immediately after sample application, the acetate strip is transferred to a petri dish containing the stain dissolved in methanol/acetic acid. Prefixation of a protein spot in methanol/acetic acid before staining results in a loss of 70-80% of the dye-binding capacity. In contrast, drying at room temperature or with a hot air stream has no influence on quantitative staining, thereby allowing for repeated sample application in the case of very dilute protein solutions.

Evaluation with Densitometry

If protein spots on acetate paper are round, any densitometer can be used for evaluation of Amido black-stained spots after the sample-containing strip is made transparent (31). Four typical densitograms of spots, each representing 2 µl of bovine serum albumin (2 µl of a solution containing 1 mg protein/ml), are shown in Fig. 1. If a sample is repeatedly measured ten times (31) and is repositioned in the equipment for measurement each time, the coefficient of variation is 0.9%. The integral of the area recorded is equivalent to the amount of protein in the spot (compare Fig. 2a). In cases in which it is difficult to obtain an ideal round spot during sample application, e.g., if the sample contains high concentrations of sucrose, glycerol, urea, etc. use of Hoechst 2495 (a fluorescent benzoxanthene derivative) is recommended for staining followed by fluorometric evaluation after elution of the stain (see below) or spot fluorometry in situ (33). For optical reasons, evaluation by densitometry across the diameter is not possible with irregular spots, and complete scanning of the spots with a

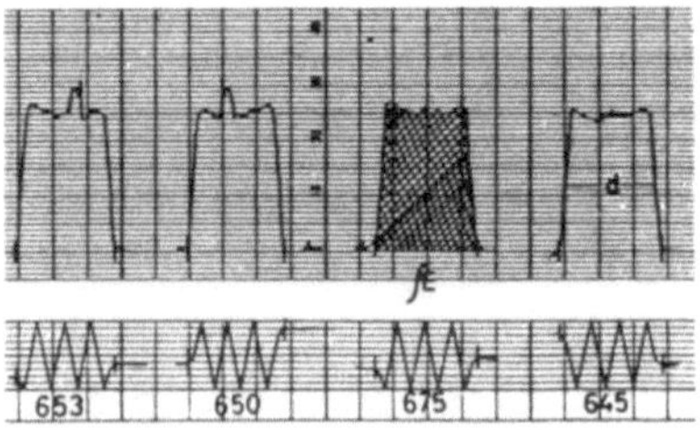

Fig. 1. Densitograms of four spots stained with amido black and representing 2 µg protein each. The densitograms were obtained with a Zeiss Gel Scanner (ZK4) and an adapter for micro-gel evaluation (123). ∫E, integrated absorption; d, spot diameter.

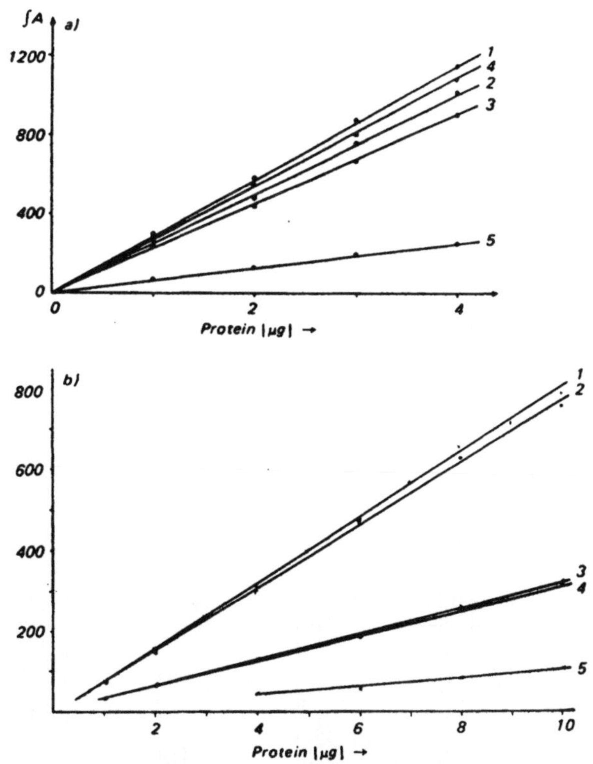

Fig. 2. Calibration curves for different proteins obtained by densitometric evaluation after amido black staining (a) and staining with Hoechst 2495 and spot fluorometry *in situ* (b). The curves are calculated according to $y = b \cdot x^a$ ($\log y = \log b + a \cdot \log x$) r, correlation coefficient; n, number of spots evaluated per curve; abscissae, μg protein; ordinate in a, integrated absorption; in b, arbitrary units. (1) Bovine serum albumin: (a) $r = 0.9989$, $n = 15$; (b) $r = 0.9928$, $n = 41$. (2) Human serum albumin: (a) $r = 0.9975$, $n = 15$; (b) $r = 0.9946$, $n = 24$. (3) Transferrin: (a) $r = 0.9989$ $n = 15$; (b) $r = 0.9725$, $n = 23$. (4) Globulin: (a) $r = 0.9985$, $n = 15$; (b) $r = 0.995$, $n = 24$. (5) Bovine myelin: (a) $r = 0.9935$, $n = 15$; (b) $r = 0.9306$, $n = 14$. Note not only the differences in stainability of the different proteins but also the differences between the two stains.

scanning microscope is prohibitively expensive and not acceptable as a routine method. Therefore, the evaluation is performed with a simple spot fluorometer (33) whereby the fluorescence of the whole spot is measured as an integral and is therefore independent of the form of the spot.

Photometric and Fluorometric Evaluation

If none of the abovementioned instruments is available for evaluation of stained spots, quantitative determination can easily be performed with a normal spectrophotometer for amido black staining after complete solubilization of the acetate strip containing the stained spot in a suitable volume of dioxane, N,N-dimethylformamide, or dimethylsulfoxide. Instead of a photometer, a fluorometer can be used after staining of the protein spots with Hoechst 2495 and solubilizing with the same three solvents; alternatively, elution of the chromophor Hoechst 2495 from the protein spot with 1% NH_4OH followed by fluorometry is possible.

Fluorescence is measured, in arbitrary units, between the peak maximum and the corresponding blank value. If the value is calculated according to the equation

$$FI = \text{arbitrary units (sample - blank) x volume/amplification}$$

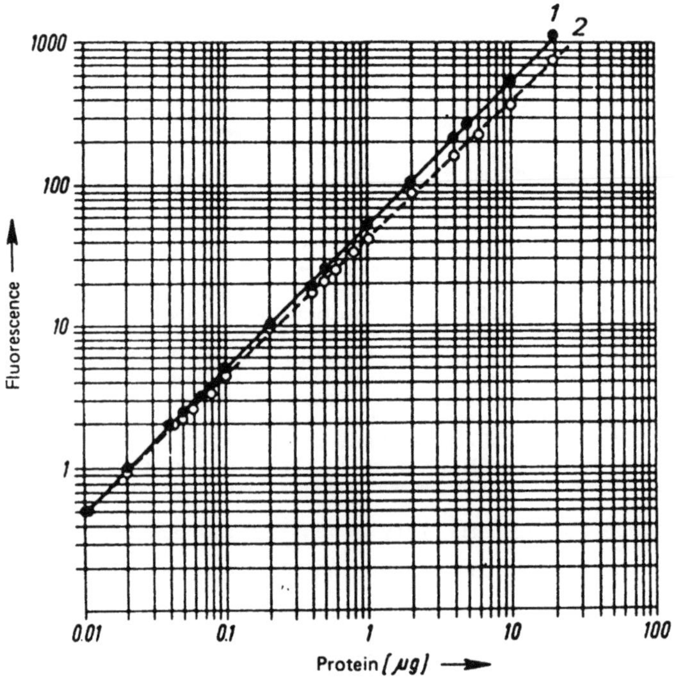

Fig. 3. Calibration curves for bovine serum albumin obtained after staining with Hoechst 2495 and either (1) elution of the stain in 15% NH_4OH or (2) solution of the acetate strip containing the sample in dimethylsulfoxide. Abscissa, μg protein; ordinate, arbitrary units. Fluorescence was measured with a Turner spectrophotometer with an excitation wavelength of 430 nm (for dimethylsulfoxide) or 425 nm (for NH_4OH) and an emission wavelength of 475 nm. The measured volumes were 2 ml for 0.01–2 μg and 5 or 15 ml for the larger amounts. The curves are calculated according to $y = b \cdot x^n$. Curve 1: $\log y = \log 51.8 + 1.01 \log x$; $r = 0.9994$; $n = 36$. Curve 2: $\log y = \log 41.9 + 0.97 \log x$; $r = 0.9995$; $n = 34$.

complete independence of volume and amplification of the instrument are reached, and a calibration curve over a wide concentration range can be prepared as shown in Fig. 3. If measurements are to be performed in the lowest concentration range possible (0.01-0.1 mg protein/ml), the precautions for fluorometric measurements described previously (34) should be followed. In this case, the lowest measuring value should be approximately five times that of the blank.

The calibration curves shown in Fig. 2 are obtained with different sample volumes in the range between 0.5 and 5 µl. Further, the dye binding of different proteins (bovine and human serum albumin, γ-globulin, transferrin, and myelin) with amido black (a) and Hoechst 2495 (b) is shown in Fig. 2. As expected, the slope for each protein is different, but there is no essential difference between the two stains. The very low binding of the myelin preparation is because of its low protein content (approximately 20%) (35). Since drying of a spot has no influence on the dye-binding capacity, several sample volumes can be applied one after another and dried between applications. The calibration curves shown in Fig. 3 were prepared in this way. Protein solutions containing 0.01, 0.1, 1.0, and 10 mg/ml were spotted repeatedly using 1-µl, 2-µl, or 5-µl capillaries. Spots were then

Table I
Compounds Tested for Their Reaction with Amido Black 10B and Hoechst 2495

Compound	Reaction with	
	Amido black	Hoechst 2495
Urea	Negative	Negative
Glucose	Negative	Negative
Sucrose	Negative	Negative
Ammonium sulfate	Negative	Negative
Chloral hydrate	Negative	Negative
Dodecyl sulfate	Negative	Negative
Triton X-100	Negative	Negative
NP 40	Negative	Negative
Mercaptoethanol	Negative	Negative
Desoxycholate	Negative	Negative
tRNA (2 mg/ml)	Negative	Positive
sRNA	Negative	—
Cholesterol	Negative	Negative
Egg kephalin (saturated solution)	Positive	Negative
Egg lecithin	Negative	Negative
Ampholine	Positive	Positive

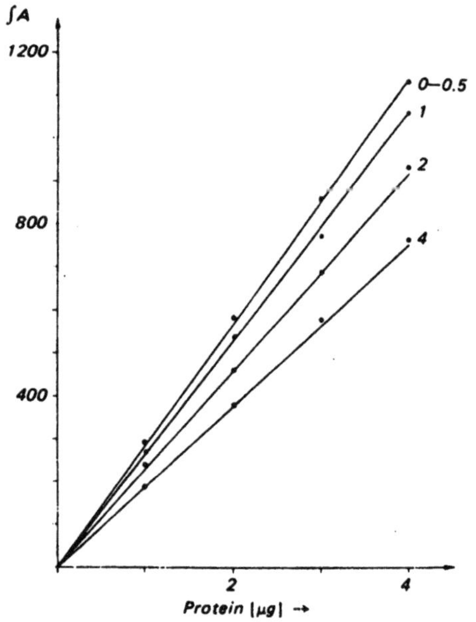

Fig. 4. Relationship between protein concentration and the integrated absorption after amido black staining for different concentrations of dodecyl sulfate. Abscissa, μg protein; ordinate, integrated absorption. Each curve was calculated according to $y = b \cdot x^{a}$ on the basis of 12 evaluated spots. The correlation coefficients are 0.9995, 0.9985, 0.9985, and 0.9979 for 0–0.5%, 1%, 2%, and 4% dodecyl sulfate present in the sample, respectively. Bovine serum albumin was the protein standard. Note that 0.5% dodecyl sulfate has no influence on the stainability with amido black; staining with Hoechst 2495 is also not affected by dodecyl sulfate (results not shown in this Figure).

stained with Hoechst 2495 and, after destaining, were solubilized in 2 ml dimethylsulfoxide (Fig. 3, curve 2) or eluted in 2 ml 1% NH_4OH and evaluated fluorometrically (Fig. 3, curve 1).

Interference in the Method

Compounds tested for their influence on the staining procedure are listed in Table I. High concentrations of egg cephalin (saturated solution) can be stained with amido black but not with Hoechst 2495. High concentrations of tRNA (2 mg/ml) are weakly stained with Hoechst 2495 but not with amido black. None of the listed compounds in a protein solution has any influence on the quantitative staining. The influence of dodecyl sulfate on amido black binding to protein is shown in Fig. 4. The presence of 0.5% dodecyl sulfate has no effect, but 1%, 2%, and 4% dodecyl sulfate decreases the staining by 6.2%, 17,4%, and 33,0%, respectively, in comparison with the same protein concentration (bovine serum albumin) without dodecyl sulfate.

A stoichiometric relationship between protein concentration and staining can be observed. This is demonstrated in the calibration curves in Fig. 4. This stoichiometry, on the other hand, may be use ful for the determination of an unknown dodecyl sulfate concentration. The presence of mercaptoethanol in a sample solution has no further effect. In contrast to amido black, the binding of Hoechst 2495 to all of the proteins studied so far is not influenced by dodecyl sulfate (0.5-2%) or dodecyl sulfate plus mercaptoethanol (1% each). Thus, turbid protein solutions containing unsolubilized particles or even protein sediments can easily be converted to clear solutions suitable for protein determination with Hoechst 2495 by the addition of a suitable volume of 4% SDS/conc. NH_4OH (1:1 v/v), e.g., 100 µl protein solution plus 10 or 20 µl SDS/NH_4OH.

Table II
Differences for Known Amounts of Different Proteins and Brain Extracts Measured According to Lowry and with Spot Analysis: Values Are Related to the Corresponding Standard Calibration Curve Prepared with Bovine Serum Albumin

Protein	Method using amido black[31]	Method of Lowry *et al.*[36]
Bovine serum albumin	0	0
Human serum albumin	− 14.9%	− 15.6%
γ-Globulin	− 3.3%	+ 15.6%
Transferrin	− 19.6%	− 14.9%
Water-soluble proteins from brain tissue	1.72 mg/ml (1.4% S.D.)	2.12 mg/ml (2.1% S.D.)
Brain total particulate fraction in 1% SDS	2.12 mg/ml (1.72% S.D.)	2.62 mg/ml (2.9% S.D.)

When solutions with a known concentration (1 mg/ml) of bovine serum albumin, human serum albumin, γ-globulin, and transferrin are assayed according to Lowry et al. (36) or with the amido black method described here, and the known amount of protein (by weight) is related to the appropriate calibration curve using bovine serum albumin as a standard, differences are observed for both methods (as shown in Table II). For human serum albumin and transferrin, the negative difference is of the same order, but for γ-globulin, there is a negative difference of 3.3% for the amido black method and a positive difference of 15.6% after determination according to the method of Lowry et al. (36) Similar types of differences are also observed for protein determinations of biological material (Table II).

Protein Determination with Unknown Volume and Unknown Concentration

For the calculation of protein concentration, the analyzed volume is usually needed, and the amount of protein in that volume is then determined. Both values can be obtained by a single determination with the amido black method. There is a linear relationship between spot size and the applied volume (1-5 µl), as shown in Fig. 5. The spot size is further determined by the protein concentration in that volume, as is shown for protein concentrations of 0.1-2.0 mg/ml. The linear relationship between the amount of protein (0.1-10 µg) and the integrated absorption for different volumes (1-5 µl) is shown in Fig. 6. As shown in Fig. 1, the densitometric evaluation of the stained spot gives both values needed for the determination of a protein concentration in an unknown volume between 1 and 5 µl the absorption as a measure of the amount of protein present in the spot and the diameter of the spot as a measure of the volume applied. The only prerequisite for the determination of protein concentration in an

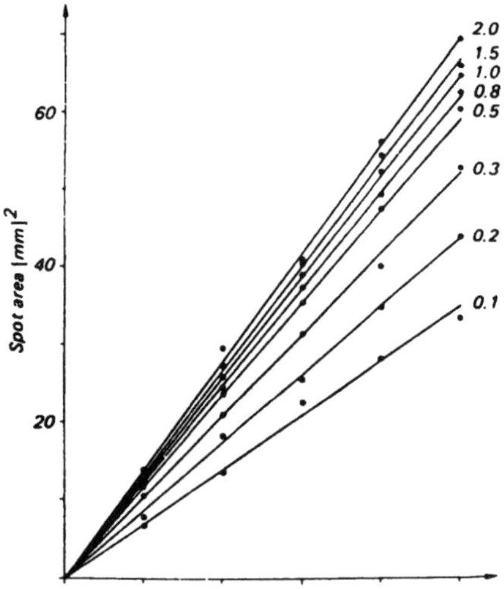

Fig. 5. Spot size on a moistened acetate strip in relation to volume for different protein concentrations evaluated after amido black staining. Bovine serum albumin was the protein standard. Abscissa, µl applied volume; ordinate, spot area (mm^2). The curves are calculated according to $y = b \cdot x^a$ on the basis of 25 evaluated spots per curve. The correlation coefficients for the different protein concentrations are: 2 mg/ml, 0.9985; 1.5 mg/ml, 0.9989; 1 mg/ml, 0.9989; 0.8 mg/ml, 0.9989; 0.5 mg/ml, 0.9989; 0.3 mg/ml, 0.9975; 0.2 mg/ml, 0.9985; 0.1 mg/ml, 0.9955.

unknown volume is an absolutely round spot on the acetate strip. This can easily be obtained if the procedures already described (31) are followed. If Triton X-100 and dodecyl sulfate are present in a sample solution, the spot remains round, but the diameter is larger, and this must then be taken into consideration. If a set of calibration curves similar to Figs. 5 and 6 is prepared, a simple iterative procedure may be used (31) to determine the protein concentration, especially if a modest computer program is available. The whole protein determination procedure with unknown volumes has a standard deviation of 4.1%.

Since protein determination are frequently related to bovine serum albumin, it is useful to calibrate the method described above with this protein. Which stain or which of the evaluation procedures is used may depend on the equipment available in the laboratory. The sensitivity could be improved if a chromophore with better fluorescence intensity is available, e.g., dansyl chloride or fluorescamine, but these compounds are not suitable for this method because they do not permit a staining technique as simple as those with amido black or Hoechst 2495.

The easy handling of this method and especially its independence of or

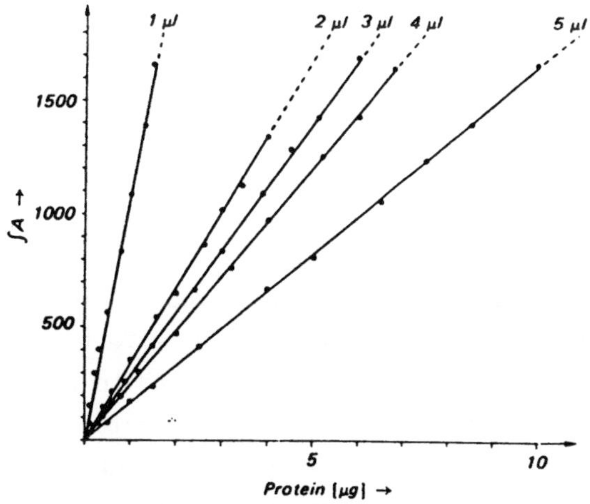

Fig. 6. Relationship between the amount of protein and the integrated absorption after amido black staining for different volumes. Bovine serum albumin was the standard. Abscissa, µg protein: ordinate, integrated absorption. Each curve is calculated according to $y = b \cdot x^a$ on the basis of 27 evaluated spots. The correlation coefficients are: 5 µl, 0.9995; 4 µl, 0.9989; 3 µl, 0.9915; 2 µl, 0.9989; 1 µl, 9.9969.

correction for other external influences are reasons for using this method in biochemical laboratories. The same principle of quantitative spot analysis can still be used if volumes smaller than 0.5 µl have to be analyzed. If suitable stains are available, the method may also be used for determination of other biological substance, e.g., glycoproteins.

GLYCOPROTEIN DETERMINATION WITH FITC-LABELED LECTINS AT THE NANOGRAM RANGE

Practical Performance

Using the same principle of spot analysis as described for protein determination, a calibration and in some respect also a characterization of glycoproteins are possible in the nanogram range (26). An aliquot of 0.5, 1, or, preferably, 2 µl of the glycoprotein solution to be analyzed is spotted as described on moistened acetate strips of the smallest possible size, e.g., 1 x 1 cm. Air drying and repeated spotting of a very dilute sample are also possible. The practical performance of the procedure (26) is given in Scheme 2 (Appendix). The binding of concanavalin A (Con A) to glycoproteins, with glucose oxidase as an example, is seen in Fig. 7 to be dependent on incubation time and temperature at the pH optimum of 7.4. For practical reasons, a standard incubation time of 80 min at room temperature is recommended, with mild shaking during incubation.

Then, any Con A not firmly bound to the glycoproteins is washed out with PBS buffer or 0.9% NaCl. Longer washing has no effect on elution of Con A from the blank. Some Con-A remains strongly bound to the blank acetate paper. To obtain reproducible results, it is therefore always necessary to use actate strips of exactly the same size for blank and sample. Finally, the Con A-glycoprotein complex is completely dissolved within approximately 45 min in a solution containing 2% SDS and 12% NH_4OH. The fluorescence is measured at 492 nm for excitation and 518 nm for emission. Care has to be taken that the slit on the excitation side is narrow enough to avoid overlapping with the emission side, since the maxima are very close together.

With this rather simple procedure, one can prepare, with FITC-labeled Con A and glucose oxidase as a glycoprotein, an acceptable standard curve ranging from 10 ng to 50 µg. The highest standard deviation is in the most sensitive range (Fig. 8). However, the use of a standard curve for quantitation in this glycoprotein method is not recommended. In addition to the problem of what to use as a reference glycoprotein, there are other reasons against the use of a calibration curve. The problem of the high blank value has already been mentioned. Further, the stability of the Con A solution is diminished by repeated freezing and thawing, and after each incubation of glycoprotein spots in the Con A solution, the Con A concentration will be reduced. This can easily be observed when several spots of high glycoprotein content are incubated successively. To counteract this, it is recommended that a fresh Con A stock solution be divided into suitable aliquots which are kept frozen until needed. The diluted Con A solution is best stored in a normal freezer at 4°C if further use in a few days is expected; otherwise, storage is at -20°C. For determination of an unknown concentration of glycoproteins, it is recommended that some reference spots with concentrations of glycoprotein in the expected range be prepared and

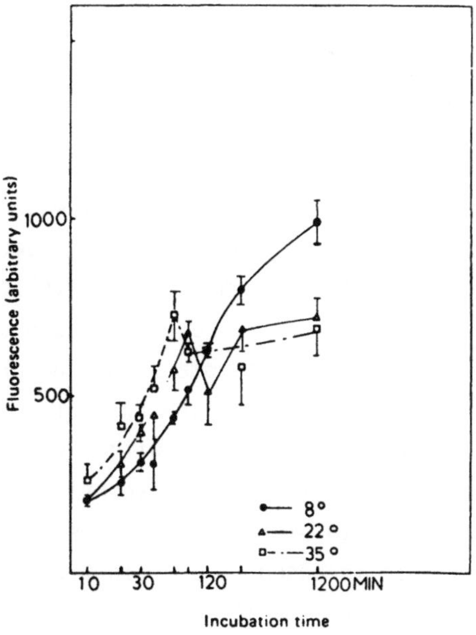

Fig. 7. Incubation in FITC-labeled concanavalin A (0.1 mg/ml in PBS buffer, pH 7.4) of 2-µl spots each representing 10 µg glucose oxidase. After incubation for different time intervals, sample and blank pieces are washed in PBS buffer (or 0.9% NaCl). Thereafter, the glycoprotein–Con A complex is solubilized in SDS/NH_4OH (2%/12% in water), and fluorescence measured in arbitrary units (bars, ±S.D.).

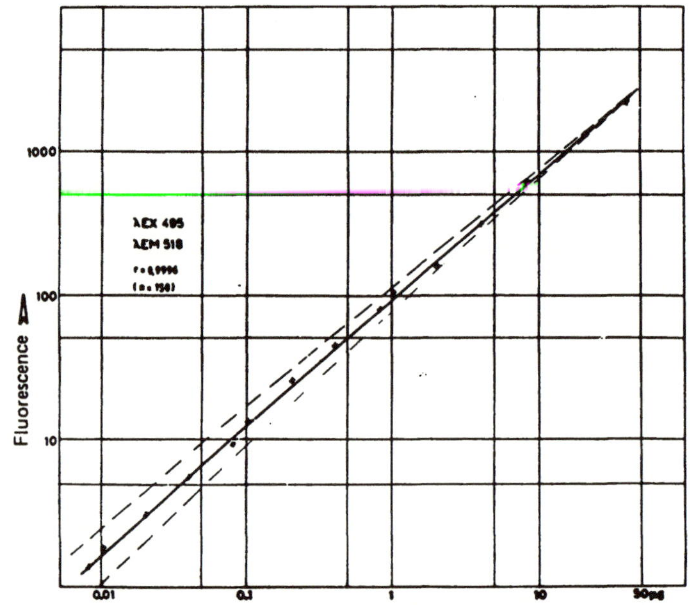

Fig. 8. Calibration curve prepared as described in Scheme 2 with glucose oxidase and FITC-labeled concanavalin A. The dashed line represents standard deviation. Fluorescence was measured in arbitrary units at 495 nm at the excitation monochromator and 518 nm at the emission monochromator; 150 sample values were evaluated, resulting in a correlation coefficient of $r = 0.996$.

the unknown sample related to these.

As with the protein determination method described above, this method also has the advantage of being independent of most common external influences as shown in Table III. Most normally interfering compounds, with the exception of urea, are completely washed out if the washing time in PBS buffer is long enough. That urea gives special problems is presumably because of its influence on glycoprotein conformation, leading to a reduced binding capacity of Con A. For extraction of glycolipids, which may also react with lectins, the sample should be pretreated twice with ether/ethanol (3:2). One must be care ful not to lose glycoproteins or proteins, which will occur if extraction with a chloroform/methanol mixture is used. Extraction of glycolipids after spotting of the sample onto the acetate layer by washing in ether/ethanol does not succeed. If a known amount of glucose oxidase is added to a brain tissue homogenate, membrane fraction, or myelin fraction, the 100% recovery demonstrates that there is no

Table III
Effect of Different Detergents on the binding of FITC-Labelled Concanavalin A to Glucoseoxidase: The Detergents Are Added to the Glucose Oxidase Solution Prior to Spotting of 2-μl Samples

Detergent	Washing time: 1 × 10 min PBS	Washing time: 4 × 10 min PBS
None	100%	100%
2% SDS	78	100
1% SDS	100	100
0.5% SDS	100	100
2% SDS/12% NH_4OH	—	100
1% SDS (5 min, 100°C)	100	100
2% Triton X-100	62	100
1% Triton X-100	68	100
0.5% Triton X-100	80	100
2% NP40	70	100
1% NP40	77	100
0,5% NP40	77	100
2% Mercaptoethanol	100	100
!% Mercaptoethanol	—	100
1% Mercaptoethanol + 1% SDS	—	100
9 M Urea	—	50
6 M Urea	—	44
1% Desoxycholate	—	100

interference with compounds in the biological material that may inhibit the binding of Con A to glycoproteins. The binding of three different lectins to three different glycoproteins using spot analysis is shown in Table IV and demonstrates that this is a very simple method for testing the binding capacities of different lectins as well as for determining the influence of different ions, pH, etc.

Table IV
Binding of Different FITC-Labeled Lectins to Different Glycoproteins[a]

FITC-lectin[b] (0.1 mg/ml)	Glucose oxidase (5 mg/ml)	Ovalbumin (5 mg/ml)	Fetuin (5 mg/ml)
Con A	802	140	40
WGA	39	39	15
RCA 120	12	—	—

[a] Experiments are performed as described in Scheme 2 with 2-μl samples. Values in arbitrary units of fluorescence.
[b] Con A, concanavalin A; WGA, wheat germ agglutinin; RCA 120: *Ricinus communis* agglutinin, mol. wt. 120,000.

Simultaneous Determination of Protein and Glycoprotein

In cases in which the material to be analyzed is very limited or rare, one can perform the protein determination and glycoprotein determination with a single spot. The protein determination is performed with the chromophor Hoechst 2495 as described above. The stain is then eluted from the protein spot with 1% NH_4OH and measured fluorometrically. The normal calibration curve can be used as for the protein determination. Thereafter, the spots are washed first with PBS buffer (twice for 5 min and once for 10 min) and then transferred to the FITC-lectin and further treated as described in Scheme 2 (Appendix). Control experiments have shown that the values obtained by this double determination with a single spot are the same as those found with separate determinations of the protein and glycoprotein concentration using two different spots.

DETERMINATION OF SUGARS IN THE PICOMOLE RANGE

For the more detailed analysis of pure glycoproteins, which are often available only in minute amounts from biological material, it may be useful to have a highly sensitive method for the resolution and quantitation of the sugar content. A microprocedure to determine sugars in the picomole range has recently been described (27). Sugars are reacted with dansyl hydrazine and amino sugars with dansyl chloride, respectively. Two-dimensional microchromatography on 3 x 4 cm micropolyamide sheets (see Appendix, Scheme 3) is used to resolve the sugar derivatives.

If the reaction is performed with a sugar mixture containing 10 mM each of fucose, ribose, arabinose, xylose, galactose, glucose, mannose, 2-deoxy-D-ribose, N-acetylgalactosamine, N-acetyglucosamine, glucuronic acid, and galacturonic acid, only 0.01-0.02 µl is needed for a microchromatogram. This small volume is applied with a very fine capillary under a stereomicroscope to a corner of the microchromato gram to give a starting point approximately 2 mm in diameter. After drying, several chromatograms are developed together using a special adapter or singly in a closed beaker of suitable size.

A UV photo (37) of such a microchromatogram, representing approximately

5 pmol sugar per single spot, is shown in Fig. 9. Dansyl l-proline is the external standard. The reaction of the sugars and uronic acids is stoichiometric, thereby allowing quantitative determination. After scraping of the spots with a small microknife (38) and complete elution with absolute ethanol, they can be measured in a microcuvette fluorometrically at 340 nm for excitation and 515 nm for emission. Alternatively, the chromatogram can be automatically evaluated by scanning fluorescence microscopy as described below.

The amino sugars glucosamine and galctosamine do not react with dansyl hydrazine under acid reaction conditions, since the amino group is protonated and not the aldehyde group as is necessary for hydrazilation.

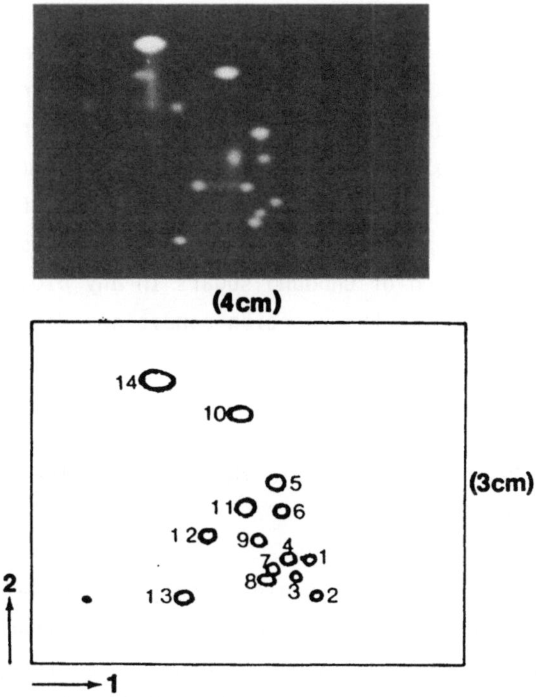

Fig. 9. Two-dimensional microchromatogram on a 3 × 4 cm micropolyamide sheet of dansyl hydrazone sugars and map for identification: 1, N-acetylgalactosamine; 2, galacturonic acid; 3, N-acetylglucosamine; 4, galactose; 5, fucose; 6, arabinose; 7, glucose; 8, mannose; 9, xylose; 10, dansyl-L-proline (internal standard); 11, ribose; 12, 2-deoxy-D-ribose; 13, glucuronic acid; 14, dansylhydrazine. Larger spots represents approximately 5×10^{-12} mol of the respective sugar, and faint spots approximately 5×10^{-13} mol sugar. Photo taken under UV illumination.[37]

Both amino sugars easily react with dansyl chloride to form a dansyl compound which can be separated chromatographically in the presence of dansyl amino acids. For the first dimension, acetone/H_2O (25:75) is used; for the second dimension, ethyl acetate/isopropanol (65:35) is used. Under these chromatographic conditions, the dansyl amino acids line up in the first dimension and do not migrate in the second dimension where the two amino sugars are detectable.

N-Acetylamino sugars react readily with dansyl hydrazine but not with dansyl chloride. This is in contrast to amino sugars and occurs because the nitrogen atom in acetylamino sugars is an amide nitrogen which, under acid conditions, is not protonated. Instead, the aldehyde group is protonated and can react with dansyl hydrazine. At the present time, fructose and neuraminic acid present problems, for neither reacts with dansyl hydrazine or with dansyl chloride since they have no amino or aldehyde group. They may, however, through their keto groups be able to react under special conditions which are being investigated with another more reactive hydrazine than dansyl hydrazine. The optimal conditions for quantitative liberation of sugars from glycoproteins or glycolipids in microscale need to be elaborated. Conditions and problems for liberation of sugars from glycoproteins using different procedures of acid hydrolysis are described elsewhere (39). Determination of unbound sugars in any biological material is without problems so long as the reaction conditions described above are used.

ONE- AND TWO-DIMENSIONAL ELECTROPHORESIS IN POLYACRYLAMIDE CAPILLARY GELS AND MICROSLAB GELS

The first application on the microscale of polyacrylamide gel electrophoresis was carried out in 1964 when Pun and Lombrozo (1) fractionated brain proteins. In 1965, Grossbach (40) used 5-µl Drummond microcaps for this technique, which was further refined in 1966 by Hydén, Bjurstam, and McEwen (41), McEwen and Hydén (42), and Hydén and Lange (43) who used 2-µl capillaries for the fractionation of brain proteins. In 1968, Neuhoff (44) introduced a gel mixture that had been specially developed for the microfractionation of water-soluble brain proteins; these gels were later found to be suitable for many different fractionation problems. Hydén and Lange (45) used microdisk electrophoresis for the analysis of the changes in proteins in different brain areas as a function of intermittent training. Griffith and La Velle (46) analyzed changes in the developmental proteins in facial nerve nuclear regions by this method. Ansorg et al. (47) used microdisk electrophoresis to study the effect of different extraction procedures on the pattern of brain proteins and on the heterogeneity of S-100 protein (48). Nir et al. (49) used the method for the analysis of the effect of light on rat pineal proteins.

Microelectrophoresis may already be, or will be in the future, the most widely used micromethod in both biochemistry and neurochemistry, since this microscale method is very flexible and is easily adaptable to almost any problem of separation. Capillary electrophoresis can be used with homogeneous polyacrylamide gels (32,40-51) as well as with gradient gels (32, 52-62), for isoelectric focusing (25,32,63-69), and in combination with immunoprecipitation (32,70,71). Furthermore, any system of capillary electrophoresis can be combined for two-dimensional separations with homogeneous or gradient microslab gels (72,73). All of these methods can be used as such or in suitable combination for enzyme analysis (for review see ref. 74).

Electrophoresis in Homogeneous Capillary Gels

The performance of these micromethods has been described in full detail previously (38) and in an abridged form by Osborne (29). Therefore, only the main features of the technique will be described in this chapter. Generally, 5- or 10-μl capillaries (Drummond Microcaps, Brand Intraend) are used, since their inner diameter (0.42 and 0.60 mm, respectively) allows for easy sample application using fine capillary pipettes. Small 2-μl, 1-μl, or even 0.5 μl capillaries can also be used freehand without special equipment, but for sample application into smaller capillaries (0.1 or 0.05 mm inner diameter (40, 63)), one needs a micromanipulator as well as a stereomicroscope. Cleaning the capillaries batchwise prior to use (32) is strongly recommended, since there should then be no problems using their capillary attraction for filling, simply by dipping them into a suitable gel mixture. For homogeneous gels, they are filled to approximately two-thirds of their total volume and are then pressed into a plasticine cushion covered with Parafilm. After careful overlayering of the remaining space in the tube with water, they are kept in a moist chamber for polymerization which normally takes place in 1 hr.

Prior to use, the water layer is aspirated off, and a few millimeters of a rapidly polymerizing 5% collecting gel is introduced and again overlayerd with water. After polymerization of the collecting gel (10-15 min), the water can be exchanged against the sample to be fractionated. The gel concentration and a suitable discontinous or continous buffer system in the gel and at the electrodes can be adapted for any problem of separation, e.g., larger or smaller protein particles, with or without SDS, Triton X-100, urea, etc. Even for the separation of sensitive enzymes, cooling during electrophoresis (routinely performed to 60-120 V at 60-120 µA per capillary gel) is normally not necessary, since the advantageous surface-to-volume ratio in capillary gels creates no heating problems during the short period of 20 to 60 min required for electrophoresis. Bromphenol blue or another suitable dye can be used as front marker to define the endpoint of separation which is normally when the front marker reaches the lower end of the gel.

Any suitable power supply may be adapted for electrophoresis with capillaries, but a power supply that allows independent control of the current for each capillary (32) is recommended (obtainable from E. Schütt, Jr., Göttingen, FRG). For electrophoresis, the single capillaries are fixed in the rubber cap of a funnel and held in position with a suitable stand. When electrophoresis is finished, gels with low acrylamide concentration can be extracted from the capillaries with water pressure (32). A well-fitting steel wire can be used for homogeneous gels, provided the lower capillary end is first filled with a small piece of plasticine to avoid the mechanical destruction of the gel with its separated protein bands.

Staining in 1% amido black 10 B in 7.5% acetic acid is done for only 10 min, whereas destaining in several baths of fresh acetic acid takes approximately 30 min. With Coomassie brilliant blue, staining and destaining take somewhat longer, even for capillary gels. The question of what dye to use for staining of proteins is often a matter of personal preference. The most commonly used dye, Coomassie brilliant blue, is not necessarily the best or most sensitive stain for proteins, compared with amido black. Coomassie blue may, however, be better for protein molecules of lower molecular weight. The optical evaluation of stained microgels is

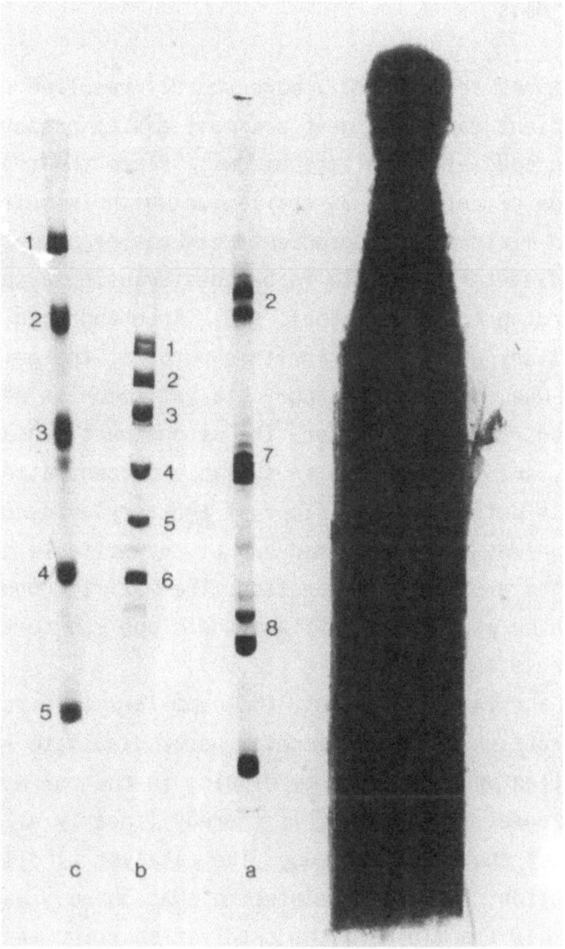

Fig. 10. Ten-microliter gels prepared in capillaries beside a normal matchstick. (a) Isoelectric focusing of a mixture of marker proteins (2 μg total protein) in a 5% polyacrylamide gel containing 9 M urea and 2% carrier ampholytes (pH 2–11); (b) 1–30% acrylamide gradient gel with 0.6 μg of marker proteins; (c) 15% homogeneous polyacrylamide gel with 0.6 μg of marker proteins. The following marker proteins were used: 1, phosphorylase *b*; 2, bovine serum albumin; 3, ovalbumin; 4, carbonic anhydrase; 5, soybeam trypsin inhibitor; 6, α-lactalbumin; 7, β-lactoglobulin; 8, myoglobin (horse); 9, cytochrome *c*.

described in Section 7. Finally, to illustrate the dimensions of microgels, a homogeneous gel, a gradient gel, and an IEF-gel, all prepared in 10-µl capillaries, are compared beside a normal matchstick in Fig. 10.

Microgradient Gels

Protein mixtures are generally more sharply resolved in gradient gels. In addition, gradient capillary gels are more easily prepared than homogeneous gels (with the special 5% collecting gel), since the top of the gradient gel can be made of only 1 or 2% acrylamide which is suitable for stacking of the protein mixture. Microgradient gels can be prepared in capillaries of different diameter or length in batches by using a special apparatus (32) or, according to Rüchel et al. (52), freehand with single capillaries by using capillary attraction. For this purpose, the gel mixture is divided in two parts. One contains only the catalyst ammonium peroxodisulfate in a suitable concentration and buffer. The second part contains the acrylamide, bisacrylamide, and TEMED, again in suitable concentration and buffer. This mixture finally determines the slope of the acrylamide concentration along the linear gradient and can be made up at any suitable concentration according to the problem of separation. The maximal concentration of acrylamide can be as high as 53%, and the slope can further be influenced by adding sucrose to this mixture (32,52).

To form the acrylamide gradient, the capillary is first half filled by capillary attraction with the ammonium peroxodisulfate solution and finally completely filled up to the rim by dipping in the correct concentrated acrylamide mixture. This mixture is thereby linearly diluted from the top to the bottom of the capillary tube. The catalyst is diluted in the opposite direction, thereby guaranteeing that on top, where the acrylamide concentration has its minimum, the catalyst starting the polymerization of the acrylamide has its maximum. This gradient formation can easily be observed if some stain is dissolved in the concentrated acrylamide mixture. The linearity of the gradient can be densitometrically checked if albumin is incorporated into the acrylamide mixture and is finally stained with amido black 10 B. After the gradient mixture is taken up, the capillary is immediately transferred into a small beaker in the bottom of which is a solution of 50% sucrose and some ammonium peroxodisulfate to insure a complete sealing of the capillary and a complete polymerization of the highly concentrated acrylamide at the lower end of the gradient. Capillary forces are again sufficient to hold the capillary with the gradient in an exactly upright position if the capillary is put onto the wall of a small glass beaker (32).

After polymerization, the capillaries can be transferred into a closable tube filled with suitable buffer for prolonged storage at 4°C. Prior to use, the unpolymerized upper phase is aspirated away, and sample application and electrophoresis can follow without further manipulations. After electrophoresis, the gels are pulled out with a well-fitting steel wire from the high-concentration end, thereby avoiding damage to the separation area. Since Triton X-100 need not be incorporated into gradient gels, they are excellently suited for all types of SDS electrophoresis and can especially be used to demonstrate the types of artifacts that can easily be produced during SDS electrophoresis (53) in micro- as well as in macro gels if the conditions of SDS electrophoresis are not carefully controlled. An

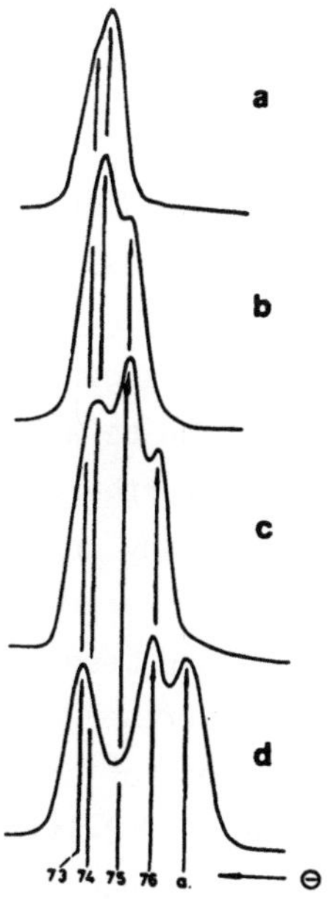

Fig. 11. **Separation of different mixtures of** $tRNA^{phe}$. **The numbers of the peaks correspond** with the numbers of the 3'- end **nucleotide of the** tRNA; *a* indicates the position of the amino **acylated tRNA; (a)** $tRNA^{phe}$ A_{73} + $tRNA^{phe}$ C_{74}; (b) $tRNA^{phe}$ C_{74} + $tRNA^{phe}C_{75}$; (c) $tRNA^{phe}C_{74}$ – $tRNA^{phe}C_{75}$ + $tRNA^{phe}$ A_{76}; (d) $tRNA^{phe}A_{73}$ + $tRNA^{phe}A_{76}$ + Phe-$tRNA^{phe}$C-C-3' NH_2A.

example of the very high resolving power of these microgradient gels is shown in Fig. 11. The difference in only one nucleotide is enough for separation of tRNA species in a 1-40% microgradient gel (75). For further examples with separated proteins from brain tissues, see refs. 2,25,32, 41-49, 56,60-62.

Isoelectric Focusing in Capillary Gels

With the development of ampholytes (76,77), it became easier to fractionate protein mixtures into single-protein components according to their isoelectric points. Isoelectric focusing (IEF) was initially carried out in saccharose gradients (78) and later in polyacrylamide gels (79,80). In a comprehensive review, Haglund (81) has described the historical development, the theoretical foundation, and the application on the macro scale of this very important and powerful method for protein fractionation. The micro version of this method, described by Riley and Coleman (82) and Catsimpoolas (83) requires 10 µg of a single protein or 0.2 to 0.4 mg of a protein mixture for a single fractionation. Quentin and Neuhoff (25), Grossbach (63), and Gainer (64), working independently, refined the IEF methods further; today, they are routinely performed in 5- or 10-µl capillaries with the same equipment described for homogeneous or gradient microgels. Even though one needs only microgram amounts or less of a protein mixture for IEF, it is sometimes advantageous, especially in combination with two-dimensional separations, to separate larger amounts of a protein mixture (up to 20 µg per 10-µl gel), and this can be achieved without danger of overloading with sample. Micro-IEF is, furthermore, excellently suited for fractionation of isoenymes followed by enzyme activity determinations (25,32,66-68,74).

Since all technical details of IEF in capillary gels have been extensively described (32,66-68), only some principles are described here. For IEF, rather soft gels are used, with only 4-7% acrylamide. For cross linking, relativly high concentrations of bisacrylamide (C = 2-4%) or diallyltartrate diamide (84) (DATD, C = 15%) were used to produce so-called nonrestrictive gels. This avoids sieving effects which are one of the main features of homogeneous and especially gradient gels.

Capillaries are filled as usual by capillary attraction with a suitable gel mixture containing 2-4% carrier ampholytes of suitable pH range and are overlayered with water. After polymerization, the water phase is removed, the sample is applied, and this is then carefully overlayered with a cushion of a suitable solution (e.g., 10% glycerine in H_2O) to avoid direct contact of sample with the acidic anolyte. Whether or not carrier ampholines have to be added to the sample has to be determined empirically, since it depends on the protein mixture to be focused. In many instances, this is not required. There is no strict rule for an optimal anolyte or catolyte - most of the published systems can be used. One of the benefits of microelectrophoresis is the very short separation time which, for IEF, is approximately 10 min. Therefore, one can find out easily by trial and error experiments which system is optimal for the separation problem in question.

The current and voltage (normally 200 V) also need to be optimized for the separation problem. Too high a voltage can easily produce artifacts, as has been found, for example, in the fractionation of LDH isoenzymes (66, 67). If well-separated and stained (66) protein separations, as shown in Fig. 10, are not achieved, this is an indication that the optimal conditions still need to be defined.

One- and Two-Dimensional Microslab Gel Electrophoresis

All three types of microelectrophoresis in capillaries are well suited for reproducible fractionation of complex protein mixtures. All three methods, however, have the same disadvantage in that a direct comparison of different sample separations is not possible in the same way that it is when slab gels are employed. Poehling and Neuhoff (72) have adapted the widely used one-dimensional (85) and two-dimensional (86,87) slab gel techniques to a very flexible and sensitive microsystem.

Preparation of Microslab Gels

The chamber in which microgels are cast is prepared from microscope slides cut to a suitable size (2.5 x 3 cm or 3 x 3,5 cm, Fig. 12). Small strips (approximately 1-2 mm wide) cut from plastic sheets of a suitable thickness are used as spacers. Gel chambers are laterally sealed by special Teflon clamps (Fig. 12.3), dental wax, or electrical tape. However, chambers sealed with wax are not suitable for long storage and require more careful handling. Sealing of the lower end of the chambers prior to filling with the gel solution is not necessary if the spacer is less than 0.25 mm thick. Capillary forces are sufficient to keep the solution between the plates. Chambers wider than 0.25 mm are sealed by pressing them into a cushion of plasticine coated with Parafilm (Fig. 12.3). For fixation of the polyacrylamide microslab gels to glass plates (a procedure that is optimal for photometric or autoradiographic evaluation after staining and drying), the glass plates are first silanized according to Radola (88). The upper buffer reservoir is constructed from Plexiglas, and the bottom has a suitably sized slit and is covered with a layer of soft rubber (approximately 5 mm thick) in order to hold the gel chamber firmly in a vertical position (Fig. 12.5). The edges between the gel chamber and the soft rubber are then sealed with agarose prior to filling with the electrode buffer. The lower buffer reservoir is a beaker of a suitable size which also supports the upper reservoir (Fig. 12.5). For two-dimensional separations, isoelectric focusing is performed in capillaries as described above.

Homogeneous Microslab Gels

Gel mixtures are often prepared, according to the method of Laemmly (85), from stock solutions of 30% acrylamide and 0.8% bisacrylamide, but, depending on the separation problems, any other gel system could be used as well. To adapt to the fractionation range required, homogeneous separating gels can be made up with 10, 15, or 20% acrylamide and 375 mM Tris HCl, pH 8.8. Stacking gels are made up from 3% or 6% acrylamide and 125 mM Tris HCl, pH 6.8. Polymerization is achieved with 0.025% TEMED and 0.025% ammonium peroxodisulfate. Chambers with spacers of 0.3 to 1 mm thickness are filled in a vertical position using fine glass pipettes until the separation gel solution fills approximately two-thirds of the chamber. This is then carefully over layered with water. Chambers with thickness less than 0.3 mm can be filled by capillary attraction and thereafter fixed to a stand. In order to fill several gel chambers simultaneously, a chamber used for preparation of gradient gels (see Fig. 14) may be utilized. After polymerization (approximately 30 min), the water layer is aspirated off, and the stacking gel solution is added up to the rim of the chamber. Into this solution a Teflon comb is introduced (Fig. 12.1) to form sample wells of 1 mm width 3 mm deep.

After removal of the Teflon comb, the sample wells are carefully cleaned with a capillary pipette (38) connected to a water vaccum pump and are then completely filled with a suitable buffer (e.g., electrophoresis buffer). The sample, containing 15-20% glycerol and 0.001% bromphenol blue, is carefully underlayered in the sample wells. This step is especially important if wells less than 1 mm in width are used; otherwise, the samples may diffuse out of the wells during application. Unevenly distributed sample solution in the wells reduces the resolution capability of the system. The end sample wells are only filled with sample buffer or marker proteins, since in these lanes the separation can be impaired if the gel chamber is not completely sealed (compare Fig. 15C). A 50 mM Tris/glycine

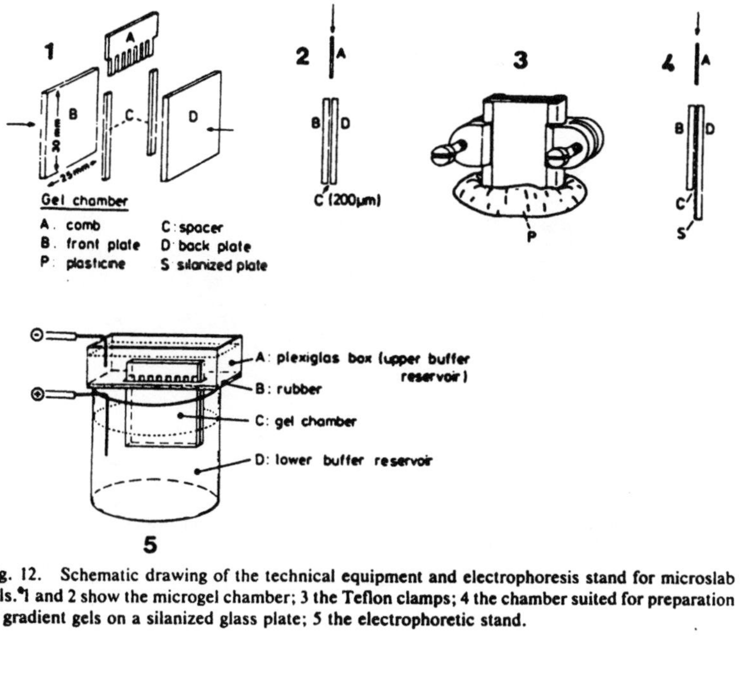

Fig. 12. Schematic drawing of the technical equipment and electrophoresis stand for microslab gels. 1 and 2 show the microgel chamber; 3 the Teflon clamps; 4 the chamber suited for preparation of gradient gels on a silanized glass plate; 5 the electrophoretic stand.

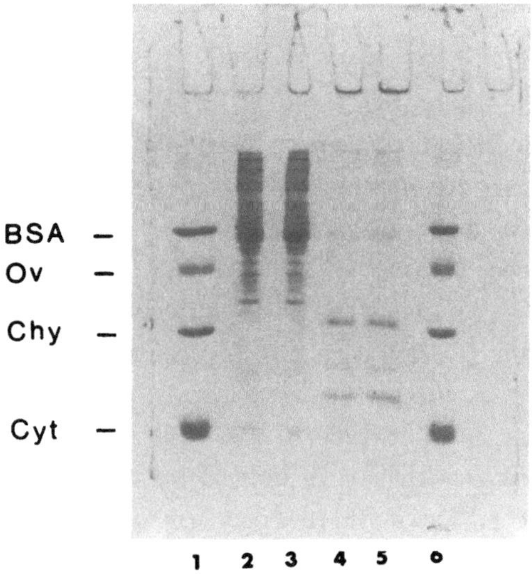

Fig. 13. SDS electrophoresis in a homogeneous polyacrylamide gel prepared in a 3 × 3.5 cm chamber and swollen after staining to 3.5 × 4 cm. Stacking gel: 6% acrylamide, 125 mM Tris HCl, pH 6.8, and 0.1% SDS; thickness of the gel 0.35 mm; electrode buffer 50 mM Tris/glycine, pH 8.4, and 0.1% SDS; electrophoresis: 60 V, 15 min, and 120 V, 50 min. Sample wells 1 and 6 are loaded with a mixture containing 0.1 μg each of bovine serum albumin (BSA), ovalbumin (Ov), chymotrysinogen A (Chy), and cytochrome *c* (Cyt). Wells 2 and 3 are each loaded with 0.6 μg soluble rat brain proteins, and wells 4 and 5 with 0.3 μg rat myelin. All samples contain 1% SDS, 15% glycerol, and 0.001% bromphenol blue.

Fig. 14. Plexiglas® chamber for simultaneous preparation of several polyacrylamide gradient gels. Not shown is the usual gradient mixer connected to a peristaltic pump. The six microgel chambers are sealed with Teflon® clamps (compare Fig. 12). By means of screws connected to Plexiglas® rods, the front plate is held in position and may subsequently be removed to facilitate removal of the polymerized acrylamide block.

buffer, pH 8.4, with 0.1% SDS is used as electrode buffer, and electrophoresis is started at low constant voltage. After the sample has stacked, and a sharp bromphenol blue front is formed, electrophoresis is continued at a higher constant voltage. An example of the resolution that can be achieved with a 15% homogeneous SDS microslab gel (3 x 3.5 x 0.03 cm) is shown in Fig. 13. This shows SDS electrophoresis of water-soluble rat brain proteins, purified rat brain myelin (0.3 µg total protein per well), and a mixture of marker proteins (bovine serum albumin, ovalbumin, chymotrypsinogen A, and cytochrome c, 0.4 µg total protein per well). Proteins of high molecular weight are very well separated into sharp bands in this system; peaks from proteins of lower molecular weight are more diffuse.

Gradient Microslab Gels

Gradient microslab gels can also be prepared, as are capillary gels, freehand by using capillary forces to form convex gradients as well as linear gradients (72). It is, however, easier and more reproducible to use the Plexiglas chamber shown in Fig. 14 for simultaneous preparation of batches of polyacrylamide gradient gels. For example, several gel chambers (3 x 3.5 cm with variable spacer width 0.1 to 1 mm) can be sealed with Teflon clamps which then rest on the Plexiglas rods in the simple Plexiglas gradient chamber (Fig. 14). Thereafter, the chamber is filled to the rim with water or 0.1% SDS to completely eliminate capillary forces within the gel chambers when small spacers are used. A suitable preformed gradient mixture is slowly pumped through a central hole in the bottom of the chamber. The ascending gradient mixture displaces the water in the gradient chamber through the overflow pipe (Fig. 14).

After the microgel chambers are filled to a suitable height with the gel mixture, a buffered 50% glycerol solution (375 mM Tris HCl, pH 8.8, containing 0.01% TEMED and 0.1% ammonium peroxodisulfate) is pumped into the bottom of the chamber to a depth of 2 mm (for suitable gel mixtures see ref. 72). Filling through the central hole in the bottom of the gradient chamber leads to a horizontal distribution of the gradient solution if the filling speed is limited so that turbulence in the gradient is minimized. Turbulence can be controlled visually if 0.01% bromphenol blue is incorporated into the acrylamide stock solution. This also provides for densitometric determination of the effective slope and linearity of the final gradient. Preparation of stacking gels, sample wells, sample application, staining, and destaining are as described above. When storage of the slab gels is required, it is recommended that a gradient-forming chamber that can be taken apart easily (Fig. 14) be used. The whole gel block with its enclosed slab gel chambers may be removed and then stored in gel buffer at 4°C. Slabs can be cut out of the block as required.

A fractionation of complex protein mixtures on microslab gels of polyacrylamide gradients is shown in Fig. 15. In the separation in Fig. 15A, a gradient gel with a convex gradient prepared by capillary attraction has been used to fractionate marker proteins, soluble rat brain proteins, and rat brain myelin (prepared according to Norton and Poduslo (89)). The protein peaks are very sharp but narrowly spaced. The same samples were also fractionated in a linear gradient prepared by capillary attraction (Fig. 15B). The reproducibility of fractionation is strongly dependent on the reproducibility of the gradient in the gels. This is difficult to achieve and requires more experience in "handmade" gradient gels.

Separations in gradient gels prepared with a gradient mixer are shown in Fig. 15C and D. In Fig. 15C, the same proteins are separated as in A and B, but to demonstrate the importance of good sealing of the gel chamber, the Teflon clamps were removed and put on again after polymerization was complete. Contact between gel and spacer was disturbed, with the result that the separations in the outer wells were disturbed. That microgradient slab gels are optimally suited for comparative studies is shown in Fig. 15D. In this gel, with a linear gradient of 6-30% acrylamide, total brain proteins of control rats of different ages (10,15, and 20 days old) are compared with those of rats with experimental phenylketonuria (90-92). The appearance of different protein bands according to age and variations caused by the experimentally induced state is clearly visible.

Homogeneous microslab gels can be used for IEF as well (Fig. 16); for technical details consult ref. 72. Recently, ultrathin slab gels for macroscale separations by IEF (88,93) have been described. These may easily be converted to the micro scale. For routine use, microslab gels 0.3 to 0.5 mm thick with sample wells of 0.7 to 1 mm width and 3 mm deep are optimal. A suitable sample volume is approximately 1 µl of a solution containing 1

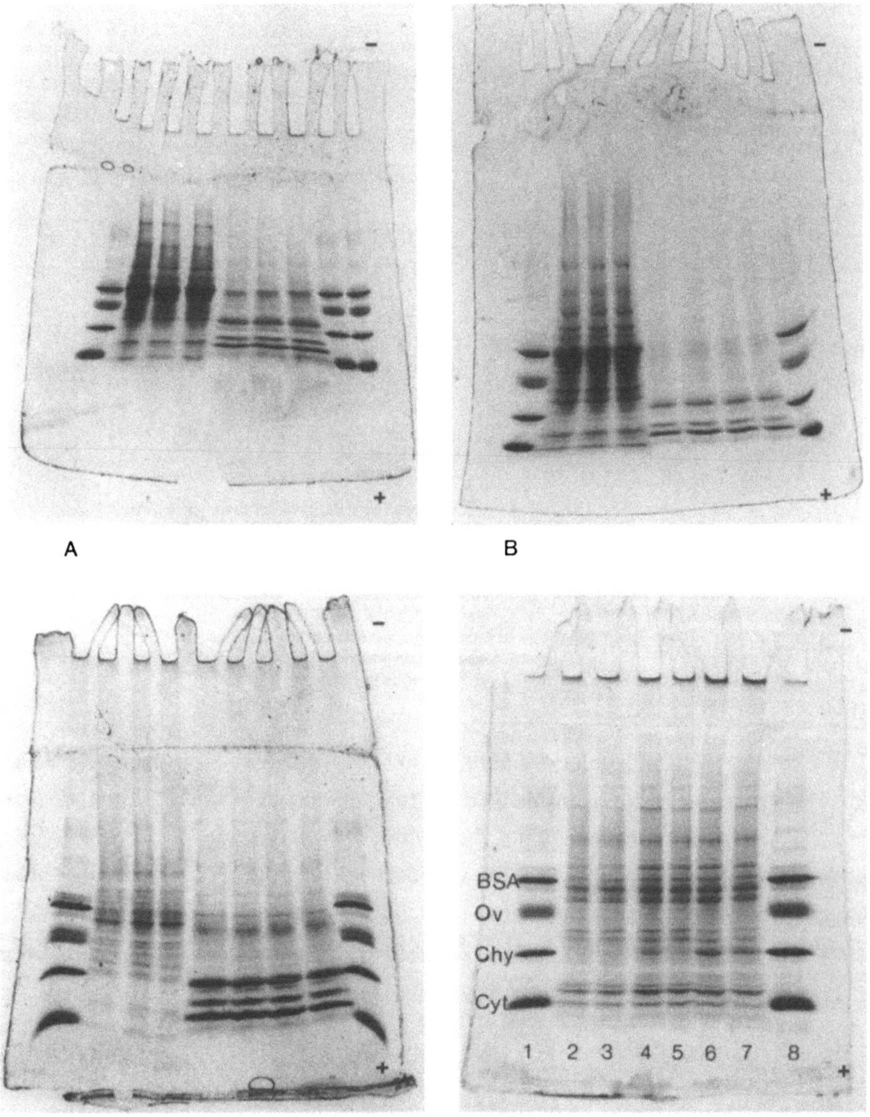

Fig. 15. SDS electrophoresis in different gradient microslab gels. In A and B, the gradients (2.5 × 3.0 × 0.02 cm) were prepared using capillary attraction to form (A) a convex 2–40% and (B) a linear 2–30% acrylamide gradient. The linear 6–25% gradient gels (C and D) (3.0 × 3.5 × 0.5 cm) are prepared with the gradient-forming chamber (Fig. 14). The proteins separated in A, B, and C are the same as in Fig. 13. In D, 0.5 µg/well total brain proteins are separated from control rats of different ages: 10 days old (lane 2), 15 days (lane 4), and 20 days (lane 6); and from rats of the same age having experimentally induced phenylketonuria, lanes 3, 5, and 7. Lanes 1 and 8: marker

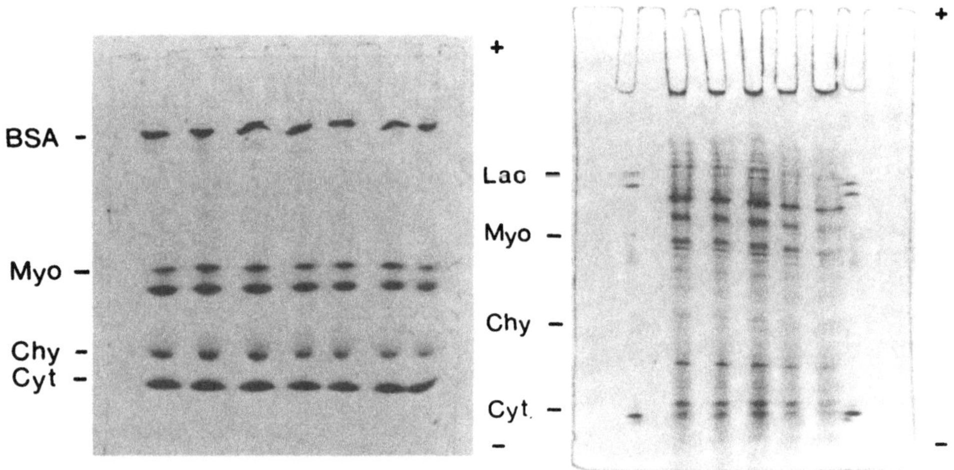

Fig. 16. One-dimensional IEF of (A) marker proteins and (B) total rat brain proteins in 2.5 × 3 × 0.02 cm homogeneous slab gels under different experimental conditions: (A) 7.5% acrylamide, 4% Servalyt®, pH 2–11, prefocusing at 60 V for 15 min, focusing after application of each 0.3 μg of bovine serum albumin (BSA), horse myoglobin (Myo), chymotrypsinogen A (Cyt) at 150 V for 15 min followed by 200 V for 15 min with 0.1% acetic acid as the anolyte and 370 mM Tris as the catholyte; (B) 7.5% acrylamide, 1% Servalyt® (2 parts pH 2–11, 4 parts pH 5–7), 9 M urea, prefocusing at 80 V for 15 min followed by sample application and focusing at 150 V, 15 min, 200 V, 15 min, and 250 V, 15 min. In 1 and 6, 0.1 μg of each marker protein [instead of BSA lactoglobulin (Lac) was used]; 2–5, 0.5 μg of soluble rat brain proteins.

mg/ml protein. An increase in sensitivity is achieved by using both thinner slabs and smaller sample wells to achieve concentration of the proteins into the smallest possible band. However, the absolute detection limit is dependent on the stainability of a protein. Both amino black and Coomassie blue can resolve in the range of 10^{-8}-10^{-9} g per single protein band. Silver staining (73, 94-97) is more sensitive than Coomassie blue staining but is not stoichiometric and is very difficult to handle (97).

Two-Dimensional Microslab Gel Electrophoresis

For the two-dimensional separation of protein mixtures in microslab gels, IEF in microliter capillaries in the first dimension is combined with electrophoresis in microslab gels for the second dimension. Either homogeneous or gradient microslab gels can be used, the most suitable size being 3 x 3.5 cm, 0.25 to 1 mm thick (for suitable gel mixtures see ref.

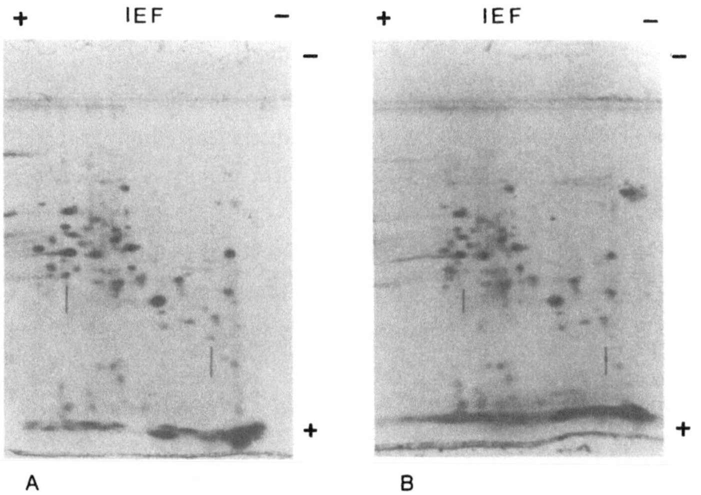

Fig. 17. Two-dimensional electrophoresis of (A) total spinal cord proteins of 30-day-old rats and (B) of rats of the same age but with experimental phenylketonuria. Spinal cords were homogenized, and urea added to a final concentration of 9 M to give a clear solution after centrifugation. Eight micrograms of protein were used for the IEF in the first dimension. The IEF gel was prepared in a 10-μl capillary, and electrophoresis performed with 0.5% phosphoric acid as the anolyte and 0.5% ethylenediamine as the catholyte. Separation in the second dimension was performed in a 3 × 3.5 × 0.075 cm, 15% homogeneous acrylamide gel containing 0.1% SDS. One of the glass plates of the gel chambers was silanized, resulting in a prolonged staining and destaining time of the gel fixed to the glass plate without interfering with the protein separation. The most obvious differences between the tissue extracts from (A) control and (B) rats with phenylketonuria are labeled with arrows. Note that in the homogeneous gel, several proteins migrating with the buffer front are not separated but that the clearly separated protein spots alone amount to approximately 400.

72). The free space on top of the stacking gel is filled with either an acrylamide gel mixture that polymerizes within 3-5 min or with 0.3% agarose (in 125 mM Tris HCl, pH 6.8, with 0.1-1% SDS). With fine forceps made from elastic spring steel (38), the IEF gel is transferred into this embedding solution and carefully pushed down onto the stacking gel. If gel chambers with spacers of 0.25 mm are used, the upper inner sides of the two glass plates have to be obliquely ground to form a V-shaped groove. When agarose is used as the embedding gel, which is imperative for the complete electrophoretic elution from IEF gels (72), the gel chamber, the foreceps, and the agarose have to be warmed to between 40°C and 50°C.

The two-dimensional microgel system has a similar resolving power to that obtained with macrosystems. As an example, the two-dimensional separation of total spinal cord proteins from (1) a 30-day-old control rat and (2) a rat with experimental phenylketonuria (90-92) is shown in Fig. 17. Independent of the differences caused by the experimental phenylketonuria (two of which are labeled with arrows), the resolving power of the homogeneous 15% acrylamide gels is good enough to separate approximately 400 spots, although, under these conditions, the low-molecular-weight proteins are not separated. It is possible that more spots would be detected if either silver staining (73,94-97) or radioactive labelling (73,98) of proteins were used.

A crucial point in a two-dimensional separation is the embedding of the IEF separation on top of the slab gel. If this is not properly achieved, there may be incomplete electrophoretic elution of the sample out of the

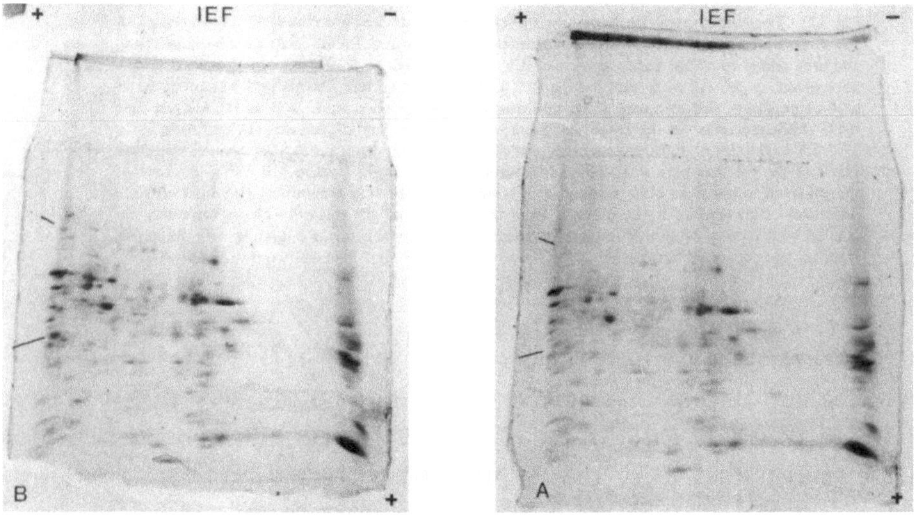

Fig. 18. Two-dimensional separation of 6 μg total proteins from rat brain under identical conditions in the two gels with the exception of embedding the IEF gel (A) in 6% acrylamide with 1% SDS and (B) 0.3% agarose with 1% SDS. For IEF in the first dimension: 10-μl capillaries, 4% acrylamide, 1% Servalyt® (2 parts pH 2–11, 2 parts pH 2–4, 2 parts pH 4–6 and 2 parts pH 5–7), 9 M urea (in gel and sample); anolyte, 0.5% phosphoric acid; catholyte, 0.5% ethylenediamine. Second dimension: 6–25% acrylamide gradient gel (3 × 3.5 × 0.075 cm) electrophoresis for 15 min at 60 V and 40 min at 120 V. Note in A that the IEF gel is not completely eluted because of embedding in fast-polymerizing acrylamide and that, in contrast, in B, the IEF gel is eluted completely because of embedding in agarose. The most obvious differences between the gels are labeled with arrows showing that several proteins are not eluted from the IEF gel embedded in acrylamide.

first-dimension gel. The most commonly used embedding medium is a fast-polymerizing acrylamide gel in order to avoid diffusion of the separated proteins in the first-dimension gel. The effect of two different embedding media on the final two-dimensional protein pattern is shown in Fig. 18. The two gels illustrated are identical in the two dimensions, but the embedding media were 6% acrylamide with 1% SDS in A and 0.3% agarose with 1% SDS in B. Note that not only is elution incomplete in the former case but also the elution of the proteins is highly variable. Spots are missing on the gel in Fig. 18A (labeled with arrow) that are clearly visible in the other slab gel (Fig. 18B).

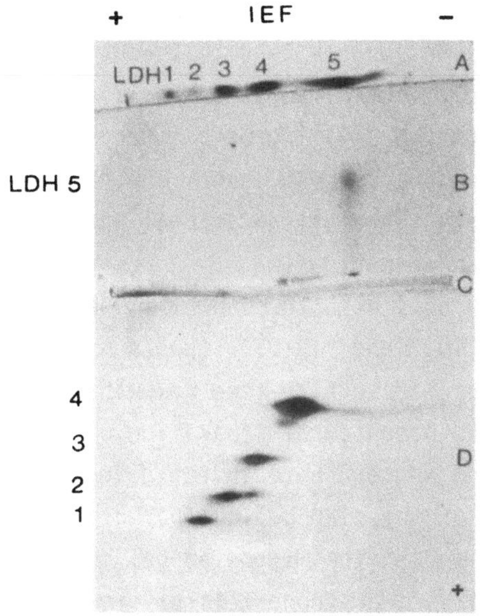

Fig. 19. Two-dimensional separation of soluble rat brain extract stained for LDH with the tetrazolium assay.[32] The IEF for the first dimension was performed in a 10-μl capillary (5% acrylamide, 4% Servalyt®, pH 2–11, 100 V for 45 min, anolyte 0.1% acetic acid, catholyte 370 mM Tris). Second dimension: 2–25% acrylamide gradient gel (3 × 3.5 × 0.075 cm), 40 V for 15 min followed by 80 V for 40 min. The IEF gel C was placed directly on top of the gradient gel D and overlayered with 1% agarose (B) in 125 mM Tris HCl, pH 6.8. The IEF gel A on top of the slab gel is a control gel with the same volume of extract as in gel C, showing the separation of the LDH isoenzymes in the first dimension after staining in the tetrazolium assay for the same time (30 min) as the staining of the slab gel.

The same observation is also made when even higher SDS concentrations are used in the embedding acrylamide gel and/or in the electrode buffer. To gain some insight into the reasons for this observation, control experiments were performed with bovine serum albumin (72). It was found that because of the speed of polymerization in the 6% embedding polyacrylamide gel, up to 33% of the total protein can be firmly bound into the gel. This effect may be partially caused by heat denaturation during rapid polymerization and also by direct polymerization of the protein with acrylamide.

That staining of enzymes (for review see ref. 74) is possible after two-dimensional separations is shown in Fig. 19. As an example, isoelectric focusing of LDH isoenzymes from rat brain extracts in 10-µl polyacrylamide gels was performed as described above. The IEF separation was then transferred directly onto the top of a 2-25% gradient slab gel, and the focused proteins were separated by SDS electrophoresis in the second dimension. Staining of isoenzymes in a corresponding capillary gel and in the slab gel was performed with a tetrazolium assay mixture as described. (25,74,99) The first-dimension gel (on top of the agarose layer) is shown in Fig. 19 together with the isoenzymes in the two-dimensional gel slab after enzyme staining. That the LDH_1 and LDH_2 isoenzymes are hardly visible in the IEF gel is in accordance with earlier observations (67). It is therefore surprising that after two-dimensional separation greater enzymatic activity of these isoenzymes was measurable than following IEF alone. The LDH_5 isoenzyme is also completely eluted from the capillary gel but under these experimental conditions (pH 8.4) migrates towards the cathode and is visible as a diffuse spot in the agarose layer. The observation that after two-dimensional separation there are additional spots below LDH_4 and beside LDH_2 requires further investigation.

Automated photometric evaluation of slab gels with the capability of measuring optical densities between 0.5×10^{-4} and 2.5 O.D. (units of optical density) combined with automated spot detection (see below) is of increasing interest, especially for the two-dimensional separation of proteins in genetic screening. Poehling and Neuhoff have investigated (72) whether the silanizing of one glass plate of the slab gel chamber according to the method of Radola (88) could be adapted to the microslab gel method. The fixation of the gel slab to an optically clear support is of great advantage in the case of photometric evaluation on a high-speed scanning stage. As has been already shown (88), silanization does not interfere at all with the protein separation. There is, however, a problem with gradient gels in that the gel can dry unevenly because of the high gel concentration at the lower end. This can easily be avoided by having the silanized glass plate approximately 0.5 cm longer than the opposing plate (see Fig. 12.4). Before staining, this extra portion of the gel is scraped off, with the result that after drying one obtains a slab gel well suited for photometric evaluation.

DETERMINATION OF AMINO ACIDS AND RELATED COMPOUNDS WITH DANSYL CHLORIDE IN THE PICOMOLE RANGE

Dansyl chloride (dans-Cl, 1-dimethylamino-naphthalene-5-chloride) was first used by Weber in 1952 for the preparation of fluorescent conjugates of albumin. This reagent has subsequently found almost as wide an application as Fischer's naphthalene sulfonyl chloride or Sanger's 2,4-dinitro-flurobenzene (for review, see refs. 101-103). Its usefulness stems from the fact that its reaction products with amino acids, amines, peptides, proteins, phenols, imidazoles, and sulfhydryl groups have an intense yellow to yellow-orange fluorescence and can be separated easily with suitable chromatographic systems (102). Woods and Wang (103) first described the fractionation of dansylated amino acids on polyamide sheets (for review see ref. 104), and Gray and Hartley (105) introduced this method of separation for the determination of end groups and in sequence analysis of proteins and peptides.

It was B.S. Hartley who, in 1969, suggested adapting this technique to the micro-scale. This was immediately successful; the normal 15 x 15 cm polyamide sheets were simply replaced by 3 x 3 cm ones, and the application of the dansylated sample was performed with a very fine capillary under the stereomicroscope. On a 15 x 15 cm polyamide layer, about 10^{-9} mol of each dansylated amino acid is detectable; using 3 x 3 cm polyamide layers, as little as 2×10^{-13} mol can be detected. However, it subsequently became obvious that, in order to obtain consistent data even when only semiquantitative results are required, a number of factors have to be taken into account. This is particularly important when working on the micro scale.

Practical Procedure

The reaction of the amino acids with dansyl chloride (dans-Cl) depends on a variety of conditions, e.g., concentration of dans-Cl, time of reaction, temperature, and pH. Neadle and Pollit (106) and Seiler (101) have shown that the carboxyl group of the N-dansyl amino acid formed can react further with excess dansyl chloride to give dansylamine with the degradation of the

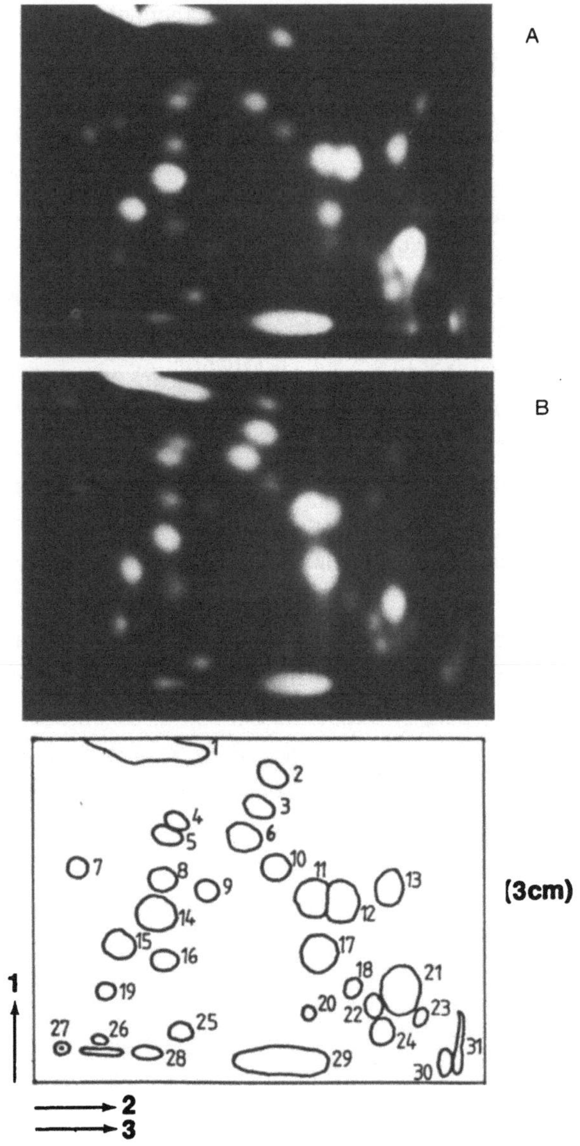

Fig. 20. Two-dimensional chromatogram on 3 × 4 cm micropolyamide layer of dansyl/amino acids from human spinal fluid (A) and human serum (B) and map for identification of the spots. 1, Dans-diethylamine; 2, dans-methylamide; 3, dans-proline; 4, dans-isoleucine; 5, dans-leucine; 6, dans-valine; 7, dans$_2$-tyrosine; 8, dans-phenylalanine; 9, dans-methionine; 10, dans-α/γ-aminobutyric acid; 11, dans-alanine; 12, dans-amine, 13, dans-ethanolamine; 14, dans-α-phenylglycine (internal standard); 15, dans$_2$-lysine; 16, dans$_2$-ornithine; 17, dans-glycine; 18, dans-hydroxyproline; 19, dans-tryptophan, 20, dans-glutamic acid; 21, dans-glutamine; 22, dans-threonine; 23, dans-asparagine; 24, dans-serine; 25, dans-tyrosine; 26, dans$_2$-cystine/cysteine; 27, START; 28, dans-taurine; 29, dans-sulfonic acid; 30, dans-amino sugars.

dansyl amino acid. The yield of dansyl amino acid is consistent for individual amino acids, and so is the quantity of dansylamine that is formed simultaneously. If a mixture of seven amino acids is allowed to react with different concentrations of dans-Cl, the best yield of dansyl amino acid is obtained with a 5.9-fold excess per reactive group. Again, differing reactions are observed for the individual amino acids when a mixture of 24 amino acids is treated with dansyl chloride (for details see ref. 107). The dependance of the dansyl reaction on the amino acid pool to be dansylated and the different reaction yield of single amino acids to form the corresponding dansyl compound makes it necessary, prior to an experiment, to find the optimal reaction conditions for the type of tissue to be analyzed. A reaction scheme suitable for practical use with most tissues is given in Scheme 4 (Appendix).

If the dansyl method is to be used for the microdetermination of amino acids from very small biological samples, e.g., isolated nerve cells, very small cores of material from particular regions of the brain, biopsy material, the usual agents for extraction, such as picric acid, ethanol, sulfosalicylic acid, acetone/hydrochloric acid, ethanol/HCl, or perchloric acid, are unsuitable. All of these substances affect the dansyl reaction and interfere with microchromatography, resulting in bad separations. After a single microhomogenization (see Section 9) of weighed pieces of tissue (14) in 0.05 M $NaHCO_3$, 85% of the free amino acids are found in the supernatant. The remaining 15% are extracted almost completely on a second homogenization with $NaHCO_3$; no further dansylamino acids can be detected on microchromatography after a third extraction (107).

The proteins contained in the supernatant after extraction and centrifugation can interfere with the dansyl reaction, since dansyl chloride also reacts with proteins (69). In addition, the proteins often remain at the origin and impair the final quality of the microchromatogram. They must therefore be removed if possible before allowing the sample to react with dansyl chloride. Heat denaturation (5 min, 85°C) using a capillary is possible, but in this very drastic method, just as in precipitation with trichloroacetic acid, many of the free amino acids coprecipitate with the denatured material, so that the final yield is reduced by ca. 30%.

Precipitation with trichloroacetic acid is also unsuitable because any excess is difficult to remove completely, and the pH for the dansylation must be adjusted accordingly. Also, trichloroacetic acid that has not been removed completely on lyophilization and which is subsequently neutralized causes very bad microchromatograms.

The best method of precipitating the proteins is to add acetone and let the mixture stand at -20°C to complete precipitation. Not all of the proteins are precipitated if a solution contains only 50% acetone (about 90% acetone is necessary to precipitate all the proteins), but enough are precipitated and can be removed by centrifugation so that good microchromatograms can be obtained after treatment of the remaining solution with dansyl chloride (Fig. 20). The use of acetone as the agent for precipitating the proteins has the additional advantage of providing the right conditions for the reaction with dansyl chloride. If the optimal concentration of dans-Cl is used, the dans-OH formed by reaction with water (easily detectable on its bright blue fluorescence under UV light and its typical position on the chromatogram) is so low that its removal by microcolumn chromatography (107) is unnecessary. If too much dans-OH is present, the chromatography and the quantitative evaluation of the chromatogram can be appreciably impaired.

Microchromatography of Dansyl Amino Acids

Microchromatography (see Appendix, Scheme 5) is best performed on micro-polyamide sheets (107,108), since these layers are homogeneous and give sharp separation of the dansyl derivatives (Fig. 20). A mixture of 30-40 dansyl derivatives can be separated on 3 x 3 cm or 3 x 4 cm micro-polyamide layers after two-dimensional chromatography. In microchromatography of dansylated amino acids on polyamide sheets, the quality of the sample applied is of critical importance. Buffers of low molarity or distilled water for extraction of amino acids from biological material are recommended, since those of higher molarity impair the chromatography significantly (107). Under the stereomicroscope, it is very easy to see if too many foreign substances are present in the sample, since the application spot appears as a microatoll with a plateau in the middle, surrounded by a yellow bank of dansylated products.

The way in which the sample is applied determines the quality of a microchromatogram. Especially fine microcapillaries should be used, and their orifices must be of such a size that the acetone/acetic acid solution can be released by applying slight pressure via the mouth piece. To apply the sample, the capillary tip is brought carefully into contact with the layer so that the latter is not damaged. If the origin is damaged, the spots of dansyl compounds show some tailing. The point of application should be at one corner of the sheet, 3 mm or, at most, 4 mm from the edges. The diameter of the application point should be no more than 0.5 mm if possible, the maximum permissible being 1 mm.

When this technique is used for the first time, it is a good idea to build up a chromatographic map of the final dansyl amino acids by systematically separating stepwise mixtures of different dansyl amino acids (107). The typical positions of the dansyl compounds in a microchromatogram are thus observed. Compounds having two reactive groups in the molecule, e.g., serotonin (-NH_2 and -OH), can easily form doubly dansylated products in addition to the two monodansyl derivatives (19). This reaction depends mostly on the pH of the mixture, and each bisdansyl compound will show specific migration during two-dimensional chromatography. Identification of compounds with more than two reactive groups (like many of the biologically active amines) as dansyl products (107) is very difficult, since each can form more than three different dansyl products.

Microchromatograms of dansyl compounds are most accurately evaluated by fluorescence scanning microscopy (see Section 7). If the necessary equipment is not available, they can be evaluated fluorometrically after elution of the isolated dansyl spot with ethanol. Complete elution is essential and is not very easy for all dansyl amino acids. If ^{14}C - labeled dansyl chloride has been used for the reaction, autoradiography can be performed as described (107). Autoradiography is also well suited for turnover studies if radioactively labeled amino acids are reacted with unlabeled dansyl chloride. Lowest accuracy with variations up to $\pm$ 10% are obtained after cumbersome scraping off of the spot area with a microknife (38) followed by scintillation counting. Many examples of the application of the micro-dansyl procedure to analyze free amino acids from many different biological materials have been described (29,107,109-129). The micro-dansyl method is also suitable for determining C- and N-terminal amino acids from minute amounts of purified proteins (107).

PHOTOMETRY AND FLUOROMETRY OF MICROGELS AND MICROCHROMATOGRAMS

Photometric evaluation of microgels creates no problem, and any densitometer or photometer fitted with a suitable device for microgels (32,130, 131) can be used, even for the evaluation of gels stained for enzyme activity measurements (32). With all the equipment commonly used for evaluating micro- or macro-gels, even under optimal conditions, errors commonly occur, since a single scan is not sufficient to correctly evaluate a whole gel. Special problems occur if the protein bands are not exactly lined up in parallel. A minute change in the relationship between the optical slit and the gel can result in quite a different pherogram. All quantitative figures given on the basis of such an evaluation are therefore doubtful. The only correct procedure is the complete scanning of both one-dimensional (as is described for two-dimensional chromatograms (132-136)) and two-dimensional microgel slabs (135). The combination of scanning photometry and image analysis (135,137-139) is a sensitive, versatile, and accurate method for the evaluation of slab gels. This method has the high sensitivity of 5×10^{-4} O.D. because it takes advantage of the noise characteristics of photometric signals. Digital image analysis is used to locate and evaluate the individual spot on the two-dimensional electropherogram. A special procedure of segmentation (140) detects the faintest protein spots hardly visible by eye and separates confluent spots reliably and automatically.

Automatic evaluation of two-dimensional thin-layer chromatograms with scanning fluorescence microscopy (132-136,140) demonstrates clearly the applicability of signal processing, image analysis, and pattern recognition to microphotometry by an enlarged dynamic range of the photometer, improved reproducibility through noise reduction, discrimination between objects and background, feature extraction, and, in addition, automatic classification (140) of the results. Using the same principles described for the evaluation of two-dimensional microchromatograms and microgels, a computer-directed scanning device suitable for macrogels as well as for microgels and a suitable program for two-dimensional automatic evaluation of one-dimensional separations are under development. It is likely that this modern type of microphotometry (137) combined with image analysis will be adapted in the near future for the quantitative evaluation of histochemical staining reactions (which are otherwise not quantifiable), and it will permit photometry on the borderline of light microscopy, e.g., a direct scan of chromosome banding.

MICROMETHODS FOR ANALYSIS AT THE CELLULAR LEVEL

The methods described so far are not well suited for analysis at the cellular level with the exception of the rather large nerve cells that can be obtained from invertebrates (for review see refs 29,141,142). A short chronological review of the development of neurochemical methods at the cellular level going back to the pioneer work of Linderstrøm-Lang has been published by Giacobini (143). Analyses at the cellular level are not as easy to perform as the methods described above, and the equipment required is generally sophisticated and expensive.

Dry mass determination at the cellular level can be performed with the famous quartz-fiber fishpole balance introduced by Lowry (144-146), using an electron microscope (147), or by X-ray absorption (13,148,149). Very many highly sensitive histochemical and radiochemical reactions can also be performed at the cellular level or even on small pieces of single cells, (74,99,142,145,150-155) A very sensitive but rather difficult method can be used for RNA-base analysis of isolated cells (3-5,17). The determination of phospholipids at the cellular level is relatively easy (22-24). Quantitative autoradiography at the cellular level is also possible (156), as are microphotometry (157, 158) and cytofluorometry (159). The cartesian diver technique for measuring enzyme activity in single neurons (6,143) is elegant but difficult to perform. By using microflame photometry (15,16), determination of K and Na is also possible for single cells.

This list is not a complete review on methods available for microanalytical approaches at the cellular level, but for all of those described, the author has at least some experience. Neurochemists seeking a specific micromethod should look to other disciplines where micromethods have been employed because of the same problem of limited material availability. Many of the above micromethods and some others are used in prenatal diagnosis, (87,160-171) in single-cell analysis of plants (172), or in connection with immunocytochemistry (170,174,175).

AUXILIARY EQUIPMENT FOR MICROANALYSIS

Ideally, anyone working with microscale methods should have available a small workshop (and handy machinist), since much of the equipment can be easily made or adapted from normal equipment. For example, capillary centrifugation can be performed in a special capillary rotor, but a normal test-tube rotor can also be adapted (176). Analytical high-speed centrifugation for determination of sedimentation coefficients with volumes less then 1 µl can be achieved with any analytical centrifuge if the usual analytical cell is adapted for capillaries (176,177). Microdialysis chambers for variable volumes between 15 µl and 500 µl (178) are easily prepared in a workshop or are otherwise commercially available (e.g., from E. Schütt, Jr., Göttingen, FRG).

Microhomogenization is an important step in microprocedures. The use of steel wire for the homogenization of isolated cells in a capillary was first described by Eichner (179). Easier to handle are nerve-canal drills driven by a dental drilling machine (28). For volumes larger then 5 or 10 µl, suitable homogenizers can easily be home made (28). The production of capillary pipettes, repeatedly used in microelectrophoresis, is easy, provided a suitable burner is available (17,180). Two-dimensional micro-immunodiffusion (70,181) is a suitable and easy way for immunoprecipitation with volumes of less then 1 µl of antigen and antibody solutions. Here there is the additional advantage that immunoprecipitates are clearly visible within minutes or hours, in contrast to days with the classical Ouchterlony test. The same procedure, for which one needs only a capillary pipette, a microscope slide, and a stereomicroscope as equipment, can also successfully be used to test complex formation other then immunoprecipitates (50,182).

CONCLUDING REMARKS

The experience of the author over more than 25 years has shown that the use of micromethods requires no more technical expertise and critical understanding than are required by other methods. The same critical appraisal of the results is needed whether they are obtained with microscale or macroscale methods. The present era in neurochemistry is that of the isolated cell. New bulk separation methods for the preparation of neurons, which maintain most of their processes (182-188), and for oligodendrocytes and astrocytes are available (187-194). These bulk-prepared cells can be maintained in culture. With such approaches, it should be possible to examine nerve tissue at the single-cell level as well as with homogeneous cell populations. There is a clear need for sensitive micromethods to obtain information from these small cell samples.

Combining currently available experimental and methodological approaches, we can now perhaps begin to tackle all the unanswered questions about the morphological appearance of a nerve cell, its function, and its inherent biochemical makeup. Characterization of these three parameters defining nerve cells in their entirety will undoubtedly reveal them as most fascinating entities.

ACKNOWLEDGMENTS. I am grateful to Dr. Martin Rumsby (York, England) for carefully reading the manuscript and to Dr. G. Huether and Dr. H.-M. Pöhling for valuable discussions.

APPENDIX

A.1. Scheme 1: Spot Analysis for Protein Determination

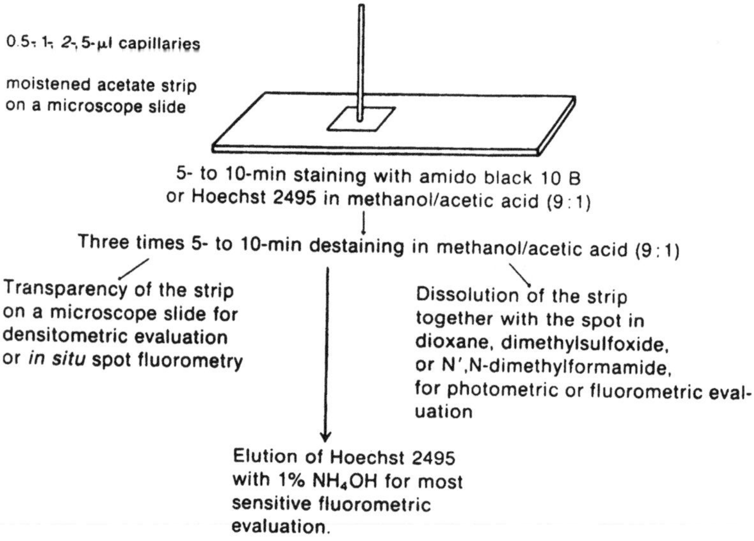

A.1.1. Reagents and Procedure

Amido black 10 B is 0.5% in methanol/glacial acetic acid (9:1); Hoechst 2495 is a fluorescent benzoxanthene derivative (Serva, Heidelberg, FRG); 500 mg are dissolved in 1 ml H_2O, and 20 μl of this solution is dissolved in 25 ml of 9:1 methanol/glacial acetic acid.

Cellulose acetate layers (Membranfilter SM 11200, Sartorius, Göttingen, FRG) are moistened prior to use in a wet chamber to facilitate sample application and to obtain ideal round protein spots in case of densitometric evaluation. Different capillary (Drummond Microcaps or Brand Intraend) volumes can be used, but most reproducible spots are obtained with 2-μl capillaries. They are filled by capillary attraction simply by dipping into the protein solution and are emptied by touching (not pressing) the acetate strip.

A.1.2. Evaluation

For quantitative evaluation of the stained spots, different possibilities exist according to the equipment available.

Densitometric evaluation is performed with any suitable densitometer after making the spots stained with amido black 10 B transparent. (31) For this procedure, the spots have to be perfectly round. For optical reasons, the slit of the measuring instrument must not be larger than one-tenth of the diameter of the spot. A suitable instrument is a photometer together with a gel scanner adapted for evaluation of microgels. (130) This can be used without further technical modification. Densitometric evaluation is also used if the protein concentration in an unknown volume has to be determined, since with a single measurement one obtains the volume applied to the acetate strip from the diameter of the spot and the concentration from the extinction of the stained protein (compare Fig. 1).

In situ spot fluorometry is used if spots stained with Hoechst 2495 are not ideally round. This may be the case, for example, if the sample contains high salt concentrations.

Photometric evaluation can be performed with any photometer if the amido black 10 B-stained spots together with the acetate strips are dissolved in dioxane, dimethylsulfoxide, or N',N-dimethylformamide. The maximum wavelength for amido black in the three solvents is 600 nm, 630 nm, and 620 nm, respectively. The lowest sensitivity in this case is 1 µg. Hoechst 2495 is not suited for photometric evaluation because its absorption is too low.

Fluorometric evaluation can be performed, as with photometric evaluation, after solubilization of the spots stained with Hoechst 2495 together with the strips. Fluorescence in dioxane (at 427 nm for excitation and 532 for emission) is tenfold less sensitive than for the two other solvents, which have the same fluorescence maxima (430 nm and 475 nm). Highest fluorescence is obtained when the chromophor Hoechst 2495 is eluted from the protein spot with 1% NH_4OH in H_2O, and the fluorescence is measured at 425 nm and 475 nm.

A.2. *Scheme 2: Spot Analysis for Glycoprotein Determination*

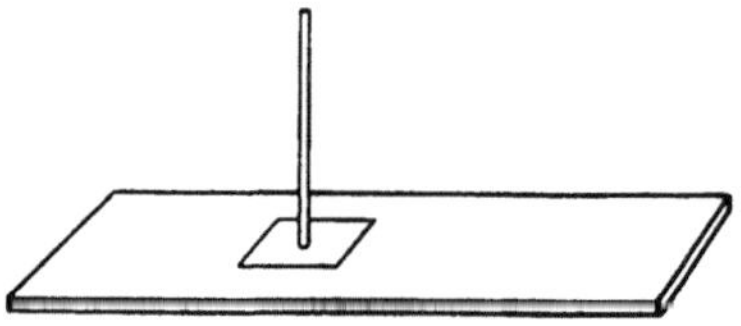

At least 5 min in methanol/glacial acetic acid (9:1)

↓

Two washes in methanol (to remove acetic acid)

↓

60-min preincubation in suitable buffer

↓

80-min (or longer) incubation in FITC-labeled lectin in buffer together with blank piece of exactly the same size as the sample piece

↓

One 5-min and two 10-min removals of unbound lectin

↓

Solubilization of the glycoprotein–lectin complex in 2% SDS in 12% NH_4OH in water

↓

Fluorometric evaluation at 492 nm for excitation and 518 nm at the emission monochromator of sample and blank piece

A.2.1. Reagents and Procedure

Pretreatment of the acetate strips and sample application is as described for spot analysis of proteins. The strips used for sample application and the blank piece have to be of exactly the same size, since some of the fluorescent-labeled lectin, even after prolonged washing in buffer, remains on the acetate sheet. Highest values of complex formation are obtained with FITC-labeled concanavalin A in PBS buffer (0.1 mg/ml Con A in PBS buffer, pH 7.4; PBS buffer: 6.46 mM Na_2HPO_4, 1.47 mM KH_2PO_4, 136 mM NaCl, 27 mM KCl, 0.68 mM $CaCl_2$, 0.49 mM $MgCl_2$).

A.3. Scheme 3: Determination and Microchromatography of Sugars in Picomole Range

100 µl sample solution
+ 300 µl glacial acetic acid
+ 500 µl dansyl hydrazine (1% in ethanol)
+ 50 µl internal standard solution

(or corresponding aliquots)

↓

10 min at 50° C in a water bath with carefully closed test tubes

↓

Transfer onto ice to stop further reaction

↓

Microchromatography on 3 × 4 cm micropolyamide sheets; first dimension (4-cm side): acetone/water (25:75 v/v) + 100 mM ammonium acetate

↓

Dry 15 min at 50°C in a vacuum oven

↓

Second dimension (3-cm side): toluene/glacial acetic acid (80:20 v/v)

Dry 15 min at 50°C in a vaccum oven

A.3.1. Reagents and Procedure

Dansyl hydrazine is solubilized in warmed ethanol and stored in the dark at 4°C. Longer incubation at 50°C or higher temperature leads to destruction of the formed dansyl hydrazine compounds. Sample application and chromatography are performed as described for dansyl amino acids (see Scheme 5). Ammonium acetate is added to the first-dimension solvent mixture to sharpen the spots during chromatography. As internal standard, a 0.5% dansyl-L-proline solution in ethanol is used as reference for the evaluation with automated scanning fluorescence microscopy. If the necessary equipment is not available, fluorometric determination can be performed after completely scraping off the spots with a microknife (38) and eluting the dansyl hydrazine sugar with ethanol. Fluorescence is measured at 340 nm for excitation and 515 nm on the emission monochromator. Evaluation using HPLC is possible, (27) but not all dansyl hydrazine sugars separated by microchromatography will be resolved in one run.

A.4. Scheme 4: Determination of Amino Acids as Dansyl Compounds in the Picomole Range

5 mg tissue in 100 µl H_2O (or aliquot)

2-min homogenization and centrifugation

100 µl supernatant ↘ ↙ 25 µl serum + 75 µl H_2O (or aliquot)

+ 25 µl internal standard solution
+ 250 µl acetone

↓

1–4 hr at −20° C

↓

5–10 min 15,000 rpm at 0–4°C

↓

250 µl supernatant
+ 25 µl 0.5 M K_2CO_3, pH 9.0
+ 50 µl dansyl chloride (1 mg/ml acetone)

↓

30-min incubation at 37° C

↓

+ 10 µl diethylamine

↓

5-min 15,000 rpm at 0–4° C (only if necessary)

↓

150 µl of the supernatant is dried *in vacuo* or in N_2 stream

↓

Dry residue + 50 µl (or less) acetone/glacial acetic acid 9:1 (v/v)

↓

stir 30 sec

↓

5-min 15,000 rpm at 0–4°C

↓

Microchromatography (see Scheme 5)

A.4.1. Reagents and Procedure

Small tissue pieces are homogenized with microhomogenizers of suitable size (see Section 9); centrifugation is performed in a capillary rotor (see Section 9) or a suitable centrifuge at 0-4°C, depending on the volume available. As internal standard, a solution of 4 x 10^{-4} M -phenylglycine dissolved in 0.01 N HCl is used. This is stored in suitable portions at 20°C until use. Prior to use 0.06 N KOH in a ratio of 1:1 (v/v) is added for neutralization. The K_2CO_3 solution is brought to pH 9.0 with gaseous CO_2 without increasing salt concentrations. Dansyl chloride of purest quality or after recrystalization is dissolved (1mg/ml) in acetone. After dansylation, any excess of dansyl chloride is converted immediately to dansyl diethylamine by adding diethylamine to the reaction mixture.

A.5. Scheme 5: Microchromatography of Dansyl Amino Acids

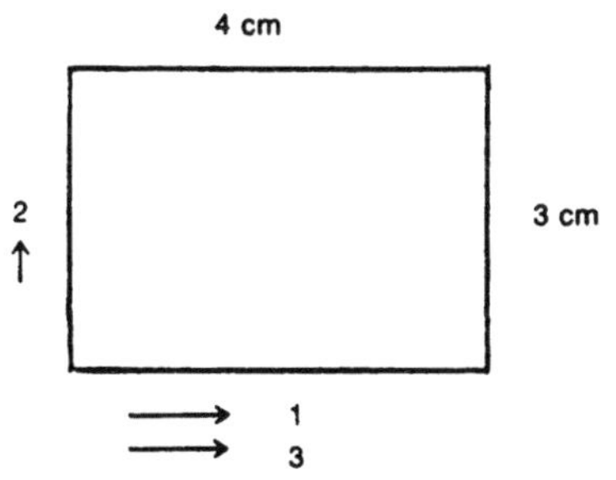

First development with distilled water; dry in vacuum over at 50–80°C for 30 min.

Second development with toluene/glacial acetic acid 8:2 (v/v); dry in vacuum oven at 50–80°C for 15 min.

Third development with formic acid/H_2O, 2:98 (v/v); dry in vacuum oven for 15 min.

Evaluation by automated fluorescence scanning microscopy

A.5.1. Reagents and Procedure

Microchromatography is performed on polyamide sheets (delivered in 15 x 15 cm size) cut to either 3 x 3 or 3 x 4 cm (Schleicher und Schüll, TLC Ready-Plastic Sheets A 1700 Micropolyamide, Dassel, FRG). The micropolyamide sheets are also available as F 1700 sheets in which the micropolyamide layer of 25 µm thickness is fexed on a plastic foil with bluish fluorescence under UV illumination. The A 1700 micropolyamide sheets are fixed on an aluminum layer, thereby increasing the contrast by a factor of ten (108) so that a spot representing 5×10^{-13} mol of a dansyl amino acid is readily visible by eye under UV light.

Great care must be taken not to contaminate the sheets with fingerprints, since the developing solvents do not flow properly over such regions, resulting in bad chromatograms. Approximately 0.5-2 µl of the reaction mixture, depending on the concentration of dansyl compounds, is carefully applied using a fine capillary under a stereomicroscope at a point 3-4mm in from the edges of the chromatogram.

A first development with water up the edge of the layer takes only 4-5 min and is necessary if the sample contains sugars or a high salt content.

Careful drying between developments is essential, since otherwise the separation will be impaired, resulting in uneven spots.

Care must be taken that the solvent mixture never reaches the sample application point when the chromatograms are dipped. Development is performed in a closed beaker. In the first dimension, it takes 4-5 min; in the second dimension, 5 min. Developed chromatograms should be stored in the dark, separately, to avoid contact between individual chromatograms. Contact may result in loss of material through fingerprinting of the spots if two sheets are packed side by side.

REFERENCES

1. Pun, J.Y., Lombrozo, K.: Anal. Biochem. 9, 9-20 (1964).
2. Hyden, H.: Acta Physiol. Scand. (Suppl.) 17(6), 1-136 (1943).
3. Edstrøm, J.-E.: Biochem. Biophys. Acta 12, 361-386 (1953).
4. Edstrøm, J.-E., Hyden, H.: Nature 174, 128-129 (1954).
5. Edstrøm, J.-E.: J. Biophys. Biochem. Cytol. 8, 39-43 (1960).
6. Lindstrøm-Lang, K.: Nature 140, 108 (1937).
7. Lowry, O.H., Roberts, N.R., Chang, M.W.: J. Biol. Chem. 222, 97-107 (1956).
8. Giacobini, E., Zajicek, J.: Nature 177, 185-186 (1956).
9. Lowry, O.H., Passonneau, J.V., Schultz, D.W., Rock, M.K.: J. Biol. Chem. 236, 2746-2755 (1961).
10. Giacobini, E., Grasso, A.: Acta Physiol. Scand. 66, 49-57 (1966).
11. Giacobini, E., Marchisio, P.C.: Acta Physiol. Scand. 66, 247-248 (1966).
12. Buckley, G., Consolo, S., Ciacobini, E., McCaman, R.: Acta Physiol. Scand. 71, 341-347 (1967).
13. Brattgard, S.-O., Hydén, H.: Acta Radiol. (Suppl.) (Stockholm) 94, 1-48 (1952).
14. Neuhoff, V.: Anal. Biochem. 41, 270-271 (1971).
15. Carlsson, B., Giacobini, E., Hovmark, S.: Acta Physiol. Scand. 71, 379-390 (1967).
16. Haljamae, H., Larsson, S.: Chem. Instrum. 1, 131-144 (1968).
17. Edstrøm, J.-E., Neuhoff, V.: Micromethods in Molecular Biology (V. Neuhoff, ed.), Springer-Verlag, Berlin, Heidelberg, New York, pp. 215-256 (1973).
18. Neuhoff, V., von der Haar, F., Schlimme, E., Weise, M.: Hoppe Seylers Z. Physiol. Chem. 350, 121-128 (1969).
19. Neuhoff, V., Weise, M.: Arzneim. Forsch. 20, 368-372 (1970).
20. Briel, G., Neuhoff, V., Osborn, N.N.: Int. J. Neurosci. 2, 129-136 (1971).
21. Briel, G., Neuhoff, V.: Hoppe Seylers Z. Physiol. Chem. 353, 540-553 (1972).
22. Schiefer, H.-G., Neuhoff, V.: Hoppe Seylers Z. Physiol. Chem. 352, 913-926 (1971).

23. Althaus, H.H., Neuhoff, V.: Hoppe Seylers Z. Physiol. Chem. 354, 1073-1976 (1973).
24. Althaus, H.H., Osborne, N.N., Neuhoff, V.: Naturwissenschaften 60, 553-554 (1973).
25. Quentin, C.-D., Neuhoff, V.: Int. J. Neurosci. 4, 17-24 (1972).
26. Neuhoff, V., Ewers, E., Huether, G.: Hoppe Seylers Z. Physiol. Chem. 362, 1427-1434 (1981).
27. Seiler, P., Neuhoff, V., Thorn, W.: Hoppe Seylers Z. Physiol. Chem. (in press) (1982).
28. Neuhoff, V.: Micromethods in Molecular Biology (V. Neuhoff, ed.), Springer-Verlag, Berlin, Heidelberg, New York, pp. 399-402 (1973).
29. Osborn, N.N.: Microchemical Analysis of Nervous Tissue. Pergamon Press, New York, Oxford (1974).
30. Feigl, F.: Tüpfelanalyse, II. Organischer Teil. Akademische Verlagsgesellschaft MBH, Frankfurt/Main (1960).
31. Neuhoff, V., Philipp, K., Zimmer, H.-G., Mesecke, S.: Hoppe Seylers Z. Physiol. Chem. 369, 1657-1670 (1979).
32. Neuhoff, V.: Micromethods in Molecular Biology, (V. Neuhoff, ed.), Springer-Verlag, Berlin, Heidelberg, New York, pp. 4-5 (1973).
33. Zimmer, H.-G., Kiehl, F., Neuhoff, V.: Hoppe Seylers Z. Physiol. Chem. 360, 1671-1672 (1979).
34. Neuhoff, V.: Micromethods in Molecular Biology, (V. Neuhoff, ed.), Springer-Verlag, Berlin, Heidelberg, New York, pp. 149-178 (1973).
35. Norton, W.T.: Myelin (P. Morell, ed.), Plenum Press, New York, pp. 161-199 (1977).
36. Lowry, O.H., Rosebrough, N.I., Farr, A.C., Randall, R.I.: J. Biol. Chem. 193, 265-275 (1951).
37. Zimmer, H.-G., Neuhoff, V.: GIT Fachz. Lab. 21, 104-105 (1977).
38. Neuhoff, V.: Micromethods in Molecular Biology (V. Neuhoff, ed.), Springer-Verlag, Berlin, Heidelberg, New York (1973).
39. Marshall, R.D., Neuberger, A.: Glycoproteins, Volume 5, Part A (A. Gottschalk, ed.), Elsevier, Amsterdam, pp. 224-290 (1972).
40. Grossbach, U.: Biochim. Biophys. Acta 107, 180-182 (1965).
41. Hydén, H., Bjurstam, K., McEwen, B.: Anal. Biochem. 17, 1-15 (1966).
42. McEwen, b., Hydén, H.: J. Neurochem. 13, 823-833 (1966).

43. Hydén, H., Lange, P.W.: J. Chromatogr. 35, 336- 351 (1968).
44. Neuhoff, V.: Arzneim. Forsch. 18, 35-39 (1968).
45. Hydén, H., Lange, P.W.: Proc. Natl. Acad. Sci. U.S.A. 69, 1980-1984 (1972).
46. Griffith, A., La Velle, A.: Exp. Neurol. 33, 360-371 (1971).
47. Ansorg, R., Dames, W., Neuhoff, V.: Arzneim. Forsch. 21, 699-710 (1971).
48. Ansorg, R., Neuhoff, V.: Int. J. Neurosci. 2, 151-160 (1971).
49. Nir, I., Dames, W., Neuhoff, V.: Arch. Int. Physiol. Biochem. 81, 607-616 (1973).
50. Neuhoff, V., Schill, W.-B., Sternbach, H., Biochem. J. 117, 623-631 (1970).
51. Cupello, A., Hydén, H.: Neurobiology 5, 129-136 (1975).
52. Rüchel, R., Mesecke, S., Wolfrum, D.I., Neuhoff, V.: Hoppe Seylers Z. Physiol. Chem. 354, 1351-1368 (1973).
53. Rüchel, R., Mesecke, S., Wolfrum, D.I., Neuhoff, V.: Hoppe Seylers Z. Physiol. Chem. 355, 997-1020 (1974).
54. Wolfrum, D.I., Rüchel, R., Mesecke, S., Neuhoff, V.: Hoppe Seylers Z. Physiol. Chem. 355, 1415-1435 (1974).
55. Rüchel, R., Richter-Landsberg, C., Neuhoff, V.: Hoppe Seylers Z. Physiol. Chem. 356: 1283-1288 (1975).
56. Endou, H., Neuhoff, V.: Hoppe Seylers Z. Physiol. Chem. 356: 1381-1396 (1975).
57. Reichel, W., Wolfrum, D., Weber, M., Scheler, F., Neuhoff, V.: Contrib. Nephrol. 1, 109-118 (1975).
58. Poehling, H.M., Wolfrum, D.I., Neuhoff, V.: Entomologia Exp. Appl. 19, 271-286 (1976).
59. Peter, R., Wolfrum, D.I., Neuhoff, V.: Comp. Biochem. Physiol 55B, 583-589 (1976).
60. Fagg, G.E., Waehneldt, T.V., Neuhoff, V.: Myelination and Demyelination (J. Palo, ed.), Plenum Press, New York, pp. 135-145 (1978).
61. Fagg, G.E., Schipper, H.I., Neuhoff, V.: Brain Res. 167, 251-258 (1979).
62. Tauber, H., Waehneldt, T.V., Neuhoff, V.: Neurosci. Lett. 16, 235-238 (1980).
63. Grossbach, U.: Biochem. Biophys. Res. Commun. 49, 667-672 (1972).

64. Gainer, H.: Anal. Biochem. 51, 646-650 (1973).
65. Bispink, G., Neuhoff, V.: Hoppe Seylers Z. Physiol. Chem. 357, 991-997 (1976).
66. Bispink, G., Neuhoff, V.: Electrofocusing and Isotachophoresis (R.J. Radola, and D. Graesslin, eds.), Walter de Gruyter, Berlin, pp. 135-146 (1977).
67. Gustke, H.H., Neuhoff, V.: Hoppe Seylers Z. Physiol. Chem. 359, 1481-1489 (1978).
68. Gustke, H.H., Neuhoff, V.: Hoppe Seylers Z. Physiol. Chem. 360, 605-608 (1979).
69. Neuhoff, V., Poehling, H.-M.: Hoppe Seylers Z. Physiol. Chem. 361, 77-78 (1980).
70. Neuhoff, V., Schill, W.-B.: Hoppe Seylers Z. Physiol. Chem. 349, 795-800 (1968).
71. Neuhoff, V., Mesecke, S.: Hoppe Seylers Z. Physiol. Chem. 358, 1623-1637 (1977).
72. Poehling, H.-M., Neuhoff, V.: Electrophoresis 1, 90-102 (1980).
73. Poehling, H.-M., Wyss, U., Neuhoff, V.: Electrophoresis 1, 198-200 (1980).
74. Huether, G., Neuhoff, V.: Histochem. J. 13, 207-225 (1981).
75. Sprinzel, M., Wolfrum, D.I., Neuhoff, V.: FEBS Lett. 50, 54-56 (1975).
76. Kolin, A.: J. Chem. Phys. 22, 1628-1629 (1954).
77. Svensson, H.: Acta Chem. Scand. 15, 325-341 (1961).
78. Vesterberg, O., Svensson, H.: Acta Chem. Scand. 20, 820-834 (1966).
79. Dale, G., Latner, A.: Lancet 1, 847-848 (1968).
80. Wrigley, C.W.: J. Chromatogr. 36, 362-365 (1968).
81. Haglund, H.: Methods in Biochemical Analysis (D. Glick, ed.), Volume 19, Wiley-Interscience, London, pp. 1-104 (1971).
82. Riley, R.F., Coleman, M.K.: J. Lab. Clin. Med. 72, 714-720 (1968).
83. Catsimpoolas, N.: Anal. Biochem. 26, 480-482 (1968).
84. Baumann, J., Chrambach, A.: Anal. Biochem. 77, 216-225 (1976).
85. Laemmly, U.K.: Nature 227, 680-685 (1970).
86. O'Farrel, P.H.: J. Biol. Chem. 250, 4007-4021 (1975).
87. Klose, J., Blohm, M., Gerner, L.: Methods in Prenatal Toxicology

(D. Neubert, H.J. Merker, and T. Kwasigroch, eds.), Georg Thieme, Stuttgart, PP. 303-313 (1977).
88. Radola, B.J.: Electrophoresis 1, 43-56 (1980).
89. Norton, W.T., Poduslo, S.E.: J. Neurochem. 21, 749-757 (1973).
90. Lane, J.D., Neuhoff, V.: Naturwissenschaften 67, 227-233 (1980).
91. Lane, J.D., Schöne, B., Langenbeck, U., Neuhoff, V.: Biochim. Biophys. Acta 627, 144-156 (1980).
92. Huether, G., Neuhoff, V.: J. Int. Metab. Dis. 4, 67-68 (1981).
93. Görg, A., Postel, W., Westermeier, R.: Anal. Biochem. 89, 60-70 (1978).
94. Switzer, R.C., Merrill, C.R., Shifrin, S.: Anal. Biochem. 98, 231-237 (1979).
95. Allen, R.C.: Electrophoresis 1, 32-37 (1980).
96. Merrill, C.R., Goldman, D., Sedman, S.A., Ebert, M.H.: Science 211, 1437-1438 (1981).
97. Poehling, H.M., Neuhoff, V.: Electrophoresis 2, 141-147 (1981).
98. O'Farrell, P.H.: J. Biol. Chem. 250, 4007-4021 (1975).
99. Cremer, T., Dames, W., Neuhoff, V.: Hoppe Seylers Z. Physiol. Chem. 353, 1317-1329 (1972).
100. Gray, W.R.: Methods Enzymol. 17, 139-151 (1967).
101. Seiler, N.: Methods Biochem. Anal. 18, 259-337 (1970).
102. Seiler, N., Wiechmann, M.: Progress in Thin-Layer Chromatography and Related Methods (A. Niederwieser and G. Pataki, eds.), Volume III, Ann Arbor Science Publications, Ann Arbor, pp.95-144 (1970).
103. Woods, K.R., Wang, K.T.: Biochim. Biophys. Acta 133, 369-370 (1967).
104. Wang, K.T., Weinstein, B.: Progress in Thin-Layer Chromatography and Related Methods (A. Niederwieser and G. Pataki, eds.), Volume III, Ann Arbor Science Publications, Ann Arbor, pp.177-231 (1972).
105. Gray, W.R., Hartley, B.S.: Biochem. J. 89, 59 (1963).
106. Neadle, D.J., Pollit, R.J.: Biochem. J. 97, 607-608 (1965).
107. Neuhoff, V.: Micromethods in Molecular Biology (V. Neuhoff, ed.), Springer Verlag, Berlin, Heidelberg, New York, pp. 85-148 (1973).
108. Zimmer, H.-G., Neuhoff, V., Schulze, E.: J. Chromatogr. 124, 120-122 (1976).
109. Leonard, B.E., Neuhoff, V., Tonge, S.R.: Z. Naturforsch. 29c,

184-186 (1974).
110. Osborne, N.N., Wu, P.H., Neuhoff, V.: Brain Res. 74, 175-181 (1974).
111. Osborne, N.N., Neuhoff, V.: Brain Res. 74, 366-369 (1974).
112. Quentin, C.-D., Behbehani, A.W., Schulte, F.J., Neuhoff, V.: Neuropediatrie 5, 138-145 (1974).
113. Behbehani, A.W., Quentin, C.-D., Schulte, F.J., Neuhoff, V.: Neuropediatrie 5, 258-270 (1974).
114. Quentin, C.-D., Behbehani, A.W., Schulte, F.J., Neuhoff, V.: Neuropediatrie 5, 271-278 (1974).
115. Neuhoff, V., Behbehani, A.W., Quentin, C.-D., Prinz, H.: Hoppe Seylers Z. Physiol. Chem. 355, 891-894 (1974).
116. Leonard, B.E., Neuhoff, V., Tonge, S.R.: Z. Naturforsch. 29c, 767-772 (1974).
117. Osborne, N.N., Neuhoff, V.: Brain Res. 80, 251-264 (1974).
118. Behbehani, A.W., Quentin, C.-D., Neuhoff, V.: Neurobiology 5, 52-59 (1975).
119. Neuhoff, V., Behbehani, A.W., Quentin, C.-D., Briel, C.: Neurobiology 5, 254-261 (1975).
120. Richter-Landsberg, C., Neuhoff, V.: Naturwissenschaften 62, 491 (1975).
121. Leonard, B.E., Neuhoff, V., Tonge, S.R.: J. Neurosci. Res. 1, 83-92 (1975).
122. Schulze, E., Neuhoff, V.: Hoppe Seylers Z. Physiol. Chem. 357, 593-600 (1976).
124. Stenzel, K., Neuhoff, V.: J. Neurosci. Res. 2, 1-9 (1976).
125. Osborne, N.N., Stahl, W.L., Neuhoff, V.: J. Chromatogr. 123, 212-215 (1976).
126. Osborne, N.N., Neuhoff, V.: J. Chromatogr. 134, 489-496 (1977).
127. Poehling, H.-M., Wyss, U., Neuhoff, V.: Physiol. Plant Pathol. 16, 59-61 (1980).
128. Ulmar, G, Neuhoff, V.: Exp. Neurol. 69, 99-109 (1980).
129. Meyer, W., Poehling, H.-M., Neuhoff, V.: Comp. Biochem. Physiol. 67C, 83-86 (1980).
130. Zimmer, H.-G., Neuhoff, V.: GIT Fachz. Lab. 19, 481-484 (1975).
131. Lane, J.D., Zimmer, H.-G., Neuhoff, V.: Hoppe Seylers Z. Physiol.

Chem. 360, 1405-1408 (1979).
132. Zimmer, H.-G., Neuhoff, V.: Informatik-Fachberichte (W. Brauer, ed.), Springer-Verlag, Berlin, Heidelberg, New York, pp. 12-20 (1977)
133. Kronberg, G., Zimmer, H.-G., Neuhoff, V.: Fresenius Z. Anal. Chem. 290, 133-134 (1978).
134. Kronberg, H., Zimmer, H.-G., Neuhoff, V.: Microsc. Acta 82, 223-228 (1979).
135. Zimmer, H.-G.: J. Microsc. 116, 365-372 (1979).
136. Kronberg, H., Zimmer, H.-G., Neuhoff, V.: Electrophoresis 1, 27-32 (1980).
137. Zimmer, H.-G., Neuhoff, V.: Naturwissenschaften 68, 464-470 (1981).
138. Zimmer, H.-G., Kronberg, H., Berstein, R., Neuhoff, V.: Pattern Recogn. 13, 79-82 (1981).
139. Zimmer, H.-G., Kronberg, H., Neuhoff, V.: Microsc. Acta (Suppl.) 4, 217-221 (1980).
140. Kronberg, H., Neuhoff, V.: Informatik-Fachberichte, Volume 17 (E. Triendle, ed.), Springer-Verlag, Berlin, Heidelberg, New York, pp. 334-337 (1978).
141. Giacobini, E.: J. Neurosci. Res. 1, 1-18 (1975).
142. Giacobini, E.: Neurosci. Res. 1, 1-202 (1968).
143. Giacobini, E.: Biochemistry of Characterised Neurons (N.N. Osborne, ed.), Pergamon Press, Oxford, New York, pp. 3-17 (1977).
144. Lowry, O.H.: J. Histochem. Cytochem. 1, 420-428 (1953).
145. Lowry, O.H., Passonneau, J.V.: A Flexible System of Enzymatic Analysis, Academic Press, New York, pp. 236-249 (1972).
146. Lehrer, G.M.: Micromethods in Molecular Biology (V. Neuhoff, ed.), Springer-Verlag, Berlin, Heidelberg, New York, pp. 285-296 (1973).
147. Bahr, G.F.: Micromethods in Molecular Biology (V. Neuhoff, ed.), Springer-Verlag, Berlin, Heidelberg, New York, pp. 257-284 (1973).
148. Hydén, H., Rosengren, B.: Biochim. Biophys. Acta 60, 638-640 (1962).

149. Hydén, H., Larsson, S.: Proceedings Second International Symposium on X-Ray Microscopy and X-Ray Microanalysis, Elsevier, Amsterdam, pp. 51-55 (1960).
150. Giacobini, E.: Biochem. Psychopharmacol. 2, 9-64 (1970).
151. Giacobini, E.: J. Neurochem. 9, 169-177 (1962).
152. Larsson, S.: Anal. Biochem. 50, 245-254 (1972).
153. Hyden, H.: Nature 4684, 433-435 (1959).
154. Cummins, J., Hydén, H.: Biochim. Biophys. Acta 60, 271-283 (1962).
155. Hydén, H., Lange, P.W., Larsson, S.: J. Neurol. Sci. 45, 303-316 (1980).
156. Dörmer, P.: Micromethods in Molecular Biology (V. Neuhoff, ed.), Springer-Verlag, Berlin, Heidelberg, New York, pp. 347-394 (1973).
157. Hydén, H., Larsson, S.: J. Neurochem. 1, 134-144 (1956).
158. Zimmer, H.-G.: Micromethods in Molecular Biology (V. Neuhoff, ed.), Springer-Verlag, Berlin, Heidelberg, New York, pp. 297-328 (1973).
159. Ruch, F., Lehmann, U.: Micromethods in Molecular Biology (V. Neuhoff, ed.), Springer-Verlag, Berlin, Heidelberg, New York, pp. 329-346 (1973).
160. Galjaard, H., Niermeijer, M.F., Hahnemann, N., Mohr, J., Sørensen, S.A.: Clin. Genet. 5, 368-377.
161. Galjaard, H., Hoogeveen, A., Keijzer, W., De Wit-Verbeek, E., Flek-Noot, C.: Histochem. J. 6, 491-509 (1974).
162. Jongkind, J.F., Ploem, J.S., Reuser, A.J.J., Galjaard, H.: Histochemistry 40, 221-229 (1974).
163. Galjaard, H., Hoogeveen, A., De Wit-Verbeek, Keijzer, W., Reuser, A.J.J.: Histochem. J. 7, 499-501 (1975).
164. Galjaard, H., Hoogeveen, A., Van der Veer, A., Kleyer, W.J.: Excerpta Med. Int. Congr. Ser. 411, 194-206 (1976).
165. Van der Veer, E., Kleijer, W.J., de Josselin de Jong, J.E., Galjaard, H.: Hum. Genet. 40, 285-292 (1978).
166. Galjaard, H.: Ann. Clin. Biochem. 16, 343-353 (1979).
167. Galjaard, H.: Trends in Enzyme Histochemistry and Cytochemistry, Excerpta Medica, Amsterdam, pp. 161-180 (1980).

168. Aitken, D.A., Kleijer, W.J., Niermeijer, M.F., Herbschleb-Voogt, E., Galjaard, H.: Clin. Genet. 17, 293-298 (1980).
169. Galjaard, H.: Trends Biochem. Sci. 5, 201-203 (1980).
170. De Josselin de Jong, J.E., Jongkind, J.F., Ywema, H.R.: Anal. Biochem. 102, 120-125 (1980).
171. Galjaard, H.: Genetic Metabolic Diseases. Early Diagnosis and Prenatal Analysis, Elsevier/North-Holland, Amsterdam (1980).
172. Outlaw, W.H., Jr.: Annu. Rev. Plant Physiol. 31, 299-311 (1980).
173. Hydén, H., Rönnbäck, L: Brain Res. 100, 615-628 (1975).
174. Sternberger, L.A.: Immunocytochemistry, John Wiley & Sons, New York (1979).
175. Chang, J.Y., Brauer, D., Wittmann-Liebold, B.: FEBS Lett. 93, 205-214 (1978).
176. Neuhoff, V.: Micromethods in Molecular Biology (V. Neuhoff, ed.), Springer-Verlag, Berlin, Heidelberg, New York, pp.205-214 (1973).
177. Neuhoff, V., Rödel, E.: Hoppe Seylers Z. Physiol. Chem. 354, 1541-1549 (1973).
178. Neuhoff, V.: Micromethods in Molecular Biology (V. Neuhoff, ed.), Springer-Verlag, Berlin, Heidelberg, New York, pp.395-398 (1973).
179. Eichner, D.: Experientia 22, 620 (1966).
180. Neuhoff, V.: Micromethods in Molecular Biology (V. Neuhoff, ed.), Springer-Verlag, Berlin, Heidelberg, New York, pp.407-409 (1973).
181. Neuhoff, V.: Micromethods in Molecular Biology (V. Neuhoff, ed.), Springer-Verlag, Berlin, Heidelberg, New York, pp.179-204 (1973).
182. Hydén, H., Rönnbäck, L: J. Neurol. Sci. 39, 241-246 (1978).
183. Althaus, H.H., Huttner, W.B., Neuhoff, V.: Hoppe Seylers Z. Physiol. Chem. 358, 773-775 (1977).
184. Althaus, H.H., Neuhoff, V., Huttner, W.B., Monzain, B., Shahar, A.: Hoppe Seylers Z. Physiol. Chem. 359, 773-775 (1978).
185. Althaus, H.H., Gebicke-Härter, P., Neuhoff, V.: Naturwissenschaften 66, 117 (1979).
186. Huttner, W.B., Meyermann, R., Neuhoff, V., Althaus, H.H.: Brain Res. 171, 225-237 (1979).
187. Fewster, M.E., Blackstone, S.: Neurobiology 5, 316-328 (1975).
188. Chao, S.W., Rumsby, M.G.: Brain Res. 124, 347-351 (1977).
189. Poduslo, S.E.: Adv. Exp. Med. Biol. 100, 71-94 (1978).

190. Szuchet, S., Arnason, B.G.W., Polak, P.E.: Biophys. J. 21, 51a (1978).
191. Kennedy, P.G.E., Lisak, R.P.: Neurosci. Lett. 16, 229-233 (1980).
192. McCarthy, K.D., de Vellis, J.: J. Cell Biol. 85, 890-902 (1980).
193. Szuchet, S., Stefansson, K., Wollmann, R.L., Dawson, G., Arnason, B.G.W.: Brain Res. 200, 151-164 (1980).
194. Gebicke-Härter, P.J., Althaus, H.H., Schwartz, P., Neuhoff, V.: Dev. Brain Res. 1, 497-518 (1981).

CHROMOGENIC PEPTIDE SUBSTRATES FOR THE DETERMINATION OF SERINE PROTEASES. A BRIEF REVIEW.

P. Friberger, L. Aurell
KabiVitrum AB, Peptide Research
S-431 33 Mölndal, Sweden

Introduction

Amino acids coupled to chromogenic groups have for many years been used for the assay of protease activities. An example is Bz-Arg-pNA for trypsin. The first chromogenic peptide substrate, Bz-Phe-Val-Arg-pNA became available more than ten years ago. The sequence of this peptide was made to mimic fibrinopeptide A, the first product formed when thrombin splits its natural substrate fibrin. Further substrates of this "second generation" of chromogenic substrates were designed in a similar way. They are characterized by a highly improved sensitivity in comparison to amino acid substrates, but still less selective and soluble than what was still to come. In cases, where natural substrates failed to serve as models, extensive screening of several hundreds of substrates and structure-activity correlations led to substrates for several other proteases. Further development included the use of non-natural amino acids as well as D-amino acids critically positioned in the peptide. This work created substrates with improved selectivity, sensitivity and solubility. These substrates, often called the "third generation", make it possible today to assay a large number of proteases with a considerable selectivity.

By combining the use of chromogenic substrates with preincubation of the sample with an excess of activator, enzyme or proenzyme, it is possible to determine the amount of proenzyme,

Modern Methods in Protein Chemistry, Volume 2

inhibitor and activator respectively in the sample. Numerous methods, such as assays for antithrombin, heparin, prothrombin, Factor X, Factor VIII, prekallikrein, kallikrein inhibitor, plasminogen, antiplasmin, plasminogen activators, α_2-macroglobulin and endotoxin, have been designed. In several cases, diagnostic kits are available today. Most of the assays have been applied on automatic clinical chemistry equipment. Chromogenic substrate assays have been found useful in intensive care units and for the diagnosis of DIC, hypercoagulability, hypertension and for drug monitoring.

In this review the development of chromogenic substrates and their use is briefly described and discussed. For the interested reader plenty of references are given for further studies.

Background

Fibrinogen is the natural substrate for the serine protease thrombin. The product first generated is fibrinopeptide A. It has been demonstrated that Phe in the 9th position N-terminally from the splitted Arg-Gly bond is of importance for the affinity between enzyme and substrate (1). When peptides

Fibrinogen $\xrightarrow{\textit{Thrombin}}$ *Fibrin + Fibrinopeptides A and B*

Fibrin ↓ *Fibrin polymer*

Fibrinopeptide A:

H-Ala-Asp-Ser-Gly-Glu-Gly-Asp-Phe-Leu-Ala-Glu-$(Gly)_3$-Val-Arg-OH

16 (Ala) · 9 (Phe) · 1 (Arg)

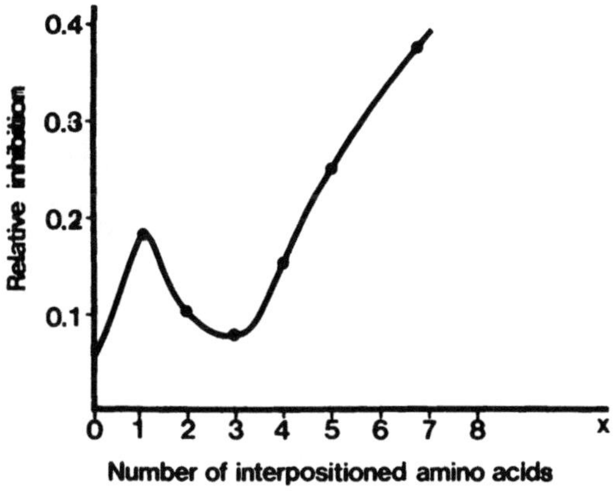

Figure 1: Affinity to thrombin as a function of the position of Phe as the N-terminal amino acid of the peptide. The preceding amino acids are those in fibrinopeptide A. From ref. 2.

of different lengths were studied as inhibitors of thrombin (Figure 1), it was found that a three-peptide was optimal to obtain a high affinity (2). The fact that Phe in position 9 as well as 3 is favourable has been explained by the fact that the distance between Phe N-terminally and Arg C-terminally in the straight three-peptide is the same as in a helix form of that part of the fibrinogen molecule. This affinity between thrombin and the peptide in question was utilized to make a substrate to be used for the assay of thrombin (3). The principle of attaching a chromogenic group to a certain peptide which mimics the sequence next to the scissible bond of the natural substrate (Figure 2) is the basis of this new branch of protein biochemistry (4). The principle for the use of a peptide-pNA is shown in Figure 3. It is based on the difference in absorbance between the intact substrate and pNA, split off from the peptide (Figure 4). At 405 nm, which is the wavelength most frequently used, the absorbance of the substrate is less than one percent of that of an equimolar amount

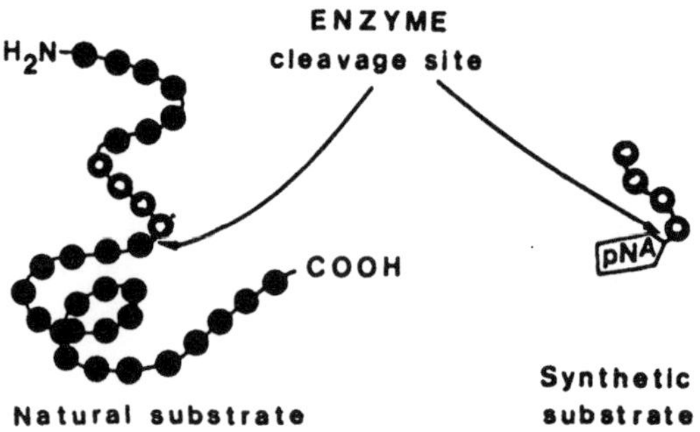

Figure 2: The synthetic substrate is made to mimic the natural substrate.

of pNA (5). As the absorbance is measured on the slope, it is important that the photometer is well calibrated. This biochemical approach to the determination of protease activities creates the possibility to standardize proteases in absolute

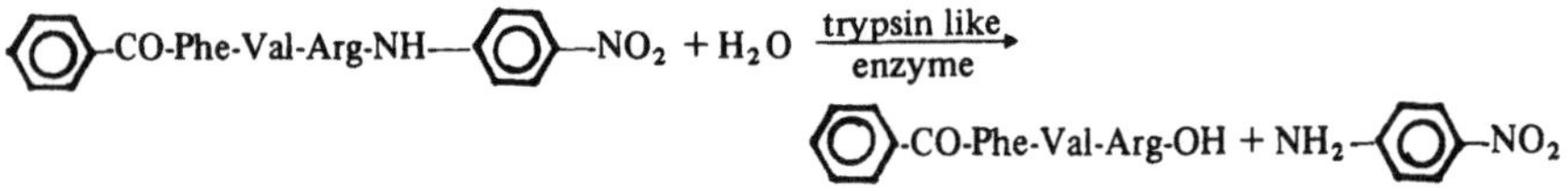

Figure 3: The rate at which pNA is released is measured photometrically at 405 nm. The change in absorbance per min is linearly proportional to the enzyme activity.

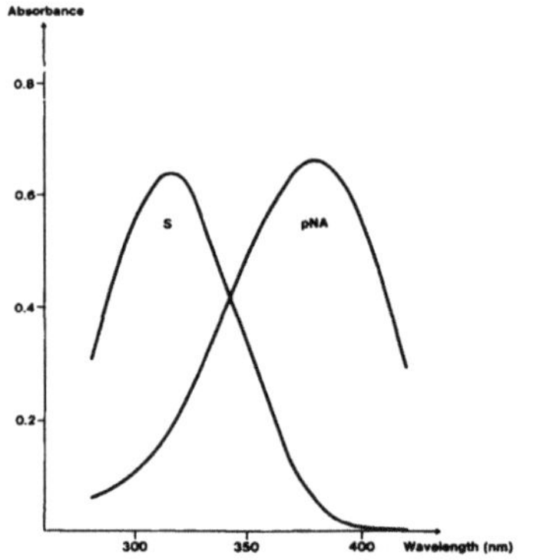

Figure 4: u.v. Absorbance curves of 50 µmol/l substrate (S) and pNA.

Table 1: Enzyme units established by using a reference preparation (IU, NIH-U) or normal plasma (PEU) and a definition (kat, U).

	1 µg of enzyme is equivalent to approx.			
Enzyme	PEU	nkat	Substrate*	Other units
Trypsin		13	S-2222	4.5 NF-U
Thrombin	0.02	5.5	S.2238	2.2 NIH-U
Factor X_a	0.12	1.9	S-2222	
Plasma kallikrein	0.01	0.94	S-2302	
Pancreatic kall.		0.20	S-2266	1.4 KU
Urinary kall.		1.0	S-2266	0-52 TAME-U**
Urokinase		0.34	S-2444	110 PU
				160 CTA-U IU
t-PA		0.26	S-2288	300 IU
Plasmin	0.006	0.20	S-2251	0.024 CU
				0.028 CTA-U
Leukocyte elastase		0.17	S-2484	
Chymotrypsin		2.8	S-2586	0.065 BTEE-U
				0.045 GPNA-mU
Cathepsin G		0.10	S-2545	

*Determined in Tris buffer 0.05 mol/l at optimal pH and I using a substrate concentration of $2 \cdot K_m$ (or at least 50 µmol/l) at 37°C.

**Without a pure enzyme, it has not been possible to estimate the specific activity. The two figures given are, however, comparable.

units like katals (Table 1). One katal is defined as the amount of enzyme, which hydrolyses one mole of substrate per second under specified conditions.

Structure-activity correlations

If the primary structure of the natural substrate is known, a synthetic small molecular weight substrate can be synthesized accordingly. It is, however, demonstrated in the case of fibrinogen, not only the primary structure that determines the affinity between the substrate and the enzyme in question. Therefore, it is important to synthesize a number of peptides, varying one amino acid at the time as illustrated in Table 2 with the thrombin substrate. This sort of comparison can also make it possible to increase the selectivity of the substrate with regard to potentially interfering enzymes. As illustrated in Tables 3-5 rather an extensive work has been carried out to find the best possible substrate for the activated coagulation Factor X, a protein with a very central position in the coagulation system. The sequence Bz-Ile-Glu-Gly-Arg-pNA mimicing the natural substrate, prothrombin, seems to be essential (6).

Table 2: Relative activities of various enzymes on a series of substrates with the formula Bz-Phe-R-Arg-pNA where R is given in the table. Optimal conditions for the respective enzymes are used.

	Relative reaction rate			
R	Thrombin	Trypsin	Plasmin	Pancreas kallikrein
-Val-	100	100	100	100
-Gly-	5	135	45	60
-Ala-	85	180	105	60
-Leu-	35	70	240	880
-Ile-	60	75	45	15
-Pro-	50	170	70	85
-Pip-	8	95	20	60

Table 3: Relative activities of Factor X_a on substrates with the formula R-Glu-Gly-Arg-pNA where R is given in the table.

R	Relative reaction rate
Bz-Ile-	100
Bz-	5
Bz-Gly-	15
Bz-Phe-	25
Bz-Tyr-	25
Bz-Val-	40
Bz-Leu-	50
Bz-Pro-	60

Table 4: Relative activities of various enzymes on substrates with the formula Bz-Ile-R-Gly-Arg-pNA where R is given in the table.

R	Relative reaction rate		
	Factor X_a	Thrombin	Trypsin
-Glu-	100	4	100
-Gln-	120	11	110
-Asp-	70	1	80
-Asn-	60	5	110
Bz-Phe-Val-Arg-pNA	2	100	25

An increased sensitivity is, however, obtained when various other groups are linked to the γ-carboxyl group of Glu (Table 4, 5). (7).

The finding that the unprotected D-amino acid can be efficient in the 3rd position was very important for making easily

Table 5: Relative Reaction Rates of substrates with the formula Bz-Ile-Glu(γ-R)-Gly-Arg-pNA where R is given in the table.

R	Relative reaction rates		
	Bovine FX_a	Human FX_a	Trypsin
OH	100	100	100
OMe	170	160	115
OEt	190	190	110
OiPr	210	200	110
NH_2	120	120	110
NHiPr	220	220	105
	250	280	90
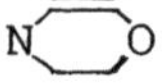	260	240	90

Me = methyl, Et = ethyl, iPr = isopropyl

soluble substrates (Table 6). This was found to be rather a general role in the three-peptides (8, 9).

Table 6: Relative reaction rates of protected and unprotected D and L amino acids of the tripeptide R-Val-Arg-pNA where R is given in the table.

R	Thrombin	Trypsin	Plasmin	Pancreas Kallikr
Bz-Phe-	100	100	100	100
H-Phe-	5	35	25	10
Bz-D-Phe-	5	35	35	55
H-D-Phe-	95	90	95	

Peptides with the sequence Pro-Phe-Arg- found in the natural substrate, kininogen, are excellent kallikrein substrates (8, 10). Different kallikreins have, however, different specificities on small peptide substrates (Table 7) even when their biological effects are in general the same. The substrates H-D-Pro-Phe-Arg-pNA (S-2302) and H-D-Val-Leu-Arg-pNA (S-2266), the activity on which was set 100, were chosen for plasma and urinary kallikrein respectively (11).

Table 7: Relative reaction rates of various kallikreins on substrates of S-2266 and S-2302 types.

Substrate	Kallikreins from			
	Pancreas	Urine	Saliva	Plasma
H-D-Val-Leu-Arg-pNA	100	100	100	35
-Phe-	145	50	50	100
-Pro-	15	25	70	15
-Gly-	5	5	30	2
-Leu-Lys-	5	40	25	<1
-Phe-	15	45	20	10
H-D-Val-Leu-Arg-pNA	100	100	100	35
-Ile-	105	-	90	50
-Leu-	100	50	65	15
-Pro-	70	15	40	20
-Val- -Lys-	5	40	25	<1
-Ile-	1	-	50	3
H-D-Pro-Phe-Arg-pNA	85	10	40	100
-Val-	145	50	50	100
-Phe-	85	35	55	55
-Pip-	85	30	45	75
-Pro- -Lys-	20	25	25	10
-Val-	15	45	20	10
Bz-Arg-Met-Arg-pNA	3	1	115	20

By screening work, the specificity of plasmin has also been determined. Lys in the 1st position is important for the selectivity. In the 2nd position Leu, Pro and Phe are favourable and in the 3rd position most D-amino acids work well both with regard to affinity (K_m) and reactivity (k_{cat}) (12).
The different enzymes for which synthetic substrates have been developed are listed in Table 8. Note that the specificity of these enzymes can be divided into three main groups, trypsin like enzymes which split bonds C-terminally to basic amino acids (Arg and Lys), chymotrypsin like enzymes which split C-terminally to aromatic amino acids (Phe, Tyr and Trp) and elastase like enzymes splitting C-terminally to small aliphatic amino acids (Ala and Val). Between these groups there is virtually no cross-reactivity, so if e.g. chymotrypsin preparation splits an Arg- substrate it certainly contains an impurity, probably trypsin.

Table 8: Proteases for which chromogenic peptide substrates have been designed.

Thrombin	$C\overline{1s}$ and $C\overline{1r}$
Trypsin	Chymotrypsin
Factor X_a	Factor XII_a
Plasma kallikrein	Factor XI_a
Tissue kallikreins	Protein C_a
Plasmin	Cathepsin G
Urokinase	Snake venoms
Tissue plasminogen activators	Limulus and crayfish lysates
Leucocyte elastase	Papainlike enzymes
Pancreas elastase	Bacterial and fungal proteases

Suitable properties of chromogenic substrates

Properties of importance for synthetic peptide substrates are: Affinity, reactivity, solubility, selectivity and stability. These factors should be considered each time a substrate for a certain enzyme is chosen as a reagent in a particular assay. High affinity means that a comparatively low substrate concentration is needed to obtain a reaction rate constant with time. High affinity certainly minimizes the risks for unwanted side reactions. High reactivity brings about a high sensitivity. The solubility has often been a limiting factor for synthetic substrates. The selectivity is specially important when pathological plasmas are analyzed because non-characterized protease activities can be encountered. The stability of amide substrates are usually quite sufficient.

Not only the peptide sequence but also the choice of blocking groups and marker groups are important for the properties of the substrates. Concerning the sensitivity, also determined by the marker group, a high dilution of plasma (1:100) which takes care of the effects of bilirubin and opaqueness is wished. The enzyme concentration after activation of such a plasma is in the order of 10 nmol/l. With a reasonable reactivity (k_{cat}=10 s^{-1}) and reaction time of 100 sec the product concentration will be around 10 µmol/l. With a molar absorptivity of 10,000, like pNA, this means an absorption of 0.1 which is suitable to measure with most equipment available.
The selectivity of an assay is beside the selectivity of the substrate influenced by several factors. By choosing conditions that are optimal for the enzyme-substrate combination in question, it is possible to reach rather a good selectivity as demonstrated in Table 9 where e.g. thrombin splits S-2238 about 60 times faster than Factor X_a and on the other hand Factor X_a splits S-2222 40 times faster than thrombin.

Table 9: By using 4 nmol/l of enzyme the following activities, ΔA/min, are obtained. The conditions used were those optimal for the enzyme above the underlined substrate. The same conditions were also used when testing this substrate with all other enzymes in the table.

	Trypsin	Thrombin	Factor X_a	Plasma kallikr.	Plasmin	Urokinase
S-2238	0.270	0.480	0.008	0.007	0.002	0.002
S-2222	0.730	0.005	0.200	0.005	0.006	0.007
S-2302	0.070	0.030	0.040	0.190	0.040	0.000
S-2251	0.030	0.002	0.001	0.007	0.040	0.000
S-2444	0.620	0.003	0.004	0.001	0.001	0.040
S-2288	0.350	0.280	0.080	0.100	0.040	0.030

Specific biochemical reactions (see below), the use of inhibitors (12) as well as a suitable blanking procedure (13) can contribute very much to the selectivity of chromogenic substrate methods.

Method principles and optimation

During ten years' work, a large number of assays have been designed. The simplest assay is the direct measurement of an enzyme in a purified system. Enzyme activities can also be measuered in biological systems like plasma. Proteins which can be determined indirectly are: inhibitors, proenzymes, enzyme activators and various types of cofactors/moderators.

One example of a direct determination of enzyme activity is the standardization of thrombin. As thrombin is an enzyme that can

$$\text{Peptide-pNA} \xrightarrow{\text{Thrombin}} \text{Peptide} + \text{pNA}$$

exist in several forms with different substrate specificities it is important to specify the substrate used for the standardization. Another example of a direct assay is the determination of enzymatic activity in plasma, usually derived from enzyme-$\alpha_2$macroglobulin complexes which are formed rather rapidly when enzymes are activated in plasma. Such complexes protect the enzyme from being inhibited by other natural "active site" inhibitors in plasma because they are sterically hindered (5). Peptide substrates can, demonstrated in the

$$(\text{Plasma} + \text{Plasmin} \longrightarrow \text{Plasmin}\cdot\alpha_2\text{antiplasmin} + \text{Plasmin}\cdot\alpha_2\text{macroglobulin})$$

$$\text{Peptide-pNA} \xrightarrow{\text{Plasmin}\cdot\alpha_2\text{macroglobulin}} \text{Peptide} + \text{pNA}$$

formula using plasmin as an example, easily reach the active site. The kinetics are, in the cases so far studied, similar to uncomplexed enzymes (15).

Inhibitors are determined by measuring the enzyme activity and subtract the activity of the enzyme after incubation with the sample (16). Below the principle of the assays of two important protease inhibitors is shown. In order to make the assays specific, the reaction between antithrombin III (AT III) and

$$\text{Plasma} + \text{Heparin (excess)} \longrightarrow \text{Antithrombin III}\cdot\text{Heparin}$$

$$\text{Thrombin} + \text{AT III}\cdot\text{Hep} \longrightarrow \text{AT III}\cdot\text{Thrombin} + \text{Thrombin(residual)}$$

$$\text{Peptide-pNA} \xrightarrow{\text{Thrombin(residual)}} \text{Peptide} + \text{pNA}$$

$$\text{Plasma} + \text{methylamine} \longrightarrow \text{Plasma with inactivated } \alpha_2\text{macrobloguline}$$

$$\text{Plasmin} + \text{Plasma without } \alpha_2\text{M} \longrightarrow \text{Plasmin}\cdot\alpha_2\text{antiPli} + \text{Pli(residual)}$$

$$\text{Peptide-pNA} \xrightarrow{\text{Plasmin(residual)}} \text{Peptide} + \text{pNA}$$

thrombin is speeded up tremendously by using heparin (Hep) as a catalysor (17). On the other hand, plasmin (Pli) reacts very rapidly with $\alpha_2$antiplasmin. $\alpha_2$macroglobulin which also is a fairly rapid plasmin inhibitor interferes, but can be selectively inactivated prior to the addition of plasmin to the sample, by using 0.12 mol/l methylamine (18).

Proenzymes can be measuered after activation. It is favourable to activate plasminogen with an excess of streptokinase because

$$\text{Plasma} + \text{Streptokinase} \longrightarrow \text{Plasminogen} \cdot \text{Streptokinase}$$

$$\text{Peptide-pNA} \xrightarrow{\text{Plasminogen} \cdot \text{Streptokinase}} \text{Peptide} + \text{pNA}$$

$$\text{Plasma} + \text{Prekallikrein Activator} \longrightarrow \text{Kallikrein}$$

$$\text{Peptide-pNA} \xrightarrow{\text{Kallikrein}} \text{Peptide} + \text{pNA}$$

the enzymatic activity generated is not inhibited by antiplasmin (19). In the case of prekallikrein, the activator itself, of phospholipid-ellagic acid type containing both Factor XII and HMW-kininogen, makes the system specific. This powerful activator also allows the high plasma dilution used in this assay. This makes the determination of kallikrein activity possible without obtaining interference from inhibitors present in the plasma (20, 21).

Activators are determined after the addition of an excess of proenzyme. The assay of tissue plasminogen activator (t-PA) has usually been set up as a single stage assay which also makes it easier to determine very small activities of enzyme (22).

$$\text{Plasma} + \text{Plasminogen} \xrightarrow{\text{Fibrin}} \text{Plasmin} + \text{Peptide-pNA} \longrightarrow \text{Peptide} + \text{pNA}$$

Moderators are determined by using all the other reactants in excess. Heparin is complexed by using an excess of anti-thrombin III and this complex inhibits Factor X_a (23, 24). The excess of Factor X_a is then determined. Factor VIII in plasma

$$\text{Plasma with Hep + AT III (excess)} \longrightarrow \text{AT III} \cdot \text{Hep}$$

$$\text{Factor } X_a + \text{AT III} \cdot \text{Hep} \longrightarrow \text{Factor } X_a \cdot \text{AT III + Hep + F } X_a \text{(residual)}$$

$$\text{Peptide-pNA} \xrightarrow{\text{Factor } X_a \text{(residual)}} \text{Peptide + pNA}$$

$$\text{Plasma + Factor } IX_a \text{ + Phospholipid + } Ca^{2+} \longrightarrow \text{Tenase}$$

$$\text{Factor X (excess)} \xrightarrow{\text{Tenase}} \text{Factor } X_a$$

$$\text{Peptide-pNA} \xrightarrow{\text{Factor } X_a} \text{Peptide + pNA}$$

determines the amount of tenase (the activator of factor ten when the other reactants are present in excess. The tenase is determined by its activation of Factor X (25).

A large number of assays have been published so far, most of them are listed in Table 10. Most necessary reagents are commercially available.

Table 10: Available chromogenic substrate assays. Preliminary methods within brackets.

Enzyme standardization:	Thrombin, plasmin, kallikreins, Factor X_a, Urokinase, $C\overline{1s}$, $C\overline{1r}$, chymotrypsin, pancreas elastase, leukocyte elastase, cathepsin G, papain and similar proteases, etc.
Enzyme activity in biological samples:	Plasmin-, kallikrein- and thrombin-like activity in plasma, urokinase in plasma, kallikrein in urine, kallikrein in saliva, bacterial proteases.
Enzyme inhibitors:	Antithrombin III, second heparin cofactor, α_2-antiplasmin, α_2-macroglobulin, C1-esterase inhibitor, t-PA inhibitor, α_1-antitrypsin, antichymotrypsin.
Proenzymes:	Prothrombin, Factor X, prekallikrein, [Factor XII], plasminogen, Factor XI, [C1s], Protein C, Factor VII.
Enzyme activators:	Factor IX_a, Factor VII_a, Factor XII_a, t-PA, Streptokinase, endotoxin.
Moderators:	Factor VIII, Factor V, Platelet factor 3, Heparin, Platelet factor 4.

Automation

Automation is often desirable in clinical chemistry and chromogenic-substrate techniques are suitable for automation. Some equipment is shown in Table 11. Certain assays have been carried out on a large number of automates (26). Such methods are AT III, Factor X and antiplasmin, methods that are rather frequently performed in some laboratories.

Table 11: Various automated instruments used in clinical chemistry.

Centrifugal analyzers:	GEMENI, CentrifiChem 300, 400 & 500, GEMSAEC, Cobas Bio, Multistat III.
Multichannel analyzers:	Olli 3000, Kone.
Random access analyzers:	Clinicon Aurora, Technicon RA-1000, Vitatron XYP, Cobas MIRA.
Other various types:	Kem-o-mat, Abbott VP, 100 & 200, LKB 2086 & 8600, Gilford 3500, Vitatron PA800 & AKES.

Lately microtiter plate techniques have come into use. This technique makes it possible to do as many as 96 determinations or even more in each series of measurement using about 250 µl as total volume.

Clinical applications

A large number of review articles on the use of chromogenic substrates have been published during the last few years (27, 28,29,30,31). Chromogenic substrates are used in investigations of coagulation defects making use of the specificity of these methods. Intensive care units have used chromogenic substrate assays to diagnose the state of the patient with e.g. septicemia (32,33). Measures like transfusions and the use of antibiotics have then been taken at an earlier stage, based on the results of the chromogenic substrate assays. It seems possible to differentiate between essential and renal hypertension by using plasma prekallikrein (34) or urinary kallikrein (35,36) assays. Some of the proteins, like prekallikrein and antithrombin III are decreased in more or less severe disorders of the liver function. Drugs like Heparin, Coumarin, Factor VIII and Streptokinase can be monitored by chromogenic substrate assays (37,38,25,39). Further clinical data are available from recently published supplements to Scandinavian journals (40,5).

The quality of plasma fractions used for substitution therapy is controlled via chromogenic substrate assays - an important application since poor quality plasma fractions could give side effects like shock.

Summary and future aspects

Several research teams, among them ours, have during the last ten years synthesized and distributed a large number of peptide substrates. Their usefulness in research and clinical practice has been documented in a huge number of articles. Diagnostic kits are also available and used in the routine in some laboratories.

Working with these things for more than ten years we can't help being very impressed by the enormous number of proteolytic enzymes present both extra- and intracellularly and the prospects to be able to study these by synthetic substrate techniques. Theoretically being able to determine the activity of most of these enzymes will make it possible to study most types of the protease regulated life processes in man and animals as well as in plants and microorganisms (43,44). After the biochemical research has been performed many diagnostic tests of practical importance for the patient will be developed. The speed and simplicity of the chromogenic substrate assays will be their characteristics.

References

1. Blombäck, B., M. Blombäck, N.J. Gröndahl, E. Holmberg. 1966. Structure of fibrinopeptides - its relation to enzyme specificity and phylogeny and classification of species. Arkiv Kemi 25, 411-416.

2. Blombäck, B., M. Blombäck, P. Olsson, L. Svendsen, G. Åberg. 1969. Synthetic peptides with anticoagulant and vasodilating activity. Scand. J. Clin. lab. Invest. 24, Suppl. 107, 59-66.

3. Svendsen L., B. Blombäck, M. Blombäck, P. Olsson. 1972. Synthetic chromogenic substrates for determination of trypsin, thrombin and thrombin-like enzymes. Thrombosis Res. 1, 267-278.

4. Claeson, G., L. Aurell. 1981. Small synthetic peptides with affinity for proteases in coagulation and fibrinolysis. An overview. In: Annals of New York Academy of Science. Vol. 370. Contribution to Haemostasis. (D.A. Walz & L.E. McCoy, eds.). pp. 798-811.

5. Friberger, P. 1982. Chromogenic Peptide Substrates. Their use for the assay of factors in the fibrinolytic and the plasma kallikrein-kinin systems. Scand. J. Clin. lab. Invest. 42, Suppl. 162, 15-16.

6. Aurell, L., P. Friberger, G. Karlsson, G. Claesson. 1977. A new sensitive and highly specific chromogenic substrate for Factor X_a. Thrombosis Res. 11, 595-609.

7. Aurell, L., R. Simonsson, S. Arielly, G. Karlsson, P. Friberger, G. Claeson. 1978. New chromogenic peptide substrates for Factor X_a. Haemostasis 7, 92-94.

8. Claeson, G., L. Aurell, P. Friberger, S. Gustavsson, G. Karlsson. 1978. Designing of peptide substrates. Different approaches exemplified by new chromogenic substrates for kallikreins and urokinase. Haemostasis 7, 62-68.

9. Friberger, P., M. Knös, S. Gustavsson, L. Aurell, G. Claeson. 1979. A new specific substrate for the determination of plasmin activity. In: Chromogenic Peptides Substrates: Chemistry and Clinical Usage. (M.F. Scully & V.V. Kakkar, eds.). Churchill Livingstone. pp. 121-127.

10. Claeson, G., P. Friberger, M. Knös, E. Eriksson. 1978. Methods for determination of prekallikrein in plasma, glandular kallikrein and urokinase. Haemostasis 7, 76-78.

11. Friberger, P., L. Aurell, G. Claeson. 1982. Chromogenic substrates for kallikreins and related enzymes. In: Agents and Actions Supplements 9. (H. Fritz et al, eds.). pp. 83-90.

12. Friberger, P. 1982. Chromogenic Peptide Substrates. Scand. J. Clin. lab. Invest. 42, Suppl. 162, 17-21.

13. Ibid. pp. 40.

14. Fässler, H., F. Duckert, G.A. Marbet. 1978. Assay with chromogenic substrates of in vivo activated proteases. Haemostasis 7, 158-163.

15. Gyzander, E., A-C. Teger-Nilsson. 1980. Activity of the α_2-macroglobulin - plasmin complex on the plasmin specific substrate H-D-Val-Leu-Lys-p-nitroanilide. Thrombosis Res. 19, 165-175.

16. Friberger, P. 1982. Chromogenic Peptide Substrates. Scand. J. Clin. lab. Invest. 42, Suppl. 162, 33-34.

17. Ødegård, O.R., M. Lie, U. Abildgaard. 1975. Heparin cofactor activity measured with an amidolytic method. Thrombosis Res. 6, 287-294.

18. Matsuda, T., M. Ogawara, R. Miura, T. Seki. 1984. Selective determination of α_2-plasmin inhibitor activity in plasma using chromogenic substrate. Thrombosis Res. 33, 379-388.

19. Friberger, P., M. Knös. 1979. Plasminogen determination in human plasma. In: Chromogenic Peptide Substrates: Chemistry and Clinical Usage. (M.F. Scully & V.V. Kakkar, eds.). Churchill Livingstone. pp. 128-140.

20. Friberger, P., E. Eriksson, S. Gustavsson, G. Claeson. 1979. Determination of prekallikrein in plasma by means of a chromogenic tripeptide substrate for plasma kallikrein. In: Kinins-II. Biochemistry, Pathophysiology and Clinical Aspects. (S. Fujii, H. Moriya & T. Suzuki, eds.). Plenum Publishing Corporation. pp. 67-82.

21. Gallimore, M.J., P. Friberger. 1982. Simpe chromogenic peptide substrate assays for determining prekallikrein, kallikrein inhibition and kallikrein "like" activity in human plasma. Thrombosis Res. 25, 293-298.

22. Rånby, M., P. Wallén. 1982. A sensitive parabolic rate assay for the tissue plasminogen activator. In: Progress in Chemical Fibrinolysis and Thrombolysis. Vol. 5. (J.F. Davidson, I.M. Nilsson & B. Åstedt. eds.). Churchill Livingstone. pp. 233-235.

23. Teien, A.N., M. Lie. 1977. Evaluation of an amidolytic heparin assay method: Increased sensitivity by adding purified antithrombin III. Thrombosis Res. 10, 399-410.

24. Friberger, P. 1982. Properties of the Coatest® Heparin assay. In: Heparin. New Biochemical and Medical Aspects (I. Witt, ed.). Walter de Gruyter. pp. 117-124.

25. Rosén, S. 1984. Assay of Factor VIII:C with a chromogenic substrate. Scand. J. Haemotol. 33, Suppl. 40, 139-145.

26. Friberger, P. 1983. Synthetic peptide substrate assays in coagulation and fibrinolysis and their application on automates. Seminars in Thrombosis and Hemostasis. 9, 281-300.

27. Blombäck, M. 1981. Chromogenic substrates in the laboratory diagnosis of clotting disorders. In: Haemostasis and Thrombosis. (A.L. Bloom & D.P. Thomas, eds.). Churchill Livingstone. pp. 809-823.

28. Fareed, J., H.L. Messmore, E.W. Bermes. 1980. New perspectives in coagulation testing. Clin. Chem. 26, 1380--1391.

29. Huseby, R.M., R.E. Smith. 1980. Synthetic oligopeptide substrates: Their diagnostic application in blood coagulation, fibrinolysis, and other pathologic states. Seminars in Thrombosis and Hemostasis VI, 173-315.

30. Latallo, Z.S. 1980. Chromogenic substrates. Chapter 11 in: Recent Advances in Blood Coagulation. Vol. 3. Churchill Livingstone.

31. Triplett, D.A. 1981. New methods in coagulation XIV-XVI. In: CRC Critical Reviews in Clinical Laboratory Sciences. pp. 55-84.

32. Aasen, A.O., N. Smith-Erichsen, M.J. Gallimore, E. Amundsen. 1980. Studies on components of the plasma kallikrein-kinin system in plasma samples from normal individuals and patients with septic shock. Adv. Shock Res. 4, 1-10.

33. Smith-Erichsen, N., A.O. Aasen. 1984. Evaluation of severity and prognosis in early stages of septicemia by means of chromogenic peptide substrate assays. Eur. surg. Res. 16, Supp. 2, 140-146.

34. Bruhn, H.D., J.P. Albert, A. Britz, M. Brüch, F. Hartmann, W. Niedermayer. 1982. Plasma-Präkallikrein-Spiegel bei Hypertonie. Deutsche Med. Wschr. 107, 797.

35. Amundsen, E., J. Pütter, P. Friberger, M. Knös, M. Larsbraten, G. Claeson. 1979. Methods for the determination of glandular kallikrein by means of a chromogenic tripeptide substrate. In: Kinins-II. Biochemistry, Pathophysiology, and Clinical Aspects. (S. Fujii, H. Moriya & T. Suzuki, eds.). Plenum Publishing Corporation. pp. 83-95.

36. Overlack, A., K.O. Stumpe, C. Ressel, F. Krück. 1979. Low urinary kallikrein excretion and elevated blood pressure normalized by orally applied kallikrein in essential hypertension. Clinical Science 57, 2635-2655.

37. Holm, H.A., U. Abildgaard. 1982. Heparin treatment of deep venous thrombosis: Correlation between clinical effects and heparin assays. In: Heparin. New Biochemical and Medical Aspects (I. Witt, ed.). Walter de Gruyter. pp. 289-297.

38. van Wijk, E.M., L.H. Kahlé, J.W. tenCate. 1981. Mechanized amidolytic technique for determination of Factor X and Factor X antigen, and its application to patients being treated with oral anticoagulants. Clin. Chem. 27, 918-921.

39. Latallo, Z.S., E. Teisseyre, S. Lopacink. 1978. Assessment of plasma fibrinolytic system with use of chromogenic substrate. Haemostasis 7, 150-154.

40. Aasen, A.O., E. Amundsen, P. Kierulf (Eds.). 1982. Significance of Chromogenic Peptide Substrate Assays in Critical Care Medicine. Acta Chir. Scand. (Suppl.) 509.

41. Alving, B.M., D.L. Tankersley, B.L. Mason, F. Rossi, D.L. Aronson, J.S. Finlayson. 1980. Contact activated factors: Contaminants of immunoglobulin preparations with coagulant and vasoactive properties. J. Lab. Clin. Med. 96, 334- 346.

42. Tölö, H., H. Suomela. 1982. An optimized assay for prekallikrein activator activity in human plasma products. Thrombosis Res. 27, 35-44.

43. Berdal, B.P., K. Bøvre, Ø. Olsvik, T. Omland. 1983. Patterns of extracellular proline-specific endopeptidases in Legionella and Flavobacterium spp. demonstrated by use of chromogenic peptides. J. Clin. Microbiol. 17, 970-974.

44. Øgaard, A.R., B.P. Berdal, K. Bøvre. 1984. Variation of endopeptidase activities in cultures of Pseudomonas aeruginosa strains. Acta path. microbiol. immunol. Scand. Sect B. 92, 31-37.

IMMOBILIZED METAL ION AFFINITY CHROMATOGRAPHY - A POWERFUL METHOD FOR PROTEIN PURIFICATION

Jerker Porath
Institute of Biochemistry, Biomedical Center, Uppsala University, Box 576, S-751 23 UPPSALA, SWEDEN

In 1961 Helfferich suggested the use of coordinative complexation between metal ions bound to ion exchangers and amines in the surrounding solution as a method to fractionate mixtures of organic amines (1). The technique has been used as "ligand exchange chromatography", the name Helfferich suggested.

Around 1974 I considered to combine the principles of metal ion complexation of proteins with the chromatographic procedures based on salt-promoted fractionation (2) and molecular sieving in hydrophilic gels such as cross-linked dextran Sephadex (3) or cross-linked agarose (4). Application of metal ion affinity as a separation parameter for proteins presents practical (and theoretical) problems of a much higher order of complexity than adsorption-desorption of small size amines from metal-loaded ion exchangers. Therefore, the published work on ligand exchange chromatography did not provide much guidance.

In the first publication (5), serum proteins were separated on a column of metal ions fixed via chelating groups (derived from imino-diacetic acid). The method was called "metal chelate chromatography". The differential affinities of the proteins for two kinds of metal ions were demonstrated. The column consisted of two consecutive beds of immobilized ions of zinc and copper, respectively. Dialyzed serum was introduced at neutral pH and at an ionic strength high enough to suppress simple charge-charge interaction. A group separation took place: some proteins became adsorbed to

Modern Methods in Protein Chemistry, Volume 2

the upper (Zn) column, others to the lower (Cu) bed and some passed through both columns unretarded. After dismantling the tandem column the adsorbed proteins were desorbed by using an eluent of lower pH. High recovery of undamaged proteins and an acceptable group separation gave impetus for the further developmental work that is still being persued.

The method has found limited application (6) but the inherent power of the protein-metal-affinity methods has not yet received the attention it deserves, a fact that prompted us to study further, in more detail, the requirements for successful and easy application. We have renamed the technique, for reasons we have given elsewhere (7), Immobilized Metal ion Affinity chromatography (IMA-chromatography).

The basic requisites for IMA-adsorption of proteins and other biopolymers are the following:

1. A suitable inert hydrophilic solid support.
2. Methods to functionalize the support.
3. Stable and selective metal-ion bonding ligands.
4. Reversible complexation of proteins with immobilized metal ions under conditions avoiding dissociation of the latter from the matrix-fixed chelating ligand.
5. Elimination or suppression of steric hindrance for the metal ion to interact with the protein(s) to be purified, effected by the use of gel of sufficiently high porosity and, if necessary, by attaching the ligand via a long, hydrophilic "spacer arm" or "leash".

IMA-adsorption is a special case of "sandwich-type" adsorption:

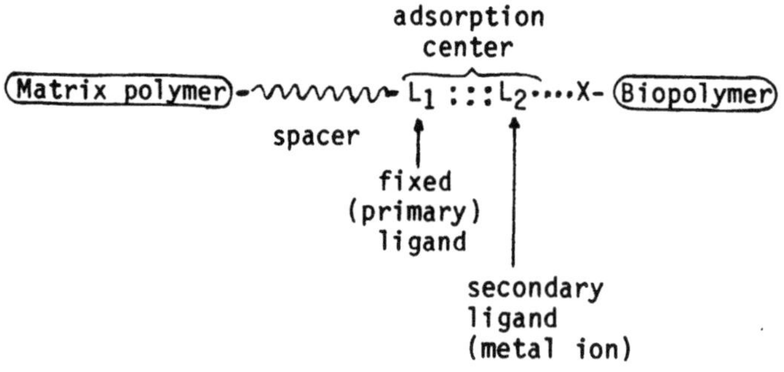

$L_1:::L_2$ bonding is much stronger than $L_2\cdots X$ to prevent metal leakage.

The simplest effective adsorbents we have used for IMA-chromatography have generalized formulae shown in Fig. 1.

"IDA"-ADSORBENT:

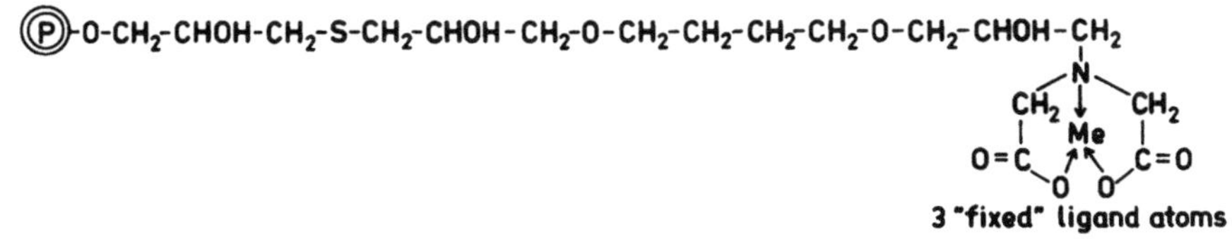

"TED"-ADSORBENT:

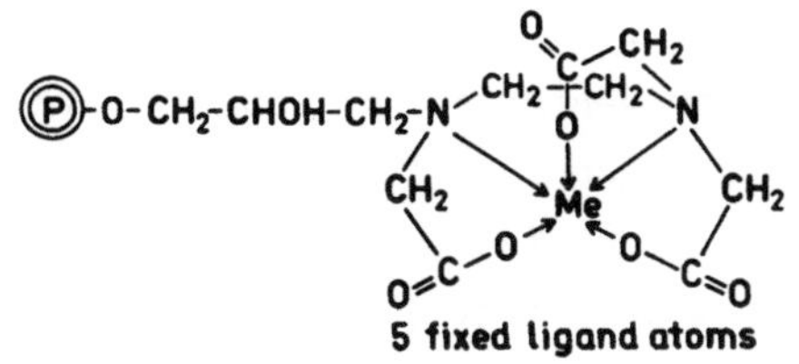

Fig. 1. Tentative structures of immobilized hexacoordinate metal complexes of IDA- and TED-adsorbents with long and a short spacer arm, respectively.

In the past only the IDA-function was used for metal chelating adsorbents. However, we found recently and not unexpectedly that other chelone-type gels (7,8) have selective adsorption properties different from those of the IDA-gels. These chelone gels may most conveniently be synthesized via amines using carboxymethylation under conditions mild enough to avoid attack on the matrix itself (9). Apart from chelones a number of other types of metal chelate forming gels have been synthesized and tested.

Metal fixation of course follows the mass-action law:

$$Me^{n+} + L \rightleftarrows Me^{n+}L \qquad 1)$$

but is complicated, kinetically and thermodynamically, by the heterogeneity of L-species due to steric hindrance and also because the metal ion interacts with more ligand species, free or bound. Thus K-values of analogous reactions in free solution can, at least, only serve as rough estimates of metal ion-ligand affinities.

Protein adsorption in the simplest terms may be visualized by a consecutive complexation reaction:

$$L:::Me + X\text{-}\text{(Protein)} \rightleftarrows L:::Me\cdots X\text{-}\text{(Protein)} \qquad 2)$$

This is an oversimplification for two reasons: both interacting species are of heterogenous nature, the immobilized metal ion for reasons just given and the protein because it may have several similar or different ligand atoms X.

IMA-chromatography, in practice, requires that reaction 1 goes virtually to completion under conditions that allows dissociation of the adsorbate to occur, i.e., quantitative elution of the protein. The metal ion must be so efficiently bound that its leakage is prevented and this usually requires a high density of the matrix-fixed ligand. On the other hand, if all coordination sites of the metal ion is occupied by stably bound ligand atoms there will be no exchange possible with potentially competent liganding solutes and the adsorption capacity will be naught. We have found that some immobilized copper cryptates constitute such "saturated" chelates. In other cases chelates, formally saturated with immobilized ligands adsorb proteins. Apparently, one or more of the matrix-fixed ligand atoms may be temporarily dissociated making competition with solute ligands possible. Histidine peptides, for example, can be retained on Cu-carboxymethylated tetraethylenepentamine-Sephadex.

Usually a fixed ligand is selected with lower "dentatity" than the coordination number of the metal ion. Then, obviously, there will exist an inverse relationship between the number of fixed atoms liganding to the metal ion and the power of the metal to bind solutes from the surrounding medium. An IDA-gel, therefore, binds nickel less strongly than does a TED-gel, whereas, on the other hand, the former gel adsorbs more proteins than does the latter (due to difference in available sites and sterical hindrance).

Ligand atoms show varying nucleophilicity. Further contribution to the adsorption site heterogeneity can be ascribed to varying extents of ligation (leading to formation of adsorption complexes, more or less saturated) where the metal ion may share one, two or possibly, but perhaps not likely, even more ligands located at different positions in the space of the matrix, the latter consequently being cross-linked by the metal ion. A decrease in heterogeneity is desirable in order to avoid leakage and at the same time to increase the selectivity in sorption and in particular in the desorption steps, which in a favorable case can make sharp isocratic elution possible.

A reduction in heterogeneity can be accomplished by washing with a solvent containing suitable chelators of moderate strength (e.g. glycine) to remove the weakest bound metal ions. In doing so the residual metal ions will remain quantitatively fixed in the gel phase in the solvent systems to be used for protein separation. This is easily checked since a gel region with such metal ions will be separated from metal-free gel by a sharp boundary (visible in case of colored ions such as Co^{2+}, Ni^{2+}, Cu^{2+} and Fe^{3+}) which will remain stationary when washing with the solvent and under proper condition also when proteins are passing.

Occasionally a protein may pick up a metal ion - an undesirable event which usually can be counteracted by using a metal free trapping bed of the same or a stronger metal ion chelating adsorbent. A metal stripping bed may be advisable whenever a sample or product of negligible metal content is desired. Incidently, a metal free chelator gel may be used to remove traces of heavy metals from a solution containing nutrients to such an extent that bacterial growth will be entirely prevented - a feature of the technique that make IMA-methods extremely promising for large scale applications.

Another unique advantage of IMA-techniques is indeed worth mentioning. The adsorption center itself can be reversibly removed from the solid support: by displacing the metal ion! However, this expedient is not commonly required to regenerate the adsorbent. In fact, for plasma fractionation we have recycled composite IMA-beds over periods of months or even years without losing much capacity or selectivity.

The access to proper hydrophilic gels is an imperative requirement for IMA-methods. Cross-linked agarose and dextran have been used most extensively, but a variety of other materials of different shapes such as polyacrylamide gels, starch grains, cellulose powder, paper, fibers, and membranes and even spongy polymers have been converted to metal adsorbents by proper matrix functionalization. Some of these products may find other kinds of applications besides analytical, e.g. as supports in cell culture and immobilization of enzymes, antigens and antibodies and possibly also for nucleic acids.

Selectivity is dependent on geometrical factors such as the spatial distribution of the matrix polymer, the length of spacer arm and the bulkiness of the ligand as is the case for other affinity methods (Fig. 1).

The metal ion acts as a super Lewis acid, albeit restricted in its ability to interact with approaching nucleophiles by the modifying influence of the matrix-fixed chelator. The latter as well as the nature of the metal governs the selectivity of the adsorbent.

IMA-adsorption of proteins also resembles hydrophobic and charge transfer (aromatic) type of adsorption in the sense that they all respond similarly towards the addition of salts at high concentration. Adsorption is thus promoted or reduced by antichaotropic or chaotropic salts, respectively. However, the adsorption onto the different types of gels differs in qualitative respects, and the variable influence of salts such as those of alkali, alkaline earth metals or of chlorides, sulfates etc. do not easily arrange themselves in the order of known Hofmeister series. Presumably, disordering of water-structure upon adsorption accounts for most of the entropic gain that drives the interaction toward enhanced IMA-complexation.

Besides salts, other additives to the buffer may change the adsorption characteristics. While ionic detergents suppress adsorption, neutral surfactants have little or no effect, which is advantageous in application of fractionation to membrane proteins. Addition of metal complexing agents, especially chelators (amino acids, for example) counteracts adsorption selectively and may therefore be used to bring about selective or complete

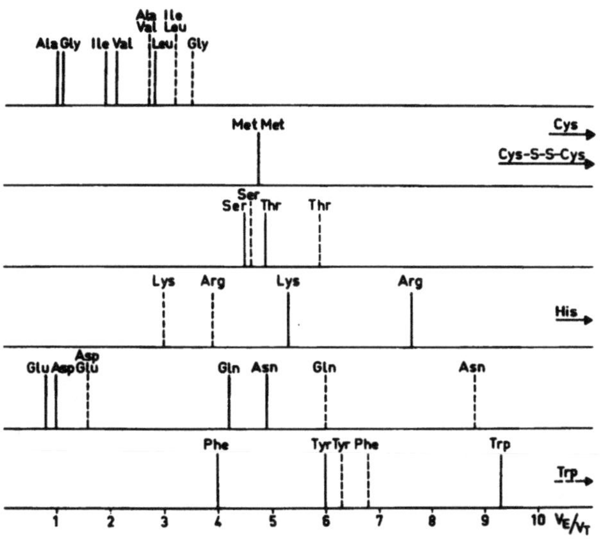

Fig. 2. Retention spectra for amino acids on Ni-IDA-Sephadex G-25 in 0.2 M ethylmorpholine-acetate buffer, pH 7.0. Solid bars and dashed bars indicate the relative retention values (V_e/V_t = elution volume/total volume) in the absence and presence of 0.5 M potassium sulphate, respectively. E. S. Hemdan and J. Porath (J. Chromatogr., in press).

desorption while keeping the pH, reduction potential etc. of the medium constant (Fig. 2).

The potential of IMAC goes far beyond fractionation of proteins but, as a prerequisite for its exploration simpler models should be studied. Chromatography of amino acids and oligopeptides has been undertaken in order to determine which amino acid side groups are chiefly involved in the IMA-adsorption (9). The observation of a strong affinity between some immobilized transition metal ions and the amino acids histidine and cysteine did not come as a surprise, but we also found that the amino acids exhibited graded affinities toward the metal ions which were dependent on the environment: pH-, salt-, and temperature-graded affinity is likely to be reflected in the IMA-properties of proteins. Figure 3 illustrates a chromatographic group fractionation of serum proteins by the use of various affinity modulating parameters.

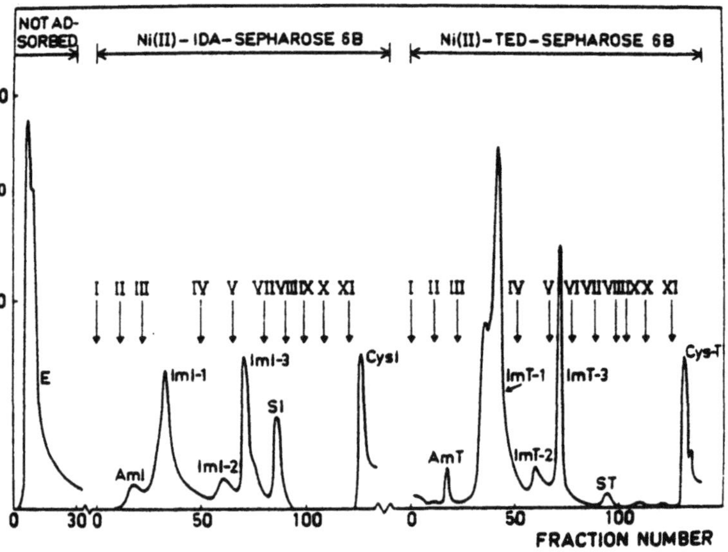

Fig. 3. Chromatographic group fractionation using various affinity modulating parameters: "affinity eluents", ionic strength and pH. The fractionation of 4 ml of serum was performed on a tandem column consisting of two 10 ml beds of Ni-TED- and Ni-IDA-Sepharose 6B coupled in series. Dialyzed serum was introduced in a 0.1 M Tris-HCl, pH 8.1, containing 1 M Na_2SO_4. After washing and collecting the proteins passing the column (far left peak), the column was dismantled and each section separately eluted. The Roman figures refer to eluent changes involving, in sequence, addition to the Tris buffer of 0.6 M $(NH_4)_2SO_4$, introduction of imidazole (0.006, 0.01 and 0.04 M), addition of histidine, removal of Na_2SO_4, change of pH to 5.5, addition of acetate buffer, and finally addition of cysteine. An efficient group separation was achieved as checked by gradient gel electrophoresis. For further details see (7).

The polymeric surrounding modifies the ability of a metal ion to interact with solute species and, likewise, more weakly interacting adjacent to imidazole or thiol contribute to the affinity of the protein-surface-located adsorption center. The cooperative interaction is revealed and is usually more pronounced in simple solutes such as in the series: histidine, histamine, imidazole or histidine oligopeptides.

These cooperative interactions as well as the minor affinity effects revealed by differences in the relative retention of solutes on IMA-columns are frequently so weak as to elude discovery by commonly used physico-chemical methods. IMA-chromatography is indeed a powerful tool for exploring weak unknown metal-biopolymer interactions, since the differences in distribution coefficients are enormously multiplied as a consequence of the continuous oscillation of the solute molecules between the moving and stationary phases.

In neutral and near neutral solutions the metal ions Co^{2+}, Ni^{2+}, Cu^{2+}, Zn^{2+} all have outstanding preferences for coordination complexation with imidazole and thiol groups on the molecular surface of proteins. Eugene Sulkowski and collaborators have shown that IMA-chromatography can be used for the study of protein topography with respect to the presence of surface located histidine residues (11,12). By the use of model proteins Sulkowski and collaborators have recently shown, in my laboratory, that it is possible to group proteins into categories according to their contents of 0, 1, 2 or 3 surface-located imidazoles. Maybe this kind of mapping can be extended to other chemical groups, for example, phosphate in phosphoproteins and sialic acid in glycoproteins (13).

Apart from IMA-chromatography on the above-mentioned metals, exploratory work with Cd^{2+}, Hg^{2+} (8), Fe^{3+} (7,13), Al^{3+}, Ga^{3+}, and In^{3+} (14) has also been published. We have also done some orientative experiments with lanthanides, Th^{4+} and UO_2^{+2} which indicate interesting possibilities to extend IMA-adsorption beyond the present boundaries. B.C. Furie and B. Furie have already made some pioneering studies on lanthanide-based IMA-chromatography (15) and have successfully purified snake venom coagulants and clostridopeptidase (16). Some of our recently developed adsorbents should be even more powerful for IMA-chromatography of Ca^{2+} and Mg^{2+} dependent proteins.

So far we have studied extracellular proteins in particular. Such proteins usually do not contain free SH-groups. We are aware that the application of IMA-methods to thiol-group containing proteins may meet problems not hitherto encountered. Thioldisulfide exchange and catalytic oxidation may have to be prevented and special modification of our present simple technique may be necessary. However, exploratory work looks promising.

No damage to serum proteins has been observed when chromatography is performed at or below room temperature. However, the IMA-behaviour of peptides and proteins at elevated temperatures remains to be studied more extensively. This is the more important in view of the risk for catalytic hydrolysis of peptides in the presence of certain heavy metal ions.

Comparing chromatography of proteins based on ion-ion interaction (ion exchange) on the one hand and coordinate covalent bonding on the other we may infer that, assuming heterogeneity in binding sites is kept at a minimum, isocratic IMA-chromatography should be more easily realized. This hypothesis has already been put to test in our laboratory and was found to hold true (unpublished results). Consequently, IMA-chromatography is one of the most powerful affinity methods for protein purification advanced so far.

Although outside the scope of this review, it may be mentioned that IMA-methods might be used to fractionate, analytically and preparatively, other biomaterials, e.g. nucleic acids. Hydrophilic IMA-adsorbents may also be used to concentrate heavy metal ions from aqueous solutions such as industrial waste water and brine, but for this purpose cheaper matrix materials have to be developed.

The reader interested in the application of IMA-chromatography or the publications of Davankov, Karger, Grushka and others on ligand exchange chromatography of amino acids should consult the reference lists given in the reviews of Sulkowski (12), Porath and Belew (17) and the doctoral thesis of Hemdan (18) and Ramadan (19).

Acknowledgements

The exploratory work on IMA-chromatography has been supported by The Swedish Natural Science Research Council, The Swedish Work Environment Fund, Alice and Knut Wallenberg Foundation, Erna and Victor Hasselblad

Foundation and The Swedish Board for Technical Development.

References

1. Helfferich, F.G.: Nature (London) 189, 1001-1002 (1961).
2. Tiselius, A.: Arkiv Kem. Min. Geol. 26B, No. 1 (1948).
3. Porath, J., Flodin, P.: Nature (London) 183, 1657-1659 (1959).
4. Porath, J., Janson, J.-C., Låås, T.: J. Chromatogr. 103, 49-62 (1975).
5. Porath, J., Carlsson, J., Olsson, I., Belfrage, G.: Nature (London) 258, 598-599 (1975).
6. Lönnerdal, B., Keen, C.L.: J. Appl. Biochem. 4, 203-208 (1982).
7. Porath, J., Olin, B.: Biochemistry 22, 1621-1630 (1983).
8. Andersson, L.: J. Chromatogr. 315, 167-174 (1984).
9. Hemdan, E.S., Porath, J.: J. Chromatogr. (1984) in press.
10. Hemdan, E.S., Porath, J.: J. Chromatogr. (1984) in press.
11. Sulkowski, E., Vastola, K., Oleszek, D., von Muenchhausen, W.: In: Affinity Chromatography and Related Techniques (Gribnau, T.C.J., Visser, J., Nivard, R.J.F., Eds.) Elsevier, Amsterdam 1982, pp. 313-322.
12. Sulkowski, E.: Trends in Biotechnology (1984) in press.
13. Ramadan, N., Porath, J.: To be published.
14. Porath, J., Olin, B., Granstrand, B.: Arch. Biochem. Biophys. 225, 543-547 (1983).
15. Furie, B.C., Furie, B.: J. Biol. Chem. 250, 601-608 (1975).
16. Evans, C.H., Mason, G.C.: Inorg. Chem. Acta 79, 274-275 (1983).
17. Porath, J., Belew, M.: In: Affinity Chromatography and Biochemical Recognition (Chaiken, I.M., Wilchek, M., Parikh, I., Eds.) Academic Press, Orlando etc. 1983, pp. 173-189.
18. Hemdan, E.S.: Acta Univ. Upsaliensis, Abstr. of Uppsala Dissertations from the Faculty of Science 759 (1984).
19. Ramadan, A.N.: Acta Univ. Upsaliensis, Abstr. of Uppsala Dissertations from the Faculty of Science 760 (1984).

USE OF SYNTHETIC DYES FOR AFFINITY CHROMATOGRAPHY OF PROTEINS

Günter A.P. Eiteljörge
Amicon GmbH, Westfalenstraße 11, D-5810 Witten (Ruhr)

Scott P. Fulton
Amicon Corporation, 25 Hartwell Avenue, Lexington, MA. 02173

Since its introduction in the late 1960's, the technique of affinity chromatography has proven to be a very successful and versatile method of purifying proteins for research purposes (1) . The high specificity of interaction between the column-immobilized substrates, cofactors, inhibitors, antigens, etc., and their corresponding enzymes, antibodies, or other binding proteins makes affinity chromatography a very useful addition or even replacement for conventional separation methods like salt precipitation, gel filtration, or ion exchange chromatography. The widespread use of affinity methods has made some one-step purifications possible maintaining a high yield of activity <u>and</u> a protein of complete homogeneity.

In spite of its many demonstrated advantages on laboratory scale, affinity chromatography has not found use to any appreciable extent in large-scale protein purifications until now. The reasons for this are primarily related to cost. Affinity media cost 10 to 100 times as much as conventional media to purchase initially because the ligand itself must first be isolated and extensively purified, then somehow covalently coupled to the gel, which can involve difficult organic synthesis. Also protein binding capacity is often low, making it necessary to use larger columns. Affinity columns are usually specialized in their applications, whereas other gel materials can be used for a wide variety of different

Modern Methods in Protein Chemistry, Volume 2

proteins and purifications. Finally, these immobilized ligands are biomolecules and so often subject to enzymatic or chemical degradation, which limits the ability of regeneration and reuse of the material, further adding to the cost.
One approach to solve these problems is to replace the biomolecule, the "natural" ligand, by synthetic chemical ligands.

Synthetic Ligands - Triazine Dyes

The first synthetic chemical ligand of widespread use in affinity chromatography was the dye Cibacron Blue F3GA (Ciba-Geigy, hereafter referred to as blue A). This material was originally developed as a textile dye and has been used extensively in the manufacture of blue jeans. It was introduced into biochemistry as a chromophore coupled to high molecular weight dextrane serving as a void volume marker in gel filtration. In 1968, while doing gel filtration of yeast pyruvate kinase, Haeckel et al. (2) attempted to determine the void fraction of the column simultaneously with the K_{av} of the pyruvate kinase by mixing the protein with blue dextrane void marker and applying the mixture to a column with Sephadex G-200. Surprisingly, the pyruvate kinase eluted in the void fraction together with blue dextrane. Further investigation showed that the enzyme was bound by the blue A and not by the dextrane. Blue dextrane and protein could be separated by gel filtration in a buffer of high ionic strength. This procedure gave a 3x purification of the enzyme.

Over the next several years a number of other enzymes were purified by use of blue dextrane. In 1973, Ryan and Vestling (3) published a method covalently coupling blue dextrane to agarose to form a true affinity medium. They used this material for the purification of various lactate dehydrogenases (LDH). In 1974, Heyns and DeMoor (4) were able to couple the blue A dye directly to agarose, without dextrane bridge.

This affinity material gave a 220fold purification of 3(17)-ß-hydroxy-steroid dehydrogenase from rat erythrocytes with 50% yield in one single step.

The use of the blue A dye-ligand for protein purification then began to grow rapidly. Now, some hundred purifications are published using this ligand, with novel uses appearing constantly (5,6). Many of these applications have involved dehydrogenase and kinase enzymes, but a wide range of other proteins have been purified too.

Table 1. Blue A Applications

NAD, NADP enzymes (>50)
ATP enzymes (>20)
Nucleic acid binding enzymes (>15)
CoA enzymes (>7)

Acyl-acyl carrier protein synthetase
Aryl sulfate sulfhydrolase
Biphosphoglycerate synthetase
Carbamyl phosphate synthetase I
Fructose diphosphatase, Glyoxalase I and II
Cyclic nucleotide phosphodiesterase
Lecithin : cholesterol acyltransferase
γ-glutamyl hydrolase, R-enzyme
Orotate phosphoribosyltransferase
Orotodylate decarboxylase
Phenol sulfotransferase, Prolyl hydrolase
Phosphoserine aminotransferase
Phosphodiesterase, ß-lactamase
Ca-sensitive ATP-dihydrophosphorylase
Phospholipase A_2, Murein transglycosylase
Uronosyl C-5 epimerase, Phosphoglycerate mutase

α-fetoprotein
Blood coagulation and complement factors
P. phosphoreum blue-fluorescent protein
Ferrocytochrome C
1,25-hydroxyvitamin D receptor
Inactive renin
Interferon (leukocyte, fibroblast, lymphoblastoid)
Myosin, serum albumin, sterol carrier protein

The apparently high specificity of binding observed between the blue A dye-ligand and certain proteins has prompted investigators to screen other reactive dyes for various applications hoping that dye-ligands of improved selectivity might be found. Baird et al. (7) found Procion Red HE-3B (ICI, hereafter referred to as red A) to be outstanding for carboxypeptidase G and NADP-requiring dehydrogenases purifications. Watson et al. (8) found that red A was much more specific in binding NADP versus NAD^+ enzymes than blue A. Later work has shown that red A is especially effective for alkaline phosphatase, dihydrofolate reductase, 5,10-methylene-tetrahydrofolate reductase, hexokinase, some tRNA synthetases, and plasminogen.

Recently the Amicon Corp., after extensive dye-ligand screening, introduced three new triazinyl dye-ligands with different properties.

Fig. 1

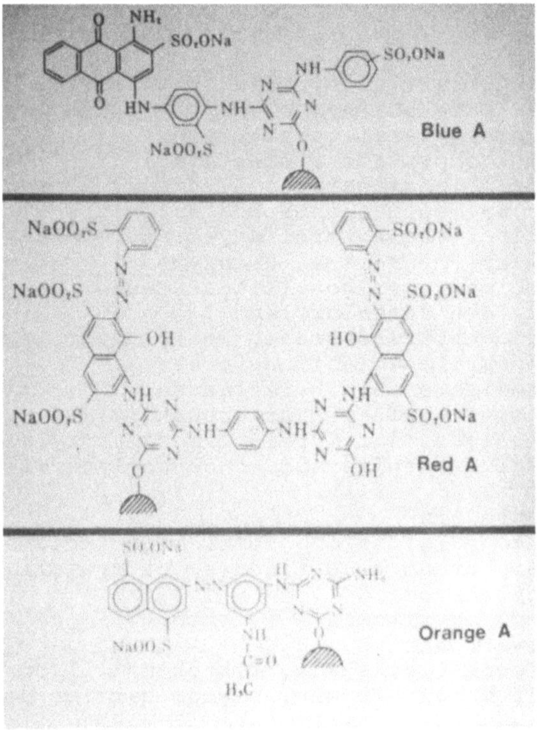

Orange A is unusual, because of the large number of proteins which do not bind in contrast with the other dye-ligands. Proteins which do bind, such as carnitine acetyltransferase, chondronectin, citrate synthetase, some complement factors, lactate dehydrogenase, and phosphoglycerate kinase, show outstanding purification factors and recoveries.
The green A dye-ligand is the opposite - more proteins typically bind, more tightly, than to the blue A. This gel material is particularly useful in removing contaminants from non-binding proteins in "negative chromatography".
The blue B dye-ligand has the bulky copper phthalocyanine chromophore and a much lower concentration of chromophore coupled to the gel. This results in weaker binding properties in general, allowing good recovery of proteins which bind too tightly to the other gels, for example pregnancy-specific β-globulin, adenylate kinase, citrate synthetase, and malate dehydrogenase.
All of the dye-ligands used to date consist of aromatic hydrocarbons substituted with sulfonates, amines, hydroxyls, and other polar groups, joined together by secondary amines or diazo bonds, with coupling to the gel matrix, crosslinked 5% agarose beads, most commonly through an ether linkage to a triazine ring, originally substituted with one or two active chlorines (9).
They are available as Matrex-Gels in a so called Dyematrex Kit for rapid screening of the five dye-ligands for any particular purification problem.

Considerable work was done studying the mechanism of action of the triazinyl dye-ligands. Electrophoresis, enzyme inhibition, bind equilibrium, absorbance and circular dichroism spectral, NMR, thermal and X-ray crystallographic studies with a variety of proteins (5) have shown that although there is considerable variation with different dye-protein systems, in many cases the dye-ligand binds to the substrate or cofactor binding site like the natural ligand. Figure 2 shows the remarkable similarity between NAD^+ (left side) and blue A. In particular, the sulfonate groups of the dye-ligand occupy similar positions to the phosphate groups of the NAD^+, and there is considerable similarity in the alignment of the π-electron planes. The structural flexibility of the dye-ligands clearly contributes the ability to bind in a wide variety of protein binding sites for small, anionic biomolecules.

Fig. 2

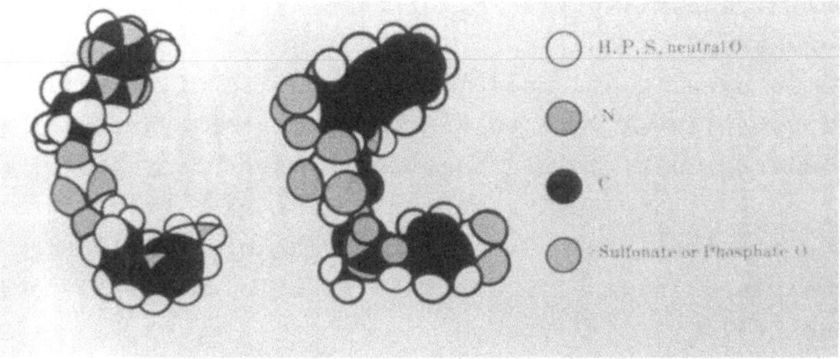

Early in the development of the blue A dye-ligand, Thompson et al. (10) proposed that the molecule was specific in binding proteins containing the so-called "dinucleotide fold", a nucleotide-binding domain found in a large number of enzymes, for example dehydrogenases and kinases. Certainly most or all of the proteins known to have this structure have been found to bind to blue A. However, so many other proteins bind that do not have the "dinucleotide fold", like serum albumin, cytochrome C, ovalbumin, that binding to the blue A is more a hint than a specific probe for this structural feature (11). The binding interaction of the dye-ligands and proteins can be very specific (sometimes a small modification of the dye molecule can prevent binding), but the degree of specificity appears to be on an individual protein-by-protein, dye-by-dye basis. As a conclusion, the use of these ligands is, and probably will continue to be, on a highly empirical basis.

In use, the dye-ligand media show great versatility and adaptability. A simple screening technique gives the best dye-ligand for the particular protein purification problem. The purification procedure may then be optimized by carefully selecting the best binding conditions and elution technique. It should give significant improvements in yield and purification factor.

Significant binding conditions include pH, ionic strength, protein load and concentration, and the presence or absence of divalent cations. Binding to the ligand normally occurs over a wide range of these parameters so that protein stability can be a more important consideration than the binding conditions. In addition, most buffer salts and additives, used to improve recovery of activity, like mercaptoethanol, DTE or glycerol, have little or no effect on the binding to the dye-ligand.

Protein binding capacity of the dye-ligand gels is in the range of 5 to 20mg protein/ml gel, depending upon the protein and the extract in which it is contained. Often competitive binding effects are important so that prior purification steps are of great influence. For reasons not entirely understood, dye-ligand gels often have binding capacities 5 to 100 times those

Fig. 3

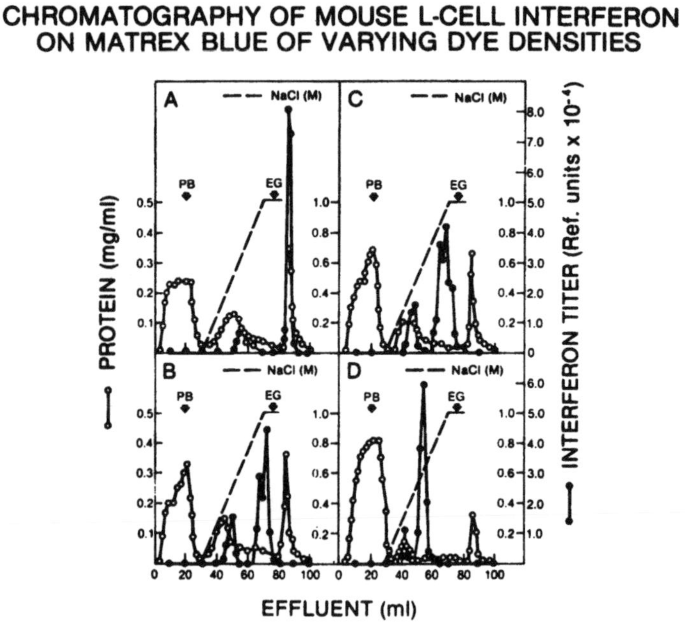

of equivalent natural ligand media. One reason may be that natural ligands, to some extent, could have lost their integrity and acceptability during binding procedures to the gel matrix or even during storage.

Elution of bound protein can be achieved by increasing the ionic strength, or introducing low (1 to 20mM) concentrations of specific eluents (substrates, cofactors, inhibitors, etc.), or using chaotropic agents, thiocyanate, guanidine-HCl or ethylene glycol. Sometimes a combination of the different types of eluents give even a better purification factor or allows the use of lower concentrations of the expensive specific eluent. Specific eluents usually give the highest purification factor, but finding the right eluent or combination of eluents can be difficult. Salt elution (0 to 2M is usually sufficient) is often more cost-effective and can give excellent results, especially with a gradient elution. Chaotropic eluents are necessary when hydrophobic interactions between the dye-ligands and protein predominate and other elution methods fail. Other useful techniques one should try include rechromatography, with different elution methods, and "negative chromatography", in which the protein of interest does not bind but the contaminants do. Tandem negative-positive column systems have also been very useful in a number of cases.

The dye-ligand gels are extremely chemically resistant and can be effectively regenerated in situ with a 6 to 8M urea/0.5M NaOH solution. Lifetimes of 30 or more runs are not uncommon and, with proper care, a column can last well over a year. If very impure protein samples are used, with high content of lipids or lipoproteins, the capacity and binding strength may drop off sharply after only one run. Usually a preliminary precipitation or solvent extraction step may prevent this. Another possibility is to restore the column by washing it with a chloroform/methanol solution.

An example of the use of dye-ligand chromatography to solve a typical problem is the separation of rabbit muscle pyruvate kinase (PK, EC. 2.7.1.40) from lactate dehydrogenase (LDH,EC. 1.1.1.27), and the simultaneous purification of both enzymes. Both enzymes are of commercial interest and are often used together in a coupled assay system. Because of the coupled reactions, it is critical to produce each enzyme substantially free of contamination by the other. Figure 4 shows the results of a screening study for the enzymes with five dye-ligand agarose media (Matrex™ gels, Amicon). Each 2ml column was equilibrated with the starting buffer (20mM Tris/HCl, pH 7.5). A 0.5ml aliquot containing the amounts of protein shown was applied to each column. The columns were then washed with

10ml of starting buffer, followed by 10ml of elution buffer (1.5M KCl in starting buffer). The sample wash and elution fractions for each column were assayed for total protein, PK, and LDH. Enzymes recovered in each fraction are expressed as percent of the sample applied. The control column contains gel with no dye-ligand.

Fig. 4

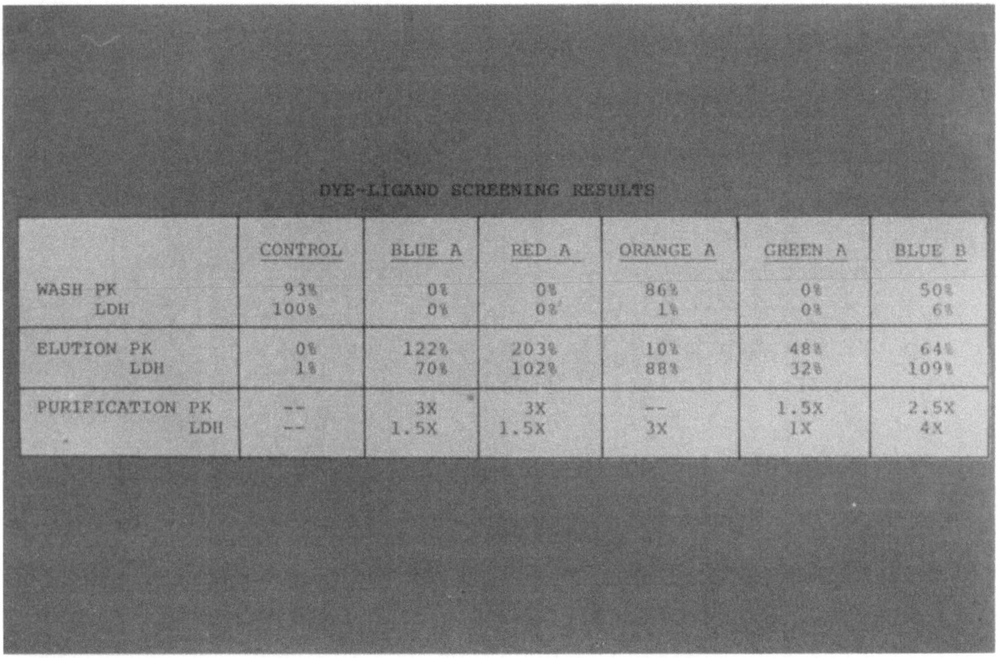

DYE-LIGAND SCREENING RESULTS

	CONTROL	BLUE A	RED A	ORANGE A	GREEN A	BLUE B
WASH PK	93%	0%	0%	86%	0%	50%
LDH	100%	0%	0%	1%	0%	6%
ELUTION PK	0%	122%	203%	10%	48%	64%
LDH	1%	70%	102%	88%	32%	109%
PURIFICATION PK	--	3X	3X	--	1.5X	2.5X
LDH	--	1.5X	1.5X	3X	1X	4X

PK bound completely to the blue A, red A, and green A columns, partially to the blue B column, and very little to the orange A column. Elution of the PK was complete only with the blue A and red A gels. The apparent increase in PK activity in the elution fraction, particularly seen with the red A material, was repeatable and may be due to removal of an inhibitor, a competing enzyme, or a protease.

The LDH bound well to all of the columns except blue B and

could be eluted with full recovery from all except the green A column. The orange A and blue B gels showed the highest purification factors for LDH. This screening results suggested an interesting approach to solve the purification problem. The orange A column can be used initially to bind the LDH, leaving the PK in the wash of the orange A column. This wash can be applied directly to a red A column, which will bind the PK. The orange A and red A columns can be eluted separately by salt gradients, producing purified LDH and PK. Appropriate column loadings were determined by frontal analysis (1, 5) to be approximately 15mg crude protein/ml gel.

<u>Fig. 5</u>

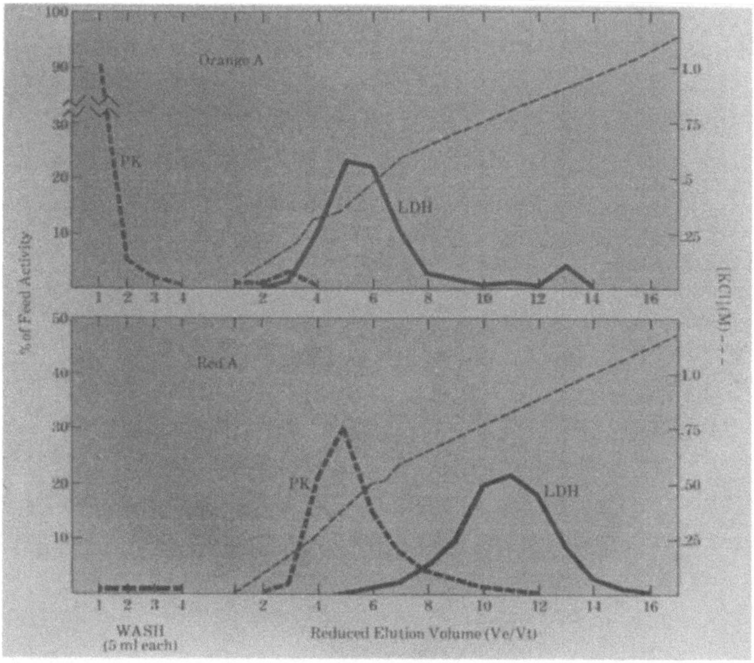

Crude extract was applied to each of the columns, and a 0-1M KCl gradient was used to elute the PK and LDH. As seen in figure 5, the two enzymes were separated by the gradient on each of the gels. The result is that in the tandem separation it will give a double separation of the two enzymes, which should effectively eliminate a crosscontamination.

Table 2 shows the results of a pilot purification on 2ml orange A and red A columns. Initially, the two columns were coupled together, with the eluate of the orange A column feeding into the top of the red A column. After washing the sample through the orange A gel with buffer, the columns were separated and the red A gel given an additional wash. The columns were then eluted by separate salt gradients. The final yield of each enzyme was high, each was purified to a considerable extent, and , most important,each was substantially free of contamination by the other.

Table 2. Final PK And LDH Purification

Orange A Eluate:	LDH -	81% Yield
		16x Purification
	PK -	0.3% of LDH
Red A Eluate:	PK -	80% Yield
		5x Purification
	LDH -	0.5% of PK

Some other applications are the binding of chondronectin only to the orange A material, from which it can be recovered in high yield or using dye-ligand media for more than simple protein purifications. These include contaminant removal, enzyme activity enhancement by separating active from inactive form, production of apoenzyme from holoenzyme, and enzyme complex and isozyme fractionation.

Summary

The remarkable ability of certain textile dyes to serve as

group-selective affinity chromatography ligands has been used to advantage in the purification of a wide variety of enzymes and other proteins. Dye-ligand chromatography often is superior to conventional natural-ligand affinity chromatography in terms of column binding capacity, biological and chemical resistance of the media, versatility of operating conditions, reusability, and cost. These advantages are especially important in large scale purifications, for which dye-ligand chromatography offers an economically feasible approach to obtaining the high performance of affinity techniques. Only some data are available for obvious reasons: Matrex blue A is used in the purification of human interferone in 50 to 100l columns. Alkaline phosphatase is purified in 20l columns with red A material. Each run gives a 200fold purification and a yield of 150 000 units of enzyme. Glucose-6-phosphate dehydrogenase, used for glucose determinations, can be 50fold purified, using the crude extract of leuconostoc mesenteroides by a combination of two dye-ligand gels.

On the laboratory scale, dye-ligand chromatography is a cost-effective and versatile improvement over conventional physical separations (gel filtration, ion exchange, salt precipitation, etc.) and can augment or often replace normal affinity techniques. To the purist, perhaps, the method has the drawback that the interactions involved are not strictly "biospecific" in the manner of conventional affinity techniques. It is true that this limits the analytical applications of the method. Synthetic ligand media, however, can be as effective as natural ligand media in providing high purification factors and high yield. Their economical advantages make them the only viable means of bringing the high selectivity of affinity chromatography to large-scale industrial purifications of enzymes and other biomolecules.

Advances in the understanding and practical applications of enzymes and other proteins are determined to a large extent by the tools available for purification. Dye-ligand chromatography is a highly colourful tool for this purpose.

References

1. Lowe, C. and Dean, P., "Affinity Chromatography", John Wiley and Sons, New York 1974.

2. Haeckel, R., Hess, B., Lauterborn, W., and Wuster, K.: Hoppe-Seyler's Z. Physiol. Chem. 349, 699 (1968).

3. Ryan, L. and Vestling, C.: Arch. Biochem.& Biophys. 160, 279 (1974).

4. Heyns, W. and DeMoor, P.: Biochim. Biophys. Acta. 358, 1 (1974).

5. Fulton, S.: "Dye-ligand Chromatography", Amicon Corp., Lexington, MA.1980.

6. Dean, P. and Watson, D.: J. Chromatog. 165, 301 (1979)

7. Baird, J., Sherwood, R., Carr, R. and Atkinson, A.: F.E.B.S. Letters 70, 61 (1976).

8. Watson, D., Harvey, M. and Dean, P.: Biochem. J. 173, 591 (1978).

9. Beech, W., "Fibre Reactive Dyes", SAF International, New York 1978.

10. Thompson, S., Cass, K., and Stellwagen, E.: Proc. Nat. Acad. Sci. USA 72, 669 (1975).

11. Stellwagen, E.: Affinity Chromatography Using Immobilized Anionic Dyes, in "Affinity Chromatography And Molecular Interactions", J.-M. Egly, ed., INSERM, Paris 1979.

FLUORESCENCE-ACTIVATED CELL SORTING: ISOLATION OF GENETIC VARIANTS FOR THE ANALYSIS OF PROTEIN STRUCTURE AND FUNCTION

Andreas Radbruch
Institut für Genetik der Universität Köln, Weyertal 121,
D-5000 Köln 41

In the last decade the techniques of flow cytometry and cell sorting could be immensely improved by the development of fluorescence-activated cell sorters, that are now commercially available (Becton-Dickinson, Ortho, Coulter, Shimadzu). Those machines allow the individual analysis of many cells for fluorescence parameters and the isolation of cells of interest. A variety of fluorescent markers has been described that detect cellular parameters like pH, ion flux, DNA content and many more (for review see 1) but the largest field of research has been opened by the use of antibodies coupled directly or indicrectly to fluorescent dyes. In this way antibodies can be used for FACS analysis and isolation of cells according to serologically defined proteins that they express. In view of protein analysis it is of particular interest for structure-function studies to obtain cells producing proteins that have gained or lost serological determinants, i.e. that have undergone defined structural changes. Other fields of application in analytical protein chemistry are the isolation of high-producing cell lines and of stable transfectants. The basic limitation for all these applications is the impossibility to apply fluorochromated antibodies

Modern Methods in Protein Chemistry, Volume 2

Figure 1:
Schematic representation of a FACS I, modified for 2 color-fluorescence analysis (Weichel et al., 1985).

(1) dichroic mirror reflecting light of less than 500 nm
(2) mirror
(3) 488 nm line filter
(4) mirror
(5) focussing lens
(6) image lens
(7) mirror
(8) and (9) filters blocking excitation light
(10) bar blocking direct laser light
(11) scatter diode
(12) photomultiplier for detection of green fluorescence light
(13) photomultiplier for detection of red fluorescence light
(14) nozzle
(15) deflection plate condensor
(16) nozzle charging wire
(17) sample vessel
(18) air pressure
(19) sheath fluid.

Not shown are the electronics required for evaluation of fluorescence signals and sorting decisions.

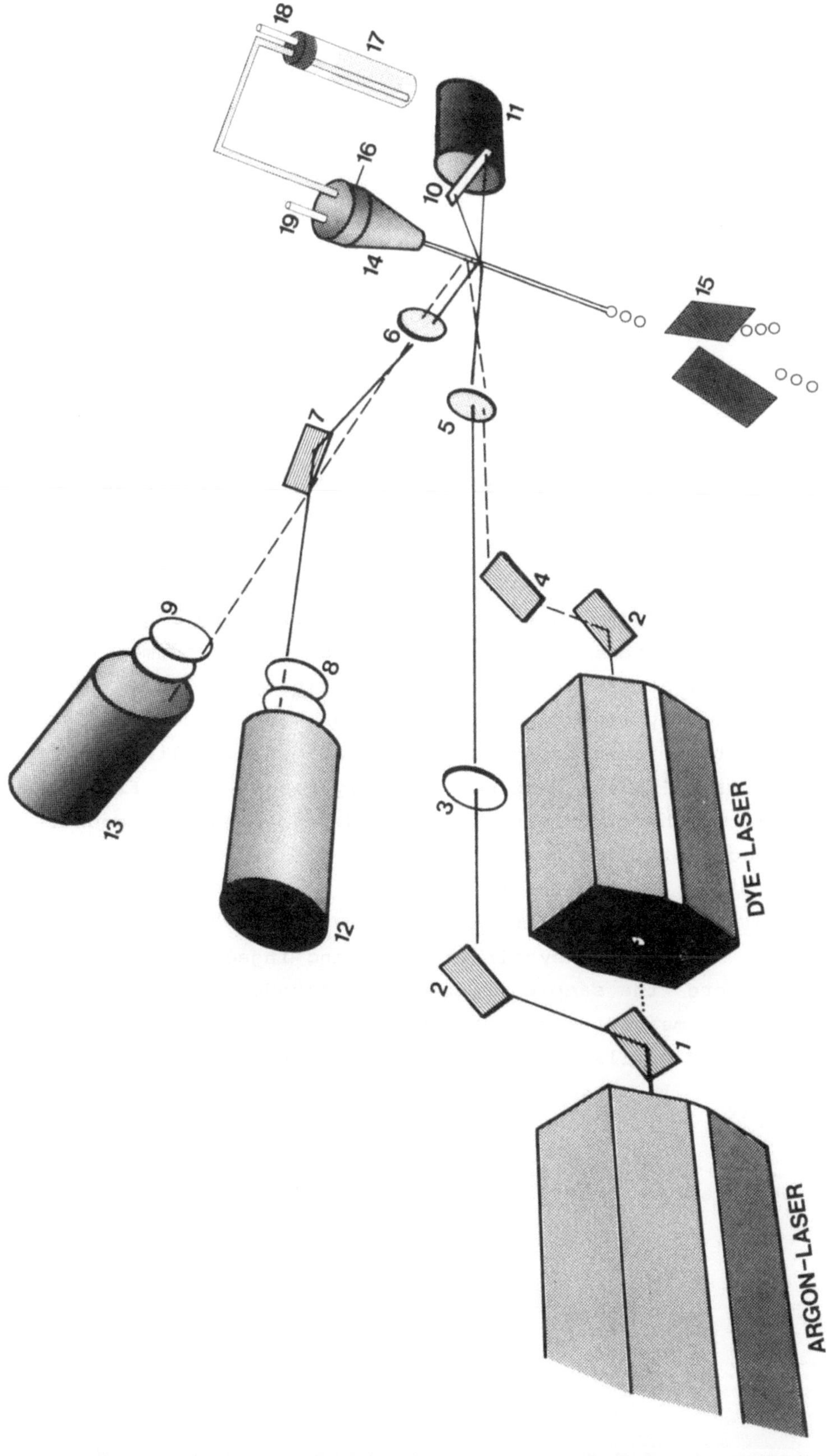
ARGON-LASER
DYE-LASER
1
2
2
3
4
5
6
7
8
9
10
11
12
13
14
15
16
17
18
19

against intracellular or secreted proteins since the fluoresceinated antibodies cannot penetrate live cell membranes. It is, however, possible to stain proteins of fixed cells in the cytoplasm and separate and analyse the fixed cells.

Fluorescence-Activated Cell Sorters

For a better understanding of limitations and possibilities of cell sorting a short introduction to the basic construction principles of these machines is necessary. The basic configuration of two-color sorters is given in Fig. 1. The principle of flow cytometry of cells in suspension is rather old and already in the early 60's a machine was developed that allowed flow cytometry and separation of analysed cells. A greatly improved machine was constructed by the Herzenberg's group in 1972 (2) that became the prototype of the first commercially available cell sorter FACS I (Becton-Dickinson). Today, sorters working by the same principle can be purchased from Ortho, Becton-Dickinson, Coulter or Shimadzu. Although the vast majority of these machines has only single-color fluorescence some of the experiments described below require double-fluorescence abilities.

1. <u>The cell path</u>. A cell population to be analysed is forced into the sample tube by air pressure and injected into the nozzle. There the sample stream is surrounded by the sheath stream that mainly serves to focus the cells for optical analysis. The cells are passing the focussed laser beam and analysed for light scatter and amount of fluorescent light emitted.

They then travel another appr. 400 μm until they arrive at the "break off point" where the solid stream is disrupted into droplets by virtue of a Piezo crystal vibrating the nozzle at high frequency (20-40 kHz). Charging all liquid within the nozzle for the time of drop formation results in a charged droplet that is deflected when passing the electrical field of the deflecting plote condensor. The task of the sorter electronics is to take in the scatter and fluorescence data of each cell, compare them to preset values and, in case they fulfil the "window" requirements, to give a stream charging impulse exactly when the cell is at the stream's "break off point". Since this charging impulse can be positive or negative it is possible to separate two populations at a time from the major population.

2. The optical path. A 2-5 Watt argon laser is today's most common light source. It is focussed on the liquid stream by the focussing lens. Direct light transmitting the stream is blocked away by a barrier. Light scattered in forward direction by particles in the stream is collected and measured by a light sensitive diode. The amount of scatter is usually correlated to size, shape and opaqueness of a particle and in practical terms serves to discriminate small and big or live and dead cells.

Fluorescent light is collected at 90° angle by a lens system that projects a picture of the cell's excitation point onto a photomultiplier tube. The exciting laser light is filtered out and the fluorescence light measured. Although single fluorescence sorters are still the standard machines, it is necessary to use double-fluorescence sorters for some of the experiments described here. We have described a simple upgrading of a single to double fluorescence machine (3, Fig.1). In our setup the argon laser light is devided by means of a dichroic mirror just in front of the argon laser into reflected light of wavelengths below 500 nm and passing light of longer wavelengths. The latter is used to pump a Rhodamine 6G dye laser that is tuned to emit yellow light of approximately 530 nm. Both the <488 nm argon laser and the 530 nm dye laser light beams are reflected and focussed on the cell stream at about 40 and 15 µm below the nozzle. Both illumination points are projected on different photomultipliers by means of a mirror (see Fig.1, (7)). The signals of these two and the scatter diode are fed into the analytical part of the sorter electronics and used for analysis and sorting.

3. Scatter. Since they are easy to obtain, are derived also from fluorescence-negative particles and are otherwise of little value, the scatter signals are usually used for triggering of the data facilities. Besides this more technical aspect they can be used for size discriminations and separation of live from dead cells (4).

4. Red fluorescence. Dead cells can also very efficiently be excluded from analysis by their intense red fluorescence, when stained with propidium iodide (1 µg/ml in the sample fluid, 5). For separation of low frequency cells this will improve results considerably. However, the red fluorescence can also or in addition be used for analysing cells tagged with specific red labels. The fluorochromes most widely used are Texas Red (6) and XRITC, both derivatives of rhodamin-iso-thiocyanate, but with higher excitation and emission wavelengths than rhodamin and thus not interfering with green fluorescence.

5. Green fluorescence. The major green fluorochrome used is fluoresceine. It used to be the only one on single-fluorescence sorters and a wide variety of antibodies coupled to fluoresceine are available. Recently, the algal fluorochrome phycoerythrin has become available that when excited at 488 nm (i.e. at the "green illumination point") emits red fluorescence which can be collected independently from the green light of fluoresceine (7). Thus a "third color" is now allowing double fluorescence on single fluorescence machines or three color analysis on two-laser sorters.

Cell Sorter Application

From the spectrum of cell sorter applications I want to describe two in some detail (table I): (a) the one-step purification of subpopulations out of heterogeneous cell mixtures according to serologically defined protein determinants and (b) the isolation of genetic variants. Both types of experiments can be examplified in the immunoglobulin system. Immunoglobulins (Ig) are produced by lymphocytes of the vertebrate B-lymphocyte lineage. They are a 2 heavy (MW appr. 50 000 D) and 2 light (MW appr. 25 000 D) chain tetramer or multiples thereoff. While the aminoterminal 100 amino acids (variable region) of each chain are highly heterogeneous among the Igs of one individual, the rest of the protein chains (constant region) is rather homogeneous in that all Igs fall into a few classes only. During its ontogeny a B cell first produces heavy chains alone that are expressed in the cytoplasm, later light and heavy chains expressed as cell surface Ig, and in the final differentiation stage of plasma cells the Ig is mainly secreted although quite a few cells also have surface Ig. The analysis of structure-function relationships in antibody variable region and isotype diversity is one reason to study B cells and the immunoglobulins they make.

Table 1: Fluorescence-activated cell sorting performance and limitations

Analysis:	scatter (90°, 180°)
	2-3 fluorescence parameters
	3000 target molecules/cell
Sterile sorting:	5000 cells/sec
	10^8 cells/day
	60-95% recovery of deflected cells
	0.1-1% background signals
Minimal size of populations for sorting	
	subpopulations sorting: 1%
	variant sorting: 10^{-7} - 10^{-6} cells

1. Selection of subpopulations. B cells that originally all make Ig of the IgM class can switch from expression of IgM to expression of any one of the other Ig classes. This class switch is most likely a consequence of antigenic stimulation of the B cell and probably plays a role for the diversity of reactions of the immune system towards the antigen. We have isolated in one step ("subpopulation sorting") mouse B cells expressing various Ig classes from LPS activated B cells. We have used these cells for analysis of their IgH gene loci by restriction endonuclease analysis (8). However, they also

could have been used for analysis of Ig proteins or biosynthesis. Subpopulation sorting can also be applied to ethanol/acetic acid (95/5%) fixed cells stained either on the surface (stained before fixation) or in the cytoplasm (stained after fixation) for serological target determinants (9). This is especially useful for the selection of plasma cells according to Ig determinants since such cells contain a lot of cytoplasmic Ig. However, many other proteins expressed in the cytoplasm could be assessed in a similar way. The structure of the proteins is apparently not changed too much because even sensitive staining antibodies can be used after fixation, e.g. antibodies recognizing complex structures made up by Ig light and heavy chains.

Another application of subpopulation sorting is the analysis and separation of live transfected cells according to the surface expression of proteins encoded by the transfected genes. This method has been used sucessfully for the isolation of the genes for the transferin receptor (10) and is the method of choice if the target molecule can best be traced by a specific staining antibody.

The distinction between subpopulation sorting and variant sorting is fading, however, in case of transfected cells since in some experiments it may be useful to purify transiently expressing cells for direct further molecular analysis of proteins, while in other experiments the purified cells are grown up and variants that have integrated the transfected genes and express them stably can be isolated by repeated sortings, growing up the deflected cells between the sorts. We have termed this technique "variant sorting" since it can only be applied for the isolation of stable genetic variants (11).

2. Variant sorting. The most general application of this method is the selection of "high-producer" variants expressing more target molecules than the "wild type" cells. This has been sucessfully used for the isolation of stably transfected high-producers of human Leu 2 antigen (12). An increase of about 100fold in staining intensity and about 30fold in gene copy number could be achieved. The selection of cells expressing high amounts of target surface molecules in a stable fashion is of general technical importance for the work not only with transfected cells but also with many other cell types like B- or T-cell hybridomas and their antigen-receptors.

While for the isolation of rare genetic variants "sequential sublining" is the method of choice for secreted target molecules (13) the selection according to surface molecules is best to be done by cell sorting. Alternative methods like "panning", "affinity column selection" or "cytotoxic selection" are far less universially applicable and have several limitations. A more detailed comparison and description of the methods is given in ref. 14.

Fluorescence activated cell sorting has been used for the isolation of murine histocompatibility antigen variants (15) and of mutant myeloma and hybridoma cells that express Ig of altered isotype (i.e. class) or idiotype, i.e. serological determinants of the constant and variable parts of Ig molecules. Isotype or class switch variants occur at low frequency ($10^{-5}/10^{-6}$) in many hybridoma cell lines (14). They can be isolated by staining of the variant cells and deflection of all cells that are not negative. Due to staining background this is usually about 0.1 to 1% of the cells. It is obvious that within the deflected cells the variant cells are only enriched but not purified. Therefore, the deflected cells are grown up and stained and sorted again 2-3 weeks later. This procedure is repeated until the variants are sufficiently enriched for isolation by cloning. This procedure has in the meantime been used for isolation of isotype variants from several hybridoma cell lines and in many cases secondary and tertiary variants have been isolated thus obtaining whole "families" of hybridoma antibodies (14). The antibodies of a family always have the same variable region but different constant regions (16). They provide a unique system to study functional differences of the various Ig classes, e.g. complement binding (17) or segmental flexibility (18).

The structural basis of antigen-binding site and idiotypic determinants (idiotype, i.e. serological determinants specific for a given Ig variable region) is of particular interest for appraisal of the unusually high rate of somatic mutation of the Ig variable region in B-cells activated by antigen (19). We have isolated idiotypic variants from the hybridoma cell line B1-8 by fluorescence-activated cell sorting with monoclonal antibodies recognizing single idiotypic determinants, i.e. idiotopes (20, 21).

The isolation of cells that have lost an idiotope is, however, complicated by the fact that in most hybridoma lines quite a few cells express no Ig at all, whereas idiotype variants are rare (10^{-6} to 10^{-8}). The non-producer cells can be excluded from the selection either by a binding inhibition method (15, 20) or by double-fluorescence sorting, using in both methods a second antibody and separating cells that have lost the first idiotypic determinant but not the second one, which could be an idiotypic or isotypic determinant. From the analysis of idiotypic variants obtained from the cell line B1-8 the existence of recombination or gene conversion for the generation of Ig variability (22) and a single amino acid exchange leading to loss of the entire idiotype while not changing the antigen binding site (21) have been demonstrated.

An even more powerful tool to analyse structure-function relations of cell surface proteins may be the combination of transfection and fluorescence-activated cell sorting and analysis. This can be done if the gene encoding the surface protein is available for transfection experiments and can be mutated in vitro. Then either all random mutations leading to a certain phenotype or the phenotype of a site-specific mutation can be evaluated by sorting or/and analysis of transfected cells.

References

1. Shapiro, H.M.: Cytometry 3, 227-234 (1983).
2. Herzenberg, L.A. Sweet, R.G., Herzenberg, L.A.: Scientific American 234, 108-117 (1976).
3. Weichel, W., Gehrke, K., Göttlinger, Chr., Holtkamp, B., Liesegang, B., Radbruch, A., Stackhouse, T.K., von Hoegen, P., Rajewsky, K., submitted.
4. Loken, M.R., Herzenberg, L.A.: Ann. N.Y. Acad. Sci. 254, 163-171 (1975).
5. Dangl, J.L., Parks, D.R., Oi, V.T., Herzenberg, L.A.: Cytometry 2, 395-401 (1982).
6. Parks, D.R., Hardy, R.R., Herzenberg, L.A.: Immunology Today 4, 145-148 (1983).
7. Oi, V.T., Glazer, A.N., Stryers, L.: J. Cell. Biol. 93, 981-992 (1982).
8. Radbruch, A., Sablitzky, F.: EMBO J. 2, 1929-1935 (1983).
9. Klein, S., Sablitzky, F., Radbruch, A.: EMBO J., in press.

10. Kamarck, M.E., Barbosa, J.A. Kühn, L., Messer-Peters, P.G., Schulman, L., Ruddle, F.H.: Cytometry 4, 99-108 (1983).

11. Liesegang, B., Radbruch, A., Rajewsky, K.: Proc. Natl. Acad. Sci. USA 75, 3901-3905 (1978).

12. Kavathas, P., Herzenberg, L.A.: Nature 306, 385-387 (1983).

13. Müller, C., Rajewsky, K.: J. Immunol. 131, 877-881 (1983).

14. Radbruch, A.: in Handbook of Experimental Immunology (Weir, D.M. and Herzenberg, L.A., eds.), Blackwell Scientific Publications Ltd., Edinburgh, 1984, in press.

15. Holtkamp, B., Cramer, M., Lemke, H., Rajewsky, K.: Nature 289, 66-67 (1981).

16. Beyreuther, K., Bovens, J., Dildrop, R., Dorff, H, Geske, T., Liesegang, L., Müller, C., Neuberger, M.S., Radbruch, A., Rajewsky, K., Sablitzky, F., Schreier, P.H., Zaiss, S.: in Immunoglobulin Idiotypes: ICN-UCLA Symposia on Molecular and Cellular Biology, Vol. XX (Janeway, C., Sercarz, E.E. and Wigzell, H., eds.), Academic Press, Inc., New York 1981, p. 169.

17. Neuberger, M.S., Rajewsky, K.: Eur. J. Immunol. 11, 1012-1016 (1981).

18. Oi, V.T.: Nature 307, 136-141 (1984).

19. McKean, D., Huppi, K., Bell. M., Strandt, L., Gerhard, W., Weigert, M.: Proc. Natl. Acad. Sci. USA 81, 3180-3184 (1984).

20. Brüggemann, M., Radbruch, A., Rajewsky, K.: EMBO J. 5, 629-634 (1982).

21. Radbruch, A., Zaiss, S., Kappen, C., Brüggemann, M., Rajewsky, K.: submitted.

22. Krawinkel, U., Zoebelein, G., Brüggemann, M., Radbruch, A., Rajewsky, K.: Proc. Natl. Acad. Sci. USA 80, 4997-5001 (1983).

THE FPLC*-SYSTEM FOR FAST AND HIGH RESOLUTION SEPARATIONS OF PROTEINS

Wilfried Duppel
Deutsche Pharmacia GmbH
D-7800 Freiburg

Introduction

Modern methods in analytical protein chemistry in most cases require pure proteins for their further analytical characterisation. Therefore the preparation of pure proteins is an essential part in analytical protein chemistry. A modern separation method should be better than older ones. In the field of protein separations this would mean: more efficiency, better resolution, more flexibility, higher yields and shorter separation times.

It is well known that the separation of complex protein mixtures of biological origin usually requires more than one step in a protein purification procedure. Therefore different properties of the proteins are used to separate them by various chromatographic methods, and in most cases a combination of several of these methods is used to achieve the required purity of the protein. These multidimensional separation techniques are common procedures to purify proteins. However, when traditional soft gels are used,these various separation steps require quite some time before the final purification stage is reached. Developments in the past years have shown, that it is difficult to improve traditional protein separation techniques, due to the specific requirements, proteins pose on such methods.

* FPLC = Fast Protein Liquid Chromatography

Modern Methods in Protein Chemistry, Volume 2

Proteins are quite often large, labile, hydrophilic biomolecules which often are sensitive to shear forces, pressure, oxidizing agents, metal ions and other environmental factors, which may cause a loss of their biological functions or enzymatic activities.

This chapter deals with a modern system which has been specifically designed for the purification and chromatographic separation of proteins, peptides and polynucleotides, the FPLC-System.

The FPLC-System

The specific design of the FPLC-System for protein purification means that the column material used for the protein separation as well as the hardware which is in contact with the buffer or sample are biocompatible and tested for high protein yields. Fig.1 shows a typical FPLC set-up with a chromatography

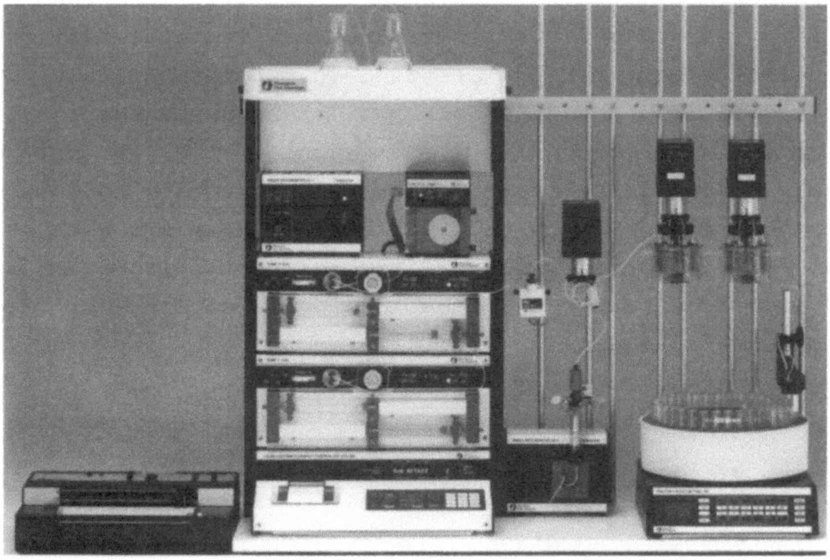

Fig.1 Automated FPLC-System for fast, high resolution protein separation.

controller, two pumps to form a gradient, a small MonoBead column, an automatic injection valve with autosampling possibilities, an UV-monitor and recorder as well as fraction collector for preparative work. As some proteins are sensitive to heavy metal ions, there is no stainless steel in contact with the sample or the separation buffers. All the materials which are in contact with these fluids are glass, fluoro polymers or titanium. To get accurate and reproducible gradient shapes, the buffers are mixed on the high pressure side in a small mixing chamber with a total volume of 600 µl. The connections between the various parts of the system are made by 0.5 mm Tefzel tubing which can stand pressures up to 10 MPa (100 bar). The column material (MonoBeads) has been specifically designed to create only very low back pressures so that the chromatographic separations can be run at rather low pressures of about 2 MPa (20 bar). With this low pressure shear forces are reduced to a minimum even for very large proteins. The system has been designed to be compatible with halide gradients which are commonly used for elution of proteins in ion exchange chromatography. This is quite important as it has been shown that stainless steel material which is regularly used in traditional HPLC systems is not resistant to halide ions. This means that corrosion takes place, which leads to corrosion deposits in the column, resulting in increased back pressure and a drastic loss of resolution and the lifetime of the column. As the columns are made of glass, one can easily see that considerable corrosion has taken place after 20-30 runs already, when the column has been used in a stainless steel HPLC system. When these columns are used in a FPLC-System, more than 2.000 runs have been performed on a single column already.

The motorized valve MV-7 is used as an automatic injection valve and is controlled by the liquid chromatography controller LCC-500. It has three positions, one for loading the sample into the loop, one for injection of the loop content onto the column and a third one which is used during wash cycles of the pump. Another type of valve, the MV-8 has a central in- or out-port and eight peripheral connections. This valve can be used as a column switching valve for up to eight different columns or as a buffer selection valve if it is put on the inlet side of the pump to choose between various buffers. In addition it can be used as an autosampling device for up to fifteen different samples with two valves as it is shown in Fig.1, or as a mini fraction collector for the collection of up to eight different fractions. These valves are resistant to pressures up to 10 MPa (100 bar), and the material in contact with the liquid is polyethyleneimine which is resistant to pH values from 2 to 13 and practically all commonly used organic solvents. When the MV-8 valve is used as an autosampling device or as a mini fraction collector, the sample holder attached to it, can be cooled with ice or with a circulating cooling bath to retain the activity of the proteins during the time when they are not in the actual separation process. As the system is designed for protein separations, the complete set-up can be run in the cold-room to preserve activity. Buffers containing urea detergents, or glycerol can be used for separations of proteins as well as various organic solvents for reversed phase chromatography.

MonoBeads for ion exchange chromatography and chromatofocusing

The hardware in a chromatographic system is essential in retaining the resolution that has been obtained within the separation column, to keep band broadening low and to preserve the activity of the proteins. But the separation itself takes place within the column and the column material is the essential part in a chromatographic system. This is also true for the FPLC-System, and the separation properties of the MonoBeads - a matrix specifically designed for protein separations - have influenced the characteristics of the hardware in the FPLC-System.

In order to get fast high resolution separations with high capacity for proteins, ion exchange chromatography has first been chosen as a suitable technique for analytical and preparative protein separation. Ion exchange chromatography is a well established method in protein purification and has proven to be a reliable technique with high yield and good recovery of enzymatic activity. This technique is particularly suited to be used in a high resolution technique, as the separation is not solely dependent on the speed of diffusion of the molecules like in gel filtration, and as resolution can be increased by varying the gradient during the elution without loosing speed. A pre-requisit, however, is that the charged groups of the ion exchanger are easily accessible and that the binding kinetics are not restricted by diffusion parameters even if you use large protein molecules.

During the development of a suitable matrix for fast high resolution ion exchange chromatography, the column size, the gradient volumes as well as the bead size distribution and pore size limit were optimized. The result of these investigations are the MonoBeads, a family of monodisperse ion exchange materials (Table I).

Table I: Properties of MonoBeads

	Mono Q	Mono S	Mono P
Type of exchanger	Anion-	Cation-	Chromatofocusing
Particle size (µm)	9.8 ±2%	9.8 ±2%	9.8 ±2%
Exclusion limit (daltons)	10 Mio	10 Mio	10 Mio
Chemical stability (pH)	2-12	2-12	2-12
Theoretical plates (m^{-1})	30-40.000	30-40.000	30-40.000
Capacity (mg per run)	25	25	25
Typical separation time (min)	10-20	10-20	20-40
Column dimensions (mm)	5x50	5x50	5x200

These MonoBeads consist of a new hydrophilic inert organic polymer with a monodisperse bead size distribution of 9.8 µm ± 2 % (Fig.2). This monodisperse bead size allows the column to be packed almost ideally, which results in a very low back

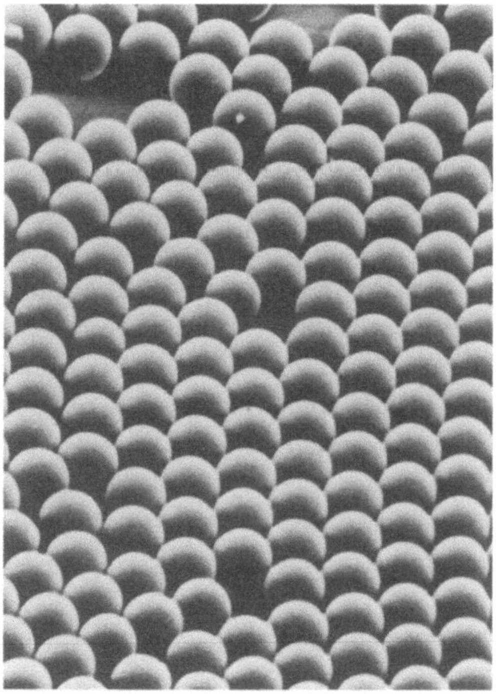

Fig.2 Monodisperse bead size of MonoBeads 9.8 µm ± 2 %.

pressure and high resolution. The typical back pressure at a flow rate of 2 ml/min is about 2-3 MPa (20-30 bar). As the beads have a macroporous structure with an exclusion limit of 10^7 Daltons, the diffusion even of large proteins is not limiting and high flow rates can be used for optimal separations. Due to the wide pore structure the resolution is not very much dependent on the flow rates in MonoBeads columns. Practically the same resolution can be obtained when a separation on a MonoBead column is performed at a flow rate of 0.25 ml/min (separation time 60 min) or with a flow rate of 3 ml/min (separation time 5 min)(1). Strong ion exchange groups are bound to the MonoBead matrix which means that the available charge is constant throughout a wide pH interval (pH 2-12). As the pores have a very high exclusion limit, these charges are available for small and large molecules. The binding capacity of a 1 ml column is about 80 mg for a medium size protein like albumin under standard conditions. Therefore 20 mg of a protein mixture can be separated in one run within 10-15 minutes with a

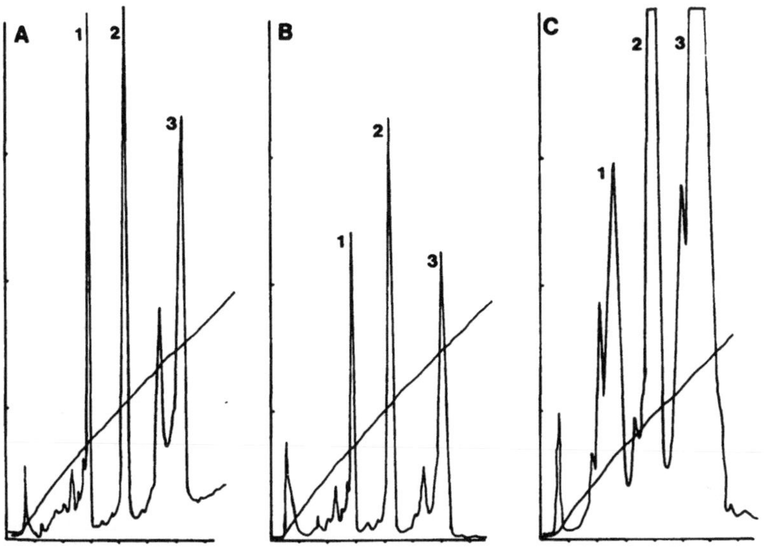

Fig.3 Separation of ribonuclease (1), chymotrypsinogen (2) and cytochrome c (3) on Mono S in 20 mmol/l Bicin pH 6.8. Elution with gradient to 0.35 mol/l NaCl. Protein load A: 12 µg, B: 1.2 mg, C: 18 mg.

very good resolution. Fig.3 shows how a protein mixture can be separated analytically with only 12 µg of protein while in the same time with the same column up to 18 mg can be separated with practically the same resolution. This separation has been performed in a Mono S HR 5/5 column (5 mm diameter, 5 cm bed height) with a bed volume of 1 ml. So, this column is ideally suited for analytical as well as semi-preparative work. If 5 cycles of a 10-12 minute separation are run after each other, up to a 100 mg can be separated per hour. If larger amounts of protein have to be purified, the same resolution is obtained in the same time in somewhat larger columns. For preparative work 8 and 20 ml columns are available in form of HR 10/10 and HR 16/10 respectively. For production purposes 200 ml columns with 5 cm diameter and 10 cm bed height are used with the same efficiency as a 1 ml column. 4 grams of protein have been separated in a single run on such a column in less than two hours.

Besides the strong anion exchanging column Mono Q and the strong cation exchanging Mono S, a third ion exchange column Mono P is available, which has the same MonoBead matrix, but which is substituted with various anion exchange groups and is optimized to be used for chromatofocusing with polybuffer. In this technique the protein is bound in its negatively charged form to the anion exchanger and then a selfforming pH-gradient is developed on the column with a solution of polybuffer adjusted to the lower pH of the gradient. This selfforming pH-gradient results in a continuous focusing of the protein bands on the column which gives sharp peaks and very high resolution. As the gradient is developed on the column, the bed height of the column is 20 cm as compared to the ion exchange columns, which have a bed height of only 5 cm. High resolution chromatofocusing separation can be obtained on such a column within 20-40 minutes while traditional chromatofocusing experiments last 6 to 12 hours. On such columns pH-gradients spanning various pH-units can be applied or only a fraction of a pH-unit can be run and resolution of the pH-profile of only 0.005 pH-units have been reported (2). Chromatofocusing in a FPLC-System is an interesting alternative to Isoelectric Focusing, as in analytical runs the same resolution can be obtained as in standard polyacrylamide gels and preparative separations can be performed at much shorter times at considerably reduced costs.

Compared to silica based ion exchange material, MonoBead columns have the distinct advantage that they can tolerate rather drastic cleaning procedures like washing steps with 2-5 mol/l NaOH as well as high concentration of urea, nonionic detergents or organic solvents, which means quite often a considerable extended lifetime of such a column.

Gel filtration with Superose

Ion exchange chromatography is a very useful method in separating proteins because the separation can be very fast, as the mechanism is not solely diffusion controlled, it is not volume dependent, the sample can be concentrated on the column and by changing the shape of the gradient the resolution can be improved. Despite these advantages, not all separations of complex protein mixtures can be resolved by this chromatographic technique. There may be molecules which are rather similar in charge, however, differ considerably in their size (molecular weight). These proteins can easily be separated by gel filtration. In addition, there is a need for the chromatographic determination of molecular weights. Therefore it is useful and necessary to have various separation mechanisms which use different physical parameters to separate the proteins in a chromatographic system. Gel filtration is a well established method for the separation of proteins since 25 years. Crucial parameters for successful gel filtration are an inert hydrophilic matrix, a highly porous matrix, low dead volume of the column, small sample sizes, even sample application and relatively low flow rates, as the separation is solely diffusion controlled. Another restriction of the gel filtration method is that the separation takes place in only one column volume which means that only a certain number of components can be separated. The number of products which can be separated depends on the performance and efficiency of the column. Despite these limitations gel filtration has a firm position in the purification of proteins and is widely used for the determination of molecular weights of proteins in their natural state.

Due to the diffusion control of gel filtration separations and the relatively low flow rates resulting in rather long separation times, there has been a need for fast high performance gel filtration separations since a couple of years. A few gel filtration media for high performance systems are available on the market based on silica or hydrophilic organic polymers. These, however, have the disadvantage that they cannot be used at pH values above 7.5 or that they show considerable adsorption phenomena under certain conditions and cannot be used under a wide variety of conditions.

The properties of two new recently introduced gel filtration media are shown on Table II. These Superose media are based on a new highly crosslinked agarose material. Agarose has been used in gel filtration for years and has proven to be a stable and inert matrix giving high protein and enzyme activity yields. With these two types of Superose, gel filtration of proteins can be performed in the molecular weight range from one thousand to 10 x 10^6 Daltons. Very high efficiencies of biocompatible glass HR 10/30 columns (10 mm Ø, 30 cm bed height) packed with Superose can be achieved due to the high percentage of pore volume. Another distinct advantage of the Superose media is that they can be used with any buffer composition in the pH range from 1 to 14. This also means that the columns can be cleaned with diluted NaOH which makes cleaning more efficient and prolongs the lifetime of such a column. Fig.4 and 5 show typical separations obtained with Superose 6 and Superose 12. With rather low back pressures (up to 3 MPa = 30 bar) separation times are usually between 30 and 60 minutes depending on the desired resolution and the flow rate used.

Table II: Chemical and physical properties of Superose columns

Properties	Superose 6	Superose 12
Separation range (Daltons)	5000 - 10^7	1000 - 3 x 10^5
Exclusion limit (globular proteins)	~4 x 10^7	~2 x 10^6
Average particle size (µm)	12 ± 2	10 ± 2
Efficiency (TP/m)	> 30.000	> 40.000
Pore volume (%)	60	55
pH-stability	1 - 14	1 - 14
Temperature stability (°C)	4 - 40	4 - 40
Protein recovery (%)	90 - 100	90 - 100
Max. loading per column (mg)	5 - 10	5 - 10

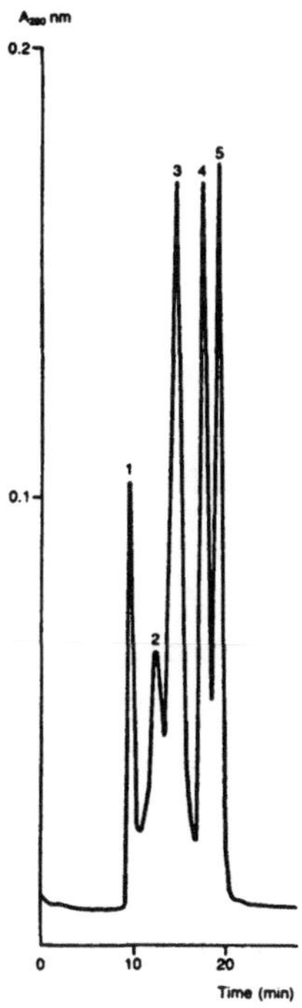

Fig.4

Fig.5

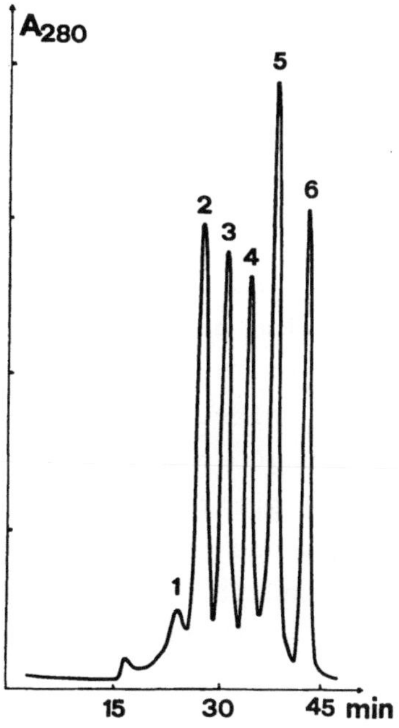

Fig.4 Separation of protein mixture on Superose 12 HR 10/30. Sample 200 µl. Flow rate 0.75 ml/min. 0.05 mol/l phosphate buffer pH 7.0 + 0.15 mol/l NaCl. 1) Void volume 2) IgG dimer 3) IgG monomer 4) β-lactoglobulin 5) lactalbumin.

Fig.5 Separation of different proteins on Superose 6 HR 10/30. Sample 100 µl. Flow rate 0.4 ml/min. Buffer as in Fig.4. 1) Thyroglobulin dimer 2) thyroglobulin monomer 3) ferritin 4) BSA 5) β-lactoglobulin 6) acetone.

Reversed phase chromatography

Reversed phase chromatography (RPC) is a well established method in the separation of small organic molecules in traditional HPLC systems. For preparative separations of biologically active proteins, RPC is not suited, as in most cases the enzymatic activity cannot be retained in the final preparation, as organic solvents are mostly used for elution. If the main aim is, however, not to isolate a biologically active protein but to purify a certain peptide with a different structure, a genetic variant or peptide fragments of a protein, then RPC is a powerful separation mechanism. It can be used in various ways in analytical protein chemistry for the determination of protein structures.

Three different reversed phase columns have been developed for protein and peptide separations in the FPLC-System. All of them are silica based, porous, microparticular beads with 5 µm in diameter. The material for peptides and other small molecules with a molecular weight up to 6.000 Daltons has an average pore size of 100 Å while the columns for protein and other polypeptide separations have an average pore size of 300 Å. The silica surface has been coated according to a new procedure which was optimized for protein and peptide separations (3). As the alkyl chain length and the ligand density affects the selectivity of reversed phase material, these parameters have been thoroughly studied by cross polarisation magic angle spinning nuclear magnetic resonance spectroscopy. The surface of the PepRPC C_{18} HR 5/5 column is coated with C_{18}/C_2 while the ProRPC C_8 HR 5/2 and HR 5/10 columns are coated with C_8/C_1. With this covalent coating of ligands a tight coating of the surface with a minimum

Fig.6 Tryptic digest of cellulase from Trichoderma viride on ProRPC HR 5/2. Eluent A: 0.05 mol/l ammonium-acetate pH 6.4. Eluent B: 0.05 mol/l ammonium-acetate in 70 % ethanol. Gradient 0-50 % B in 30 min.

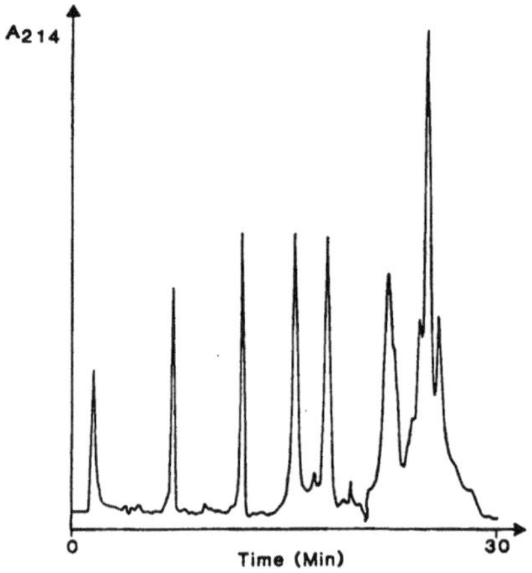

of free silanol groups was achieved, which results in a special selectivity, different from endcapped silica material. By this synthesis procedure the RPC material can be manufactured very reproducible. The ProRPC C_8 HR 5/2 and HR 5/10 columns contain the same material. They are, however, used for different purposes. The HR 5/2 column with a diameter of 5 mm and a bed height of 25 mm is used for screening purposes as well as for protein and polypeptide separations according to an on-off mechanism. The HR 5/10 column with a diameter of 5 mm and a bed height of 100 mm, however, is used for the separation of polypeptides and proteins showing only small differences in hydrophobicity and are therefore separated according to an equilibrium partioning mechanism with shallow gradients and low flow rates. Therefore the bed height of this column is longer. Fig.6 shows the separation of a tryptic digest of bacterial cellulase on a ProRPC HR 5/2 column while the high resolution of large proteins is shown in Fig.7 by the separation of various proteins of the ribosomal 30 S subunit. A separation of synthetic enkephalins and related fragments on a PepRPC HR 5/5 is shown in Fig.8. Peptides differing by only one amino acid can easily be separated on such a 5 cm column within 20 to 30 minutes.

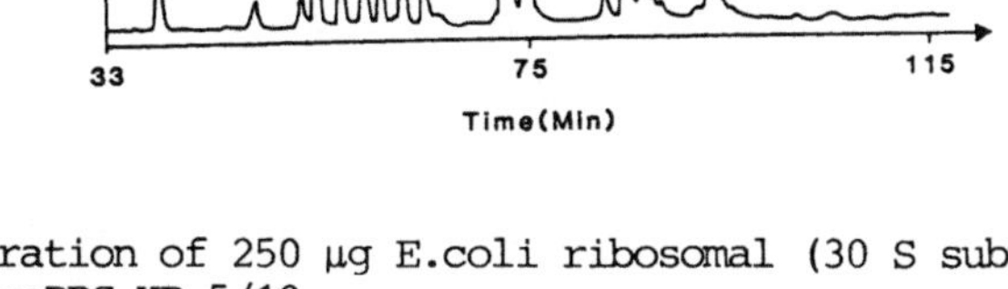

Fig.7 Separation of 250 µg E.coli ribosomal (30 S subunit) proteins on ProRPC HR 5/10.
Flow rate 0.2 ml/min. Eluent A: 0.1 % TFA in water pH 2.0.
Eluent B: 0.1 % TFA in water/acetonitrile 50:50.
Linear gradient 40-100 % B in 120 min.

Fig.8 Separation of a mixture of synthetic enkephalins and related fragments on PepRPC HR 5/5. Sample see figure. Flow rate 0.7 ml/min. Eluent A: 10 mmol/l KH_2PO_4 in H_2O pH 4.5. Eluent B: 5 mmol/l KH_2PO_4 in H_2O/acetonitrile 50:50. Linear gradient 0-50 % B in 30 min.

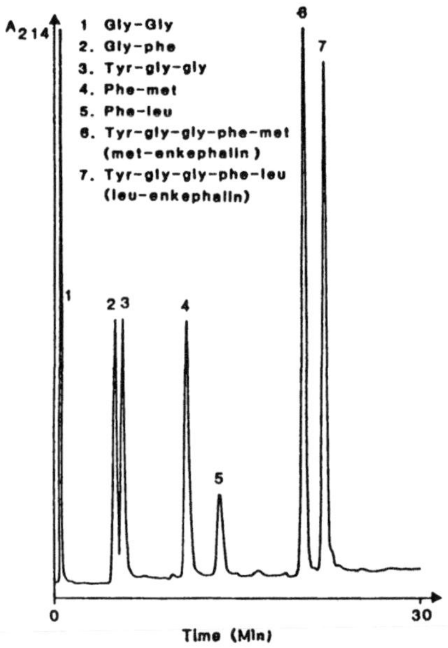

The high resolution and the specific selectivity of the FPLC reversed phase columns add considerably to the analytical separation power of the FPLC-System especially when multidimensional separations are required for detailed structure analysis.

Selection of chromatographic conditions

If one wants to separate a protein in the FPLC-System, one has the choice of various techniques. If published data or own experiences on the purification of a certain protein are available, these techniques and conditions can be used as start conditions and should then be further optimized. One will readily realize that a lot of the published methods on traditional gels have not been optimized. As one step in such a method could require 10 to 20 hours, one can understand that this has not always been done. With the FPLC-System, however, when a normal separation step will last less than half an hour, it is easy to optimize a separation in less than half a day. And it is worth to invest this time because one can get much purer preparation with higher activities in an optimized separation system.

Starting with a crude preparation, it is generally useful to use a concentrating step in the beginning of the purification scheme. Ion exchange chromatography is a suitable method with high capacity, good resolution and an often important concentrating mechanism as a first step. If the isoelectric point and the pH stability range of desired protein is known, it is easy to decide whether an anion or a cation exchanger should be used and which pH of the startbuffer should be chosen. If such data are not available it has been proven to be very valuable to run an electrophoretic titration curve analysis to get this kind of information. The principle of this analysis is outlined in Fig.9. In the first step, an isoelectric focusing with a wide pH gradient is run in a polyacrylamide or Agarose IEF gel. This gel should have a long slot into which the protein sample is added after the formation of the pH gradient in the first dimension. In the second dimension the proteins in the slot are exposed to different pH values and the gel is turned by 90 degrees and an electrophoretic separation is performed. In this second dimension the proteins migrate according to their charge and form a so-called titration curve. After staining or identifying the desired protein bands, one can easily see the iso-

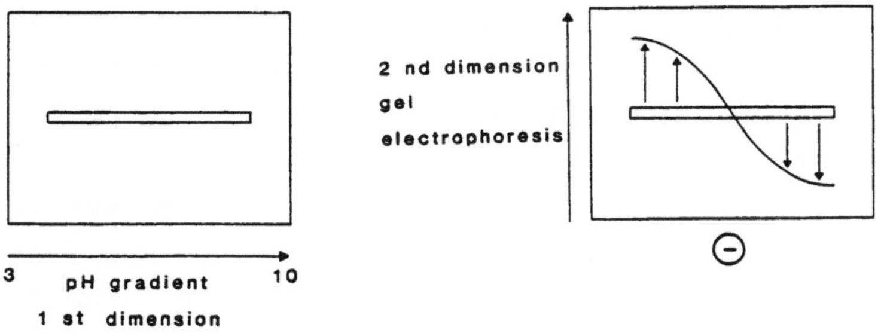

Fig.9 Principle of electrophoretic titration curves.

Fig.10 Use of electrophoretic titration curve analysis to decide which ion exchange to use. For details see text.

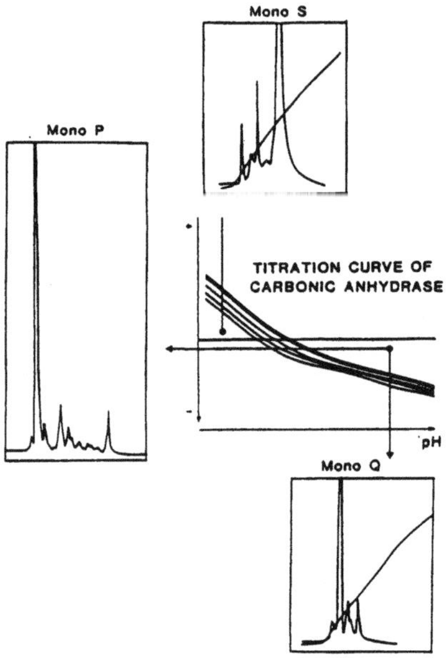

electric point and at which pH the desired component is best separated from contaminating proteins. How this information is used to decide which exchanger to use in the chromatographic separation is outlined in Fig.10. The central part shows the titration curve analysis of carbonic anhydrase where the upper thick curve represents the major protein part of this protein. As proteins are positively charged below their isoelectric point, one would use a cation exchanger Mono S for the separation in this pH range. From the titration curve one can see that the main protein component has the highest charge at this pH value and is therefore eluted last in a salt gradient from a Mono S cation exchange column. If we go to high pH values the main component has the lowest charge and is therefore eluted early in the salt gradient on a Mono Q anion exchange column. If the separation of the protein band is best along the pH axis of the titration curve, chromatofocusing may be the method of choice. When we start at a high pH value, the main carbonic anhydrase band will be eluted first when we apply the descendent pH gradient by elution with polybuffer. Looking at these chromatograms, one would decide to use chromatofocusing on a Mono P column because the resolution seems to be best in this system. For some other reasons (speed, capacity, etc.) one could also decide to use a Mono S column instead.

If conditions from an electrophoretic titration curve are transferred into a chromatographic system this usually results in optimal starting conditions and the separation can then be further optimized by adjusting sample size, flow rate and gradient shape to get the required resolution. In a high performance separation with MonoBeads it is essential that the buffering conditions are kept constant during the run and that a suitable buffering ion which has its pk value close to the start pH value is used.

As gel filtration is a volume dependent technique it is usually used for analytical purposes with small sample sizes or as a second step after concentration of the crude protein solution on an ion exchange column. In this technique the most important parameters to be optimized are sample size and flow rate if the gel with the correct fractionation range has been chosen. Due to the stability of Superose buffering conditions can be chosen as required by the protein.

The conditions in reversed phase chromatography depend on the hydrophobicity of the proteins to be separated. If no information is available from published material, the best way to proceed is to choose the corresponding reversed phase material with respect to its pore size and the molecular weight of the compounds to be separated. The elution pattern of a separation starting in water with a linear gradient to acetonitrile will give information in which range of the gradient the system has to be further optimized. To eliminate unspecific effects of charges and to increase hydrophobicity, ion pairing substances like 0.1 % TFA are added to the system. Depending on the separation pattern, one can get optimized resolution by modifying the gradient shape or by choosing certain isocratic conditions.

Automated and multidimensional FPLC-Systems

The FPLC-System is in a unique way an analytical as well as a preparative tool for the separation of proteins and polypeptides. The automation of the FPLC-System increases reproducibility, saves working time and increases the number of analysis per time or the total output in preparative systems. With the liquid chromatography controller LCC-500 the FPLC-System can be automated in various ways with the motor-driven valves MV-7 and MV-8.

1) Multisample injection

If several samples have to be separated in a FPLC-System - these can either be samples of similar nature which are separated on the same column according to the same separation program, or it can be different samples which are separated on different columns with different programs - one or two motor-driven valves MV-8 are connected to the inlet of the sample loop at the motor-driven sample valve MV-7. A peristaltic pump P-1 connected to the MV-7 and controlled by the liquid chromatography

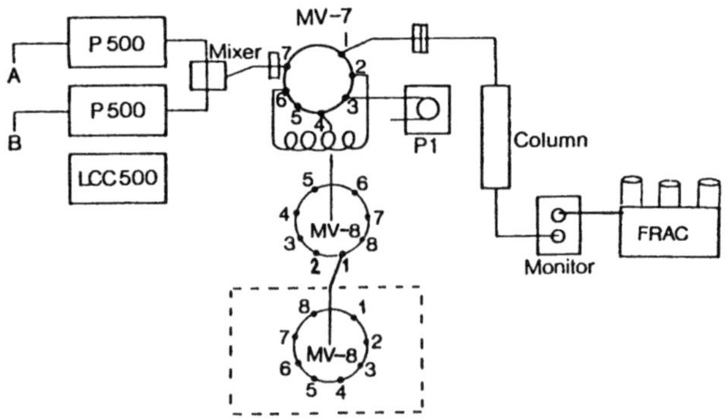

Fig.11 Multisample injection with two motorized valves MV-8 for up to 15 samples.

controller LCC-500 pumps the different samples into the sample loop. The set-up shown in Fig.11 is suitable for up to 15 different samples. By suitable programming of the LCC-500, single or duplicate runs as well as washing cycles between each sample can be performed automatically with this set-up. If a larger number of samples had to be processed in an automatic sequence, more MV-8 valves can be added to the sample application system, or an autosampler can be attached to the system which will also be controlled by the LCC. This arrangement for automatic sample injection is used for clinical or other routine applications like the separation of HbA_{1c} in diabetes mellitus. HbA_{1c} is a glycosylated hemoglobin. Fig.12 shows a baseline separation of HbA_{1c} from hemoglobin on a Mono S column. The optimal separation conditions have been determined by electrophoretic titration curves starting at a pH of 5.7 with a salt gradient, which has been optimized to give a baseline separation of the HbA_{1c} peak (4). As the HbA_{1c} is well resolved from the main HbA and other glycosylated hemoglobins, this peak can be quantitatively evaluated by the LCC and the actual and relative percentage of the HbA_{1c} can be printed out in the printer/plotter.

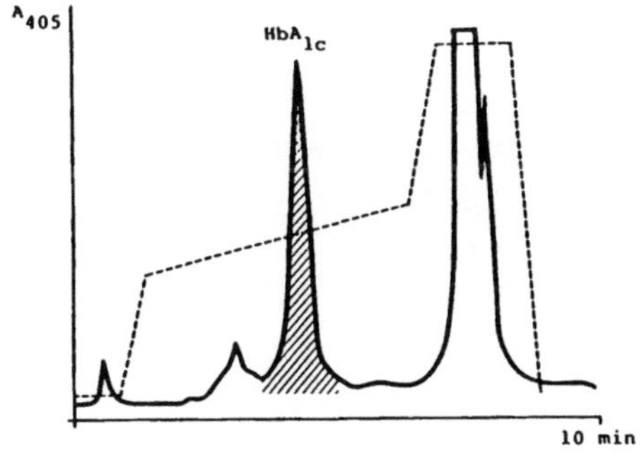

Fig.12 Separation of glycosylated hemoglobins on Mono S HR 5/5. 50 µl hemolysate are injected after equilibration with 0.01 mol/l malonate buffer pH 5.7 and eluted with a gradient of up to 0.3 mol/l LiCl (dashed line).

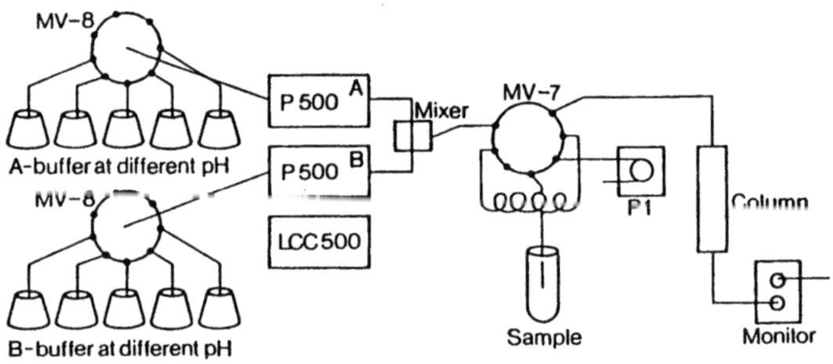

Fig.13 Use of motorized valves MV-8 for method development and optimization.

2) Method development

If electrophoretic titration curve data are not available, the separation conditions can be optimized by the automated FPLC-System. In this case a set-up as shown in Fig.13 is used. Motor-driven 8-port-valves MV-8 are connected to the one or both pump inlets while the peripheral ports of the MV-8 valves are connected to reservoirs containing different buffer salts, different pH values or different salt concentrations. With a suitable method development program one can automatically run the same sample repeatedly under different buffer conditions and check which pH or salt concentration is optimal for the required separation. Before the separation is performed with a new buffer composition, the total system is purged automatically with the next buffer solution.

3) Multicolumn system

The MV-8 valve can also be used if various separation techniques are used in the FPLC-System. In this case different columns like such for Gel Filtration, Ion Exchange Chromatography, RPC or Chromatofocusing are directly connected to the system via two MV-8 valves. If another separation mechanism is used the

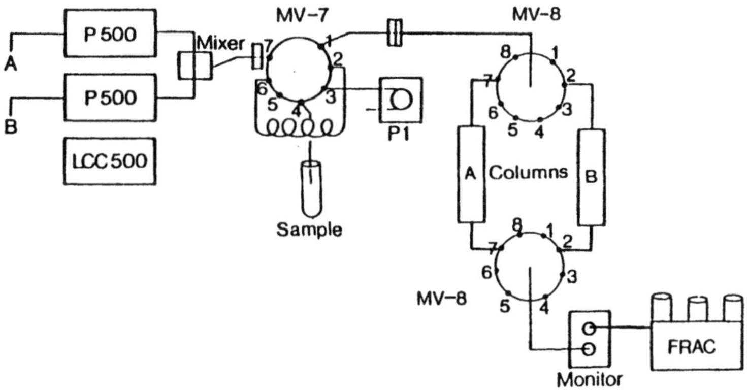

Fig.14 Two to eight different columns can be connected to MV-8 valves for column switching.

pumps are flushed with the new buffer, the MV-8 is switched to the correct column, and the column is equilibrated with the new starting buffer. By using this set-up shown in Fig.14 the new column must not be connected and disconnected mechanically to the system each time when it is used. Such a set-up can also be used for quality control runs after a preparative separation. In those cases two different types of column material may be used. For small molecular weight components up to 30.000 Daltons ion exchange chromatography can be performed on Polyanion SI material (8 µm). This silica based ion exchange material is covalently coated by polyethyleneimino groups and gives excellent yields for small molecules. Preparative separations can be performed on a material with 17 µm diameter which has the same surface coverage of charged groups. The separation on such a preparative column can be performed with the same program if a higher flow rate is used and much more material can be separated in a single run. If in Fig.14 column A would be a column with analytical 8 µm Polyanion SI material and column B would be the preparative 17 µm material, the purification of NADP could be optimized on column A with small amounts of material (µg-amounts). Then the same program could be used to run the separation with a higher flow rate on column

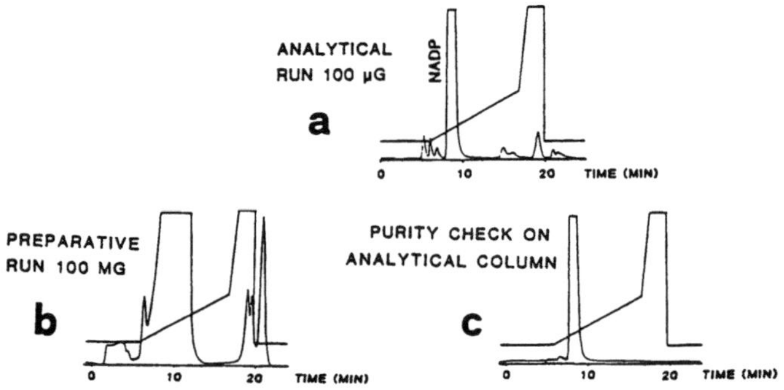

Fig.15 Purification of a commercial NADP on Polyanion SI 8 µm (a+c) and 17 µm (b).
Eluent 0.01 mol/l phosphate buffer pH 3.3 with linear gradient to 0.4 mol/l. Flow rate a+c: 1 ml/min, b: 4 ml/min.

B with up to a 100 mg, giving exactly the same resolution. Our fraction of the major NADP peak shown in Fig.15 can be re-run after switching the MV-8 valves back to the analytical column A to check the purity of the material.

4) Twodimensional separations

When twodimensional separations are performed, the system can be automated as shown in Fig.16. In this arrangement in renal diagnosis a desalting step on a Sephadex column is combined with an ion exchange step on a Mono Q column. In order to run both dimensions at the same time, the Sephadex G-25 column is connected to a P-1 peristaltic pump which is controlled by the LCC, while the Mono Q column is connected to a regular gradient system also controlled by the same LCC. If the number of samples that have to be processed per day do not require that both dimensions are run simultaneously, the FPLC-System offers the unique opportunity that Sephadex columns and other conventional gel filtration or affinity material can be run directly in the FPLC-System with the P-500 pumps. This unique combination with the use of traditional gels, which no other high performance liquid chromatography system can offer, is due to the fact that the P-500 pumps work at a constant flow rate even if the back pressure is zero. In the analysis of urine proteins it is necessary to remove UV-absorbing low molecular weight materials in a desalting step first. This is done on a Sephadex G-25 Superfine column, the proteins are eluted in the void volume of the column and collected in a superloop. In a superloop sample volumes up to 50 ml can be collected without zone spreading or dilution and afterwards injected on to the next column. While the Sephadex column is washed free from low molecular weight components and the liquid stream is directed to waste by a three-way-valve PSV-100, the protein fraction of the urine is pumped onto the Mono Q column by switching the MV-7 valve. The proteins are then eluted by a salt gradient, the next sample is automatically applied to the Sephadex column and the next protein fraction is ready in the superloop when the Mono Q column is reequilibrated again. Fig.17 shows the elution profile of urine proteins from Mono Q. This profile is depending on the function of the kidney as increased levels of

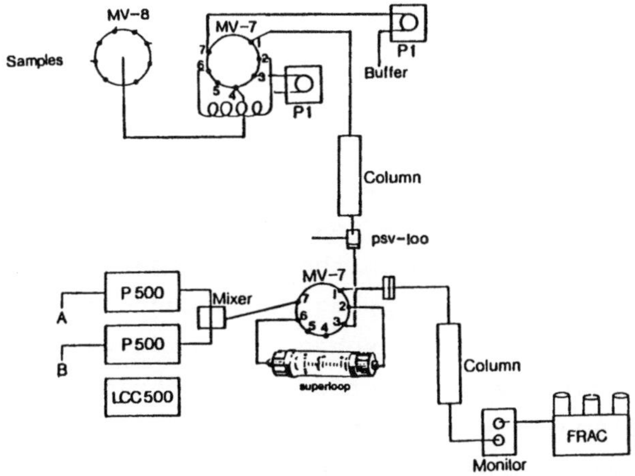

Fig.16 Automated two-dimensional chromatography, in which the interesting fraction from the first column is collected in a superloop and then further purified on a second column.

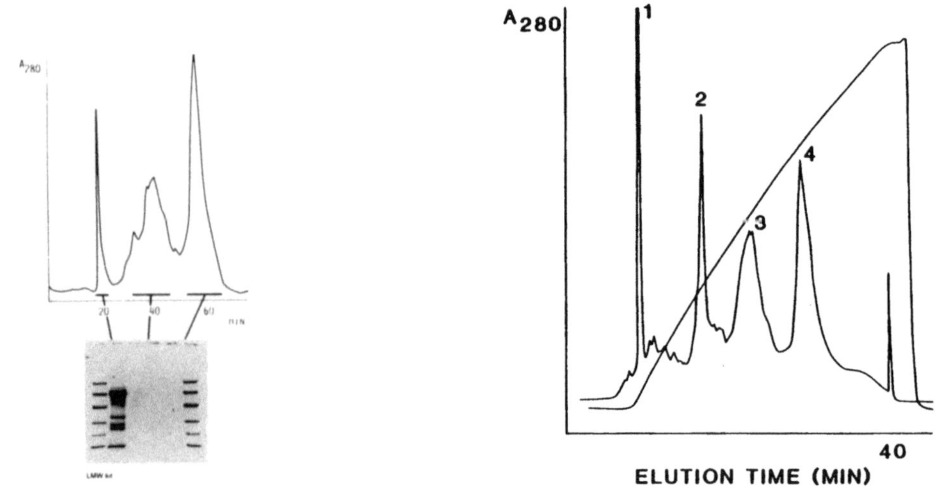

Fig.17 Separation of urine proteins.
Left: Separation of 0.5 ml urine on Sephadex G-25 Superfine (1.6 x 40 cm) in BisTris propane 10 mmol/l pH 7.5. Void volume peak (6 ml) at 20 min is collected in superloop and transferred to a Mono Q column equilibrated in the same buffer.
Right: Elution from Mono Q with a gradient to 10 mmol/l BisTris propane pH 9.5 with 0.35 mol/l NaCl. 1) β_2-microglobulin 2) retinol-binding protein 3) acid glycoprotein 4) albumin.

high molecular weight components like albumin indicate a glomerular misfunction of the kidney while increased levels of low molecular weight components like β_2-microglobulin indicate a tubular misfunction (5,6).

The automation of the FPLC-System in multidimensional separation is beneficial not only in preparative separation or complicated routine analysis but also in the investigation of genetic variants as well as peptide mapping and sequencing work. In those cases combinations of ion exchange chromatography and reversed phase chromatography or reversed phase chromatography on two different columns are quite often used. Fig.18 shows a preparation of human hemoglobin on a ProRPC HR 5/10 column. The high resolving power of this column shows that in this specific sample both normal and abnormal beta chains are present. The two fractions of the beta chains, β and β^*, have been isolated,

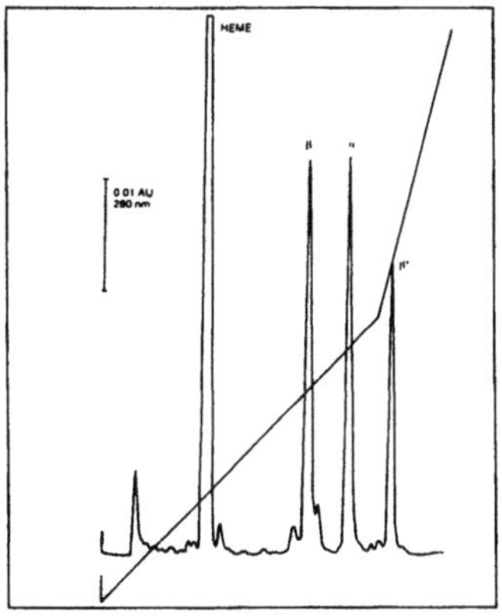

Fig.18 Isolation of normal and mutant human hemoglobin on ProRPC HR 5/10. Eluent A: 0.3 % TFA in H_2O/acetonitrile 61:39. Eluent B: water/acetonitrile 50:50. Linear gradient 0–50 % B in 75 min.

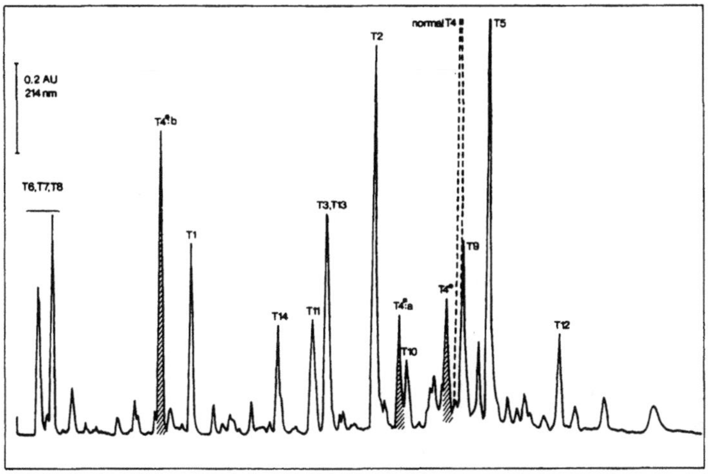

Fig.19 Fractionation on PepRPC HR 5/5 of tryptic peptides of β-chains of Fig.18. Eluent A: 49 mmol/l KH_2PO_4 in H_2O pH 2.9. Eluent B: eluent A/acetonitrile 50:50. Linear gradient 0–75 % B in 80 min.

subjected to a tryptic digestion and compared to one another. Fig.19 shows the separation of these tryptic fragments of the two beta chain on a PepRPC HR 5/5 column. The superimposed chromatograms reveal that the T4-peptide is not present in the β*-mutant, but that three new peaks (hatched) are present. Amino acid analysis of the two fragments T4*:a and T4*:b shows them to be cleavage products of T4*.

5) Automated Chromatofocusing

Chromatofocusing in the FPLC-System has been shown to be a very high resolution technique. Fig.20 shows a separation of serum on a Mono P column in a narrow pH interval from pH 5 to 4. In this region the genetic variants of G_c-globulin and α_1-antitrypsin are resolved. The results of a fused rocket immunelectrophoresis which followed the separation on the Mono P column, are shown on the lower part of Fig.20. Further investigations on a pH gradient with an interval of only 0.5 pH units showed that it is possible to differentiate mutants with Z and S variants of α_1-antitrypsin which are associated with plasma deficiency. If a lot of separations have to be performed on a Mono P column and time is the rate limiting step, the FPLC-System can be automated in a way shown in Fig.21. Here two Mono P columns are connected via two MV-7 motor valves to the two P-500 pumps, one of which delivers start buffer, the other one the

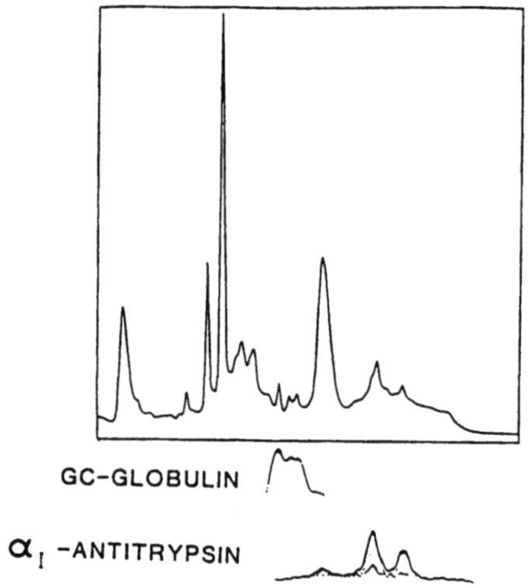

Fig.20 Chromatofocusing of albumin depleted (by Blue Sepharose CL-6B) plasma on Mono P.
Start buffer: 25 mmol/l piperazine/HCl pH 5.4
Eluent: Polybuffer 74/HCl diluted 1:15 pH 4.0
Lower part: Fused rocket precipitates in agarose containing specific antisera.

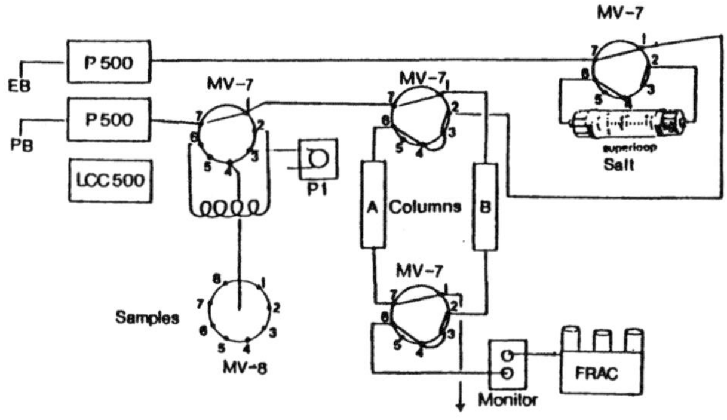

Fig.21 Chromatofocusing with two Mono P columns running in parallel.

polybuffer for elution. This set-up offers the possibility that the separation is run on one Mono P column, while the other one is equilibrated with start buffer. When the separation is finished both MV-7 valves are switched and the next sample can be applied to the second regenerated column. The first column is then reequilibrated with start buffer. An additional MV-7 valve may be included to wash the column to be regenerated with a high salt solution before equilibration to remove all bound substances. Before the next sample is applied to the reequilibrated column, the pH in the effluent has to reach the pH of the start buffer. This can be controlled by a flow-through pH-monitor. A flow-through electrode with a volume of only 30 µl suitable for such high resolution techniques as chromatofocusing in a FPLC-System is available. The signal of the pH-monitor is fed into the second monitor input of the LCC and can be used to control the pH level and to start the next sample application when the required pH-value is reached.

6) Multidimensional separations

The LCC-500 can also be used to control more complex multidimensional chromatography systems as they may occur during the routine purification process of one or a few selected proteins from a crude homogenate. Fig.22 shows the automated purification process for various enzymes from chicken skeletal muscle. The first step is a desalting on Sephadex G-25 Superfine from which the protein peak is transferred in the correct buffer composition to a cation exchange chromatography step. Under these conditions some enzymes are eluted in the void volume, others are fractionated as sharp peaks during gradient elution and can be collected in a fraction collector FRAC-100. The peak which was eluted in the void volume is transferred to a second small Sephadex G-25 Superfine column to perform a buffer exchange which makes this protein suitable to be chromatographed on a fourth step, which is anion exchange chromatography.

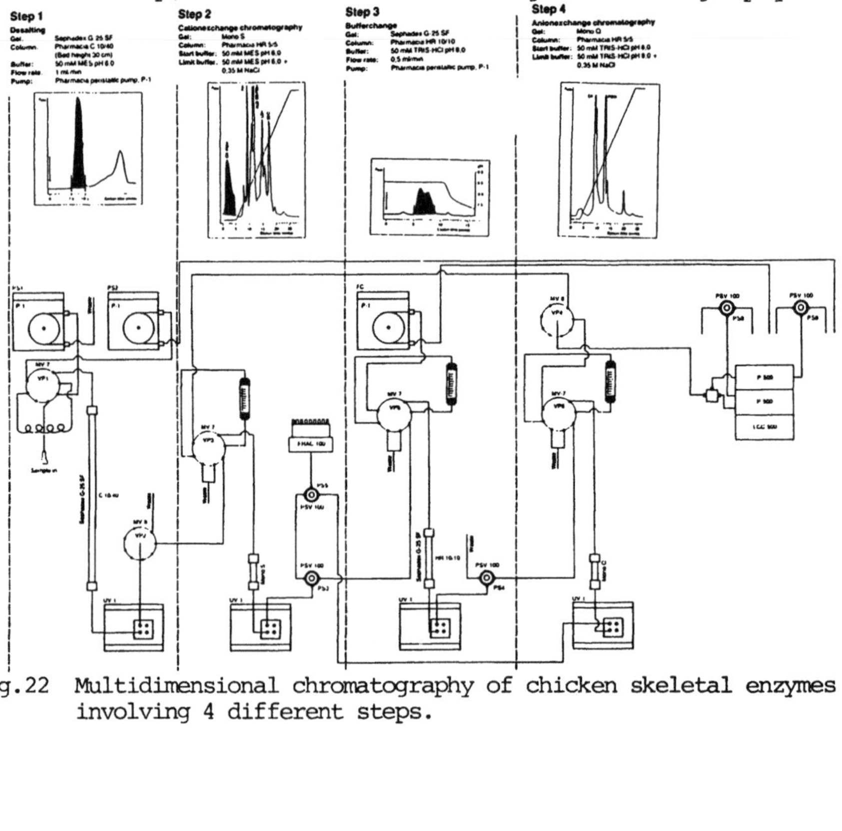

Fig.22 Multidimensional chromatography of chicken skeletal enzymes involving 4 different steps.

On this column two of the interesting enzymes, creatine kinase and phosphoglucomutase, are eluted as sharp peaks in the salt gradient. The interesting protein fractions between the various chromatographic steps are collected in a superloop and from there transferred to the next chromatographic column. The LCC controls all the necessary equipment including two P-500 pumps for gradient formation of the MonoBead columns as well as the various valves and pumps involved in the selection of different buffers and columns. This set-up enables the maximum throughput of samples, as most steps are running in parallel all the time.

These few examples should demonstrate how the high resolution of the MonoBead columns as well as the other special FPLC media in combination with the biocompatible hardware of the FPLC-System and the flexible automation possibilities can be used to solve problems in the analytical and preparative separation of proteins, polypeptides and polynucleotides.

References

1. Söderberg, L., Wahlström, L., Bergström, J.: Protides Biol. Fluids, Proc. 3, Colloq. 30, 635-640 (1982).
2. Fägerstam, L.G., Lizana, J., Axio-Fredriksson, U., Wahlström, L.: J. Chromatog. 266, 523-532 (1983).
3. Lindgren, G., Källman, J., Lundström, B.: 3rd Int. Symp. on HPLC of Proteins, Peptides and Polynucleotides, Monaco 1983, Abstract 105.
4. Jeppson, J.-O., Englund, H., Nylund, V.: 5th European Congress of Clinical Chemistry, Budapest 1983, Abstract 289.
5. Cooper, E.H., Turner, R., Johns, E.A., Lindblom, H., Britton, V.J.: Clin. Chem. 29, 1635-1640 (1983).
6. Lindblom, H., Axio-Fredriksson, U., Cooper, E.H., Turner, R.: J. Chromatog. 273, 107-116 (1983).

AMINO ACID ANALYSIS BY HIGH PERFORMANCE LIQUID CHROMATOGRAPHY AND PRECOLUMN DERIVATISATION

Keith Ashman and Armin Bosserhoff

Max-Planck-Institut für Molekulare Genetik
Abteilung Wittmann
Ihnestrasse 63-73
D-1000 Berlin 33 (Dahlem), West Germany

Introduction

Automatic amino acid analysis is a well established routine technique, however the sensitivity and analysis time of the traditional methods have not improved as rapidly as have those of protein isolation and microsequencing (1). For example on-line separation and detection of the phenylthiohydantoin (PTH) derivatives released during protein degradation in a fully automated protein sequencer already functions at a sensitivity of 25 pmol (2). Typically the classical automatic amino acid analysis methods make use of ion exchange chromatography to separate the free amino acids, followed by post column derivatisation either with ninhydrin (detection level 1 nmol to 500 pmol) or a fluorogenic reagent, e.g. fluorescamine (3) (detection level about 500 pmol) (4,5). The advent of high performance liquid chromatography (HPLC), and the development of precolumn derivatisation techniques offers an alternative to the established methods. There are several reagents which may be used to modify free amino acids, e.g.:

1. 9-Fluorenylmethylchloroformate (FMOC) (6)
2. Phenylisothiocyanate (PITC) - to detect the thiocarbamyl derivatives of the amino acids (7)
3. Dabsyl chloride (8)
4. Dansyl chloride (9,10)

Modern Methods in Protein Chemistry, Volume 2

5. 7-Fluoro-4-nitrobenzo-2-oxa-1,3-diazole (NBD-F) (11)
6. 7-Chloro-4-nitrobenzo-2-oxa-1,3-diazole (NBD-Cl) (12)
7. Ortho-phthaldialdehyde/mercaptoethanol (OPA) (13)

Of these different reagents OPA is the one we will be largely concerned with in the following pages. OPA offers a number of advantages as a derivatising reagent, particularly for an automatic system. Firstly the reagent is easy to handle, stable and can be applied to pre-column derivatisation techniques; the reaction with primary amines is both very fast, of high yield and easy to perform. The amino acid derivatives formed are detectable either by their U.V. absorption (330 nm) or by their fluorescence (λ_{ex}330 nm, λ_{em}455 nm). The latter permits a detection sensitivity of at least 1 pmol to be achieved; the introduction of microbore HPLC together with the appropriate equipment may well lower this to 10 femtomoles, i.e. the method is very sensitive. Further, fluorescence detection has the advantage that the derivatising reagent itself does not fluoresce, and hence produces no peak on the chromatogram. In general it produces a lower level of background noise compared to other fluorescent reagents, such as fluorescamine.

Methods

Apparatus

The following equipment, a Waters HPLC-system, was used: two M6000A HPLC pumps; Model 720 system controller; Model 710B WISP; Model M730 data module. The fluorescence detector was a model RF530 from Shimadzu ($_{ex}$330 nm, $_{em}$455 nm). The column (empty steel-column, 250 x 4.6 mm from Knauer, Berlin) was packed with Shandon Hypersil ODS 5 µm using a Shandon HPLC packing pump. Alternatively, columns obtained from Dr. I. Molnar, Berlin, filled with the same support were used. The column oven consisted of an aluminium block (made in the institute's workshop) through which water from a thermostatically controlled water bath (Haake) set at 30°C, is passed. A guard column was also included in the system and was filled with glass beads (reflex beads no. 31/18 from Dragon Werk, Bayreuth).

Reagents

Buffer A*	12.5 mM di-sodium hydrogen phosphate pH 7.2
Buffer B	methanol/tetrahydrofuran (1000:30, v/v) both UVASOL grade from Merck
Reaction Buffer	1 M potassium borate pH 10.4
Stock OPA Solution	56 mg OPA dissolved in 1 ml methanol + 50 µl mercaptoethanol
Amino Acid Standards	Standard amino acid mixture from Pierce diluted to a concentration of 1 pmol μl^{-1} or 10 pmol μl^{-1}, respectively
Sample Buffer	0.2 M citrate buffer pH 2.2

*Buffer A was made from phosphate as this buffers more strongly than acetate and contains less contaminants.

The OPA reaction mixture consisted of the stock solution, which was stored at 4°C for a month, diluted 1:10 with the reaction buffer. This mixture was also stored frozen and freshly prepared weekly.

The automatic derivatisation was achieved essentially as described in (14,15). The sample processor makes 2 injections, one of OPA reagent and one of sample with the solvent flow set at zero. The flow rate is then increased from zero to

0.1 ml min^{-1} over the next 2 minutes, before increasing it to 1.5 ml min^{-1} over a 30 seconds period. Finally, the flow rate is held constant for the rest of the run and is reset to zero before the next set of injections (OPA plus sample) occurs. The amino acid derivatives were eluted from the column with a gradient of buffers A and B, as shown in Fig. 1, and with the following sequence of eluent changes:

GRADIENT TABLE

Time min.	Flow ml min^{-1}	%A	%B
0	0	75	25
2	0.1	75	25
2.5	1.5	75	25
7.5	1.5	75	25
13	1.5	55	45
14	1.5	47	53
16	1.5	47	53
17	1.5	45	55
20	1.5	25	75
24	1.5	25	75
25	1.5	75	25
29	1.5	75	25
30	0	75	25

RESULTS AND DISCUSSION

The amino acid separation on C18-reversed phase HPLC after OPA-precolumn derivatisation is presented in Fig. 1. The analytical system detailed above can be described as a micro system; it permits amino acid analyses of low picomole amounts of material as a simple routine technique. However, a number

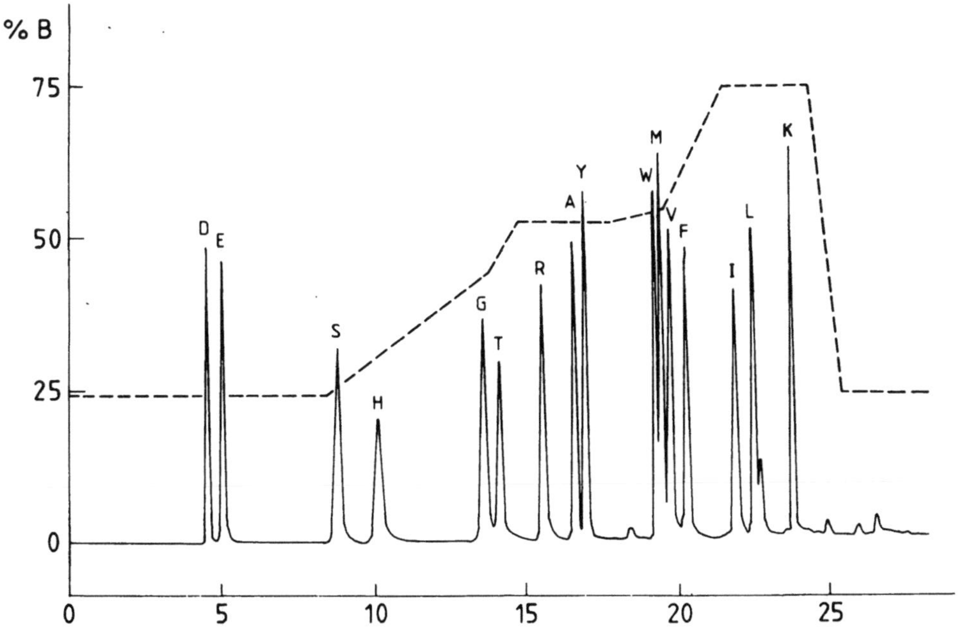

Fig. 1: Shown is the separation of 100 pmol of a standard amino acid mixture by reversed phase HPLC, using a 250 x 4.5 mm column, filled with Shandon Hypersil ODS 5µ reversed phase material. Buffer A was 12.5 mM Na_2HPO_4 pH 7.2 and buffer B 1 litre of methanol to which 30 ml of tetrahydrofuran were added.

Fig. 2: Shown is a blank HPLC-trace of ortho-phthaldialdehyde solution and citrate buffer pH 2.2 which were injected into the system under the same conditions as given in Fig. 1.

of criteria should be taken into account with such a system which are discussed below:

Specificity

This requires that the applied method yields information only about those substances that are of interest, but not about contaminants even if at low substance levels the substance to contaminant ratio becomes increasingly adverse.

Resolution

Separation efficiency in any chromatographic technique bears the relationship of selectivity versus band broadening for two adjacent peaks; the relationship is expressed as

$$R = \frac{V_e}{(W_a - W_b)/2}$$

that is the distance between two adjacent peak maxima (V_e) divided by the sum of the peak widths (W_a and W_b) at half height. Resolution greater than 1.0 must be obtained to ensure accurate quantization.

Accuracy

The method of quantization is important. In general, in an automatic analysis system an integrater is used and this equipment must be correctly calibrated if accurate results are to be obtained. This involves careful choice of the parameters used for the integration, e.g., the peak width, level of background suppression and selection of a peak window that ensures that the machine correctly identifies the various peaks on a chromatogram. A further constraint is accurate retension times (see below).

Sensitivity

The absolute sensitivity of an analytical method is often considered to be the detection limit, i.e., the smallest signal that can be distinguished from the background noise. However, for most practical purposes reliable results can only be obtained at signal levels at least twice as high as the detection limit. Another important factor is the level of contaminants in the samples: the background noise increases with decreasing amino acid concentrations.

Linearity

The relationship between the concentration and the measured property of the substance should, where possible, be linear over the working concentration range of the analytical method. Although the Lambert-Beer law holds true over a wide range of concentrations it is necessary to check the linearity experimentally where spectroscopic techniques are applied, particularly fluorescence spectroscopy.

Practicability

For use as a routine method it is desirable that an analytical technique be straightforward, with as few manipulation steps as possible, and that the retension times and sensitivity levels can be maintained.

OPA Pre-column Derivatisation as an Analytical System

Does the system described meet the criteria laid down above. OPA reacts with primary amines to form fluorescent derivatives (16) (scheme 1). As far as amino acid analysis is concerned the reagent is specific for primary amino acids,

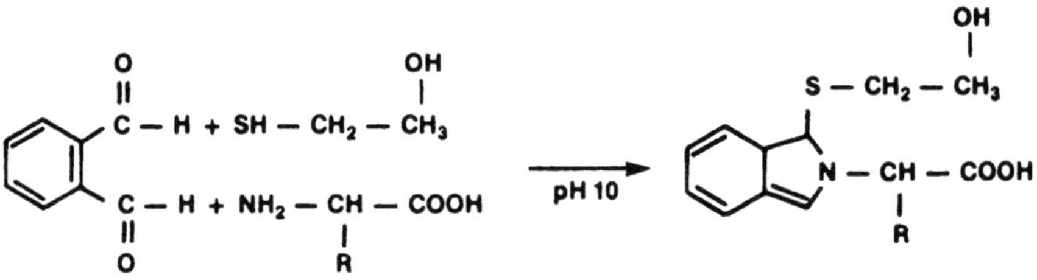

Scheme 1: OPA-reaction with primary amino acids (16)
A substituted iso-indole-derivative rapidly forms between a primary amino acid and ortho-phthaldialdehyde in the presence of a reducing agent (ß-mercaptoethanol).

as long as the sample contains no other primary amines. It is therefore of importance that the samples are carefully prepared. The tubes used for the hydrolysis must be clean: tempering at 400°C or soaking in chromic acid overnight followed by rinsing with double distilled water and methanol removes all traces of contaminating primary amines. The hydrochloric acid used for the hydrolysis must be free of chlorine and contaminants (twice-distilled), or gas phase hydrolysis can be used. The same holds true for the citrate buffer used to dissolve the hydrolysed samples. Figure 2 shows a blank analysis using the same gradient system as above, where only 10 µl citrate buffer were derivatised and recorded at the same sensitivity range. There are clearly amino acids and other fluorescent compounds present, at the 100 pmol level. This result shows that such a contamination is not of significance for routine analyses at this level, but at higher sensitivity it becomes more problematic. The same is true for the elution buffers. Procedures to purify the buffers are being developed. They should be filtered through a 0.2 µm membrane filter (0.2 µm nitrocellulose filter from Sartorius Membranfilter, Göttingen), as this removes any dust particles and microorganisms that they may contain, and has the further advantage of thoroughly degassing them. In the case of buffer B the tetrahydrofuran should be added after filtering, as vacuum filtration of the mixture leaves an undefined composition.

Reproducibility

The data in Table 1 show that the method gives reproducible results. But there are a number of pitfalls that should be mentioned. It is clear that standard solutions of known concentrations are required for the calibration runs. However as extreme dilutions have to be made accuracy is difficult to achieve. Further the highly diluted standards tend to degrade more rapidly than concentrated ones. The glassware used for the standard preparation must also be contamination free, as must be the buffers used for the dilution. Another crucial point is the precision of the gradient forming system used. This must provide reproducible gradients as constant retention times have to be obtained for the final automatic quantization. The merits and disadvantages of the different gradient mixing methods available (high pressure mixing with 2 pumps versus low pressure mixing with one pump and a control valve) are beyond the scope of this paper. The results depend largely on the HPLC-system employed. The gradients used should be kept as simple as possible (e.g., linear slopes as opposed to curves), since the more complex a system the harder it is to reproduce it accurately. Routine analyses demand stable columns, a desire which is difficult to meet with present day HPLC-columns and supports, especially at high back pressures. We found the use of an automatic system for the OPA-derivatisation advantageous as far as reproducibility is

Table 1: Results from 10 successive standard runs with the HPLC amino acid analysis system given in Fig. 1

The retention times for each of the amino acid peaks are given in minutes (upper value) and the amino acid amounts in picomoles (lower value), as calculated by the Waters Modul M70 data module (parameters: peak width at half height, 11 sec; noise reaction, 2).

	1	2	3	4	5	6	7	8	9	10
Asp	4.87 107.7	4.89 101.1	4.87 98.7	4.91 97.5	4.93 99.0	4.95 101.5	4.93 110.7	4.91 102.0	4.85 100.8	4.89 101.6
Glu	5.44 105.6	5.46 101.3	5.40 98.6	5.46 98.0	5.50 98.5	5.51 100.8	5.46 112.2	5.44 103.2	5.42 101.7	5.42 101.2
Ser	9.40 106.5	9.46 103.5	9.33 101.2	9.46 101.8	9.46 100.5	9.53 102.4	9.44 114.2	9.22 104.1	9.27 101.2	9.20 99.4
His	10.89 104.8	10.92 102.3	10.76 101.0	10.92 100.8	10.90 100.6	10.98 102.6	10.83 113.1	10.59 104.1	10.64 101.5	10.51 99.0
Gly	14.17 100.5	14.22 105.8	14.13 106.1	14.24 106.9	14.26 107.3	14.31 106.4	14.50 113.6	14.02 107.6	13.92 101.8	13.79 98.5
Thr	14.74 109.9	14.79 105.2	14.70 100.8	14.81 100.8	14.83 102.1	14.86 102.7	15.01 113.8	14.63 101.2	14.54 100.7	14.43 98.6
Arg	16.35 103.8	16.40 103.0	16.37 102.7	16.42 103.0	16.48 102.8	16.48 102.7	16.51 112.2	16.37 104.7	16.08 100.6	16.01 100.0
Ala	17.49 107.1	17.54 104.0	17.56 104.5	17.56 104.1	17.63 105.6	17.60 103.0	17.60 114.7	17.56 105.7	17.34 102.5	17.29 101.2
Tyr	17.94 106.6	18.00 102.6	18.04 103.1	18.02 101.8	18.07 101.7	18.04 102.9	18.00 114.6	18.07 103.8	17.80 103.7	17.73 100.7
Trp	20.46 102.2	20.47 101.6	20.57 101.0	20.51 101.5	20.57 103.4	20.49 102.6	20.42 111.9	20.69 105.7	20.32 110.2	20.25 104.1
Met	20.64 101.8	20.68 99.2	20.77 101.1	20.71 99.2	20.77 98.6	20.69 99.1	20.62 109.9	20.90 104.0	20.50 101.1	20.45 101.7
Val	20.97 104.3	20.99 99.9	21.10 98.3	21.02 97.8	21.08 97.4	21.01 98.5	20.91 108.2	21.24 103.4	20.82 105.2	20.75 102.5
Phe	21.61 107.0	21.61 104.4	21.74 106.2	21.65 104.5	21.70 104.1	21.61 108.2	21.48 115.8	22.01 104.2	21.42 107.8	21.31 107.4
Ile	23.46 102.2	23.44 97.4	23.63 98.8	23.48 101.1	23.57 96.5	23.41 100.1	23.22 107.2	24.03 100.6	27.21 105.8	23.01 106.5
Leu	24.31 104.2	24.27 101.5	24.45 107.7	24.31 104.6	24.40 103.7	24.23 105.3	24.03 107.0	24.78 109.8	24.03 109.4	23.81 109.0
Lys	26.10 99.0	26.10 93.8	26.16 102.5	26.12 97.8	26.18 99.8	26.12 96.3	26.08 101.6	26.16 102.3	26.02 103.4	25.99 102.5

concerned, because the reaction time and conditions can be held constant, thus removing one source of error.

Sensitivity

Pre-column derivatisation with OPA is a sensitive technique as it avoids the additional peak-broadening of the post-column technique. The method described above can be used on a routine basis with concentrations as low as 10 pmol per amino acid.

Fluorescence quenching effect: Dissolved oxygen in the elution buffers can lead to a serious reduction in the fluorescence levels. This difficulty can be overcome by bubbling either helium or nitrogen through the buffers which allows their continuous degassing. We chose nitrogen because it is considerably cheaper.

Reagent to sample volume: We have observed that the serial injection of reagent and sample affects the derivatisation yield. A sample to reagent ratio of, e.g., 5 - 10 µl to 20 µl was satisfactory for analytical runs. It was found necessary to keep the respective volumes constant.

Linearity: In Table 2 data to support the linearity of the method are listed.

Practicability

The HPLC amino acid analysis method described above is practical, and can be used routinely. As an example the amino acid analysis of glucagon, performed with the OPA system, is shown in Fig. 3. The peptide was hydrolysed with thioglycolic acid, in order to prevent the destruction of tryptophan. The chromatogram shows the good yield of this amino acid obtainable by the OPA method.

Table 2:

Shown are the results of successive runs with 10, 100, 200 and 1000 pmol. The data module was calibrated at the 100 pmol level. The results show that the method is linear over this concentration range.

	10 pM	10 pM	100 pM	100 pM	200 pM	200 pM	1000 pM	1000 pM
Asp	10.9	10.4	101.1	98.7	225.1	229.9	955	936
Glu	10.6	9.8	101.3	98.6	225.2	228.0	972	958
Ser	9.9	9.6	103.5	101.2	223.8	225.0	836	962
His	9.0	8.7	102.3	101.1	243.5	251.7	1230	1221
Gly	10.0	9.4	105.8	106.1	219.4	223.9	1011	1010
Thr	9.5	9.8	105.2	100.8	220.3	217.5	972	951
Arg	9.7	9.4	103.0	102.7	235.1	226.2	961	962
Ala	10.8	10.7	104.0	104.5	225.1	226.2	1028	1009
Tyr	10.3	10.2	102.6	103.1	247.7	230.9	982	976
Trp	11.0	11.2	101.6	101.0	240.6	229.2	968	970
Met	10.6	10.0	99.2	101.1	237.8	235.9	1009	1001
Val	10.3	10.5	99.9	98.3	227.2	226.7	1003	986
Phe	9.9	9.6	104.4	106.2	215.2	211.7	863	985
Ile	10.9	10.3	97.4	98.8	237.9	243.2	968	947
Leu	9.2	9.1	101.5	107.7	238.3	243.0	964	949
Lys	10.6	10.4	93.8	102.5	222.4	222.1	867	874

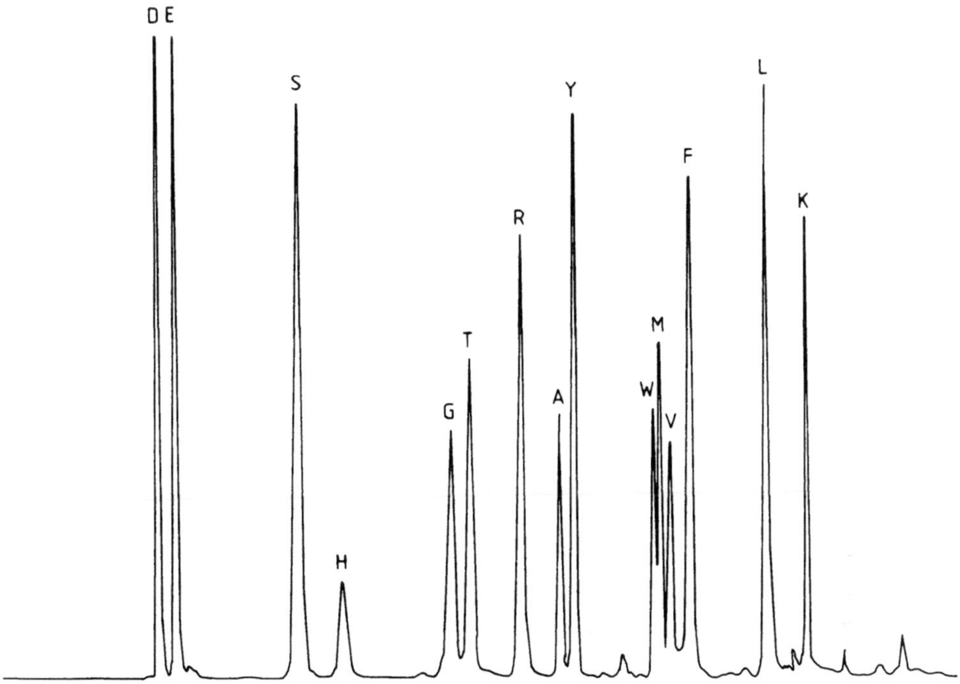

Fig. 3: Analysis of glucagon hydrolysate

Shown is the analysis of 500 pmol of glucagon, hydrolysed in 4% thioglycolic acid in 5.7 M HCl for 22 h at 110°C under vacuum. One fifth of the hydrolysed sample was injected. The tryptophan recovered after this hydrolysis is clearly visible.

Automation with OPA is easy: due to the speed and simplicity of the reaction no lengthy reaction procedures or extractions prior to the analysis are necessary. Further, OPA itself, does not fluoresce and hence there is no peak of excess reagent on the chromatogram if the derivatives are detected with a fluorescence spectrophotometer. This is an advantage over the HPLC-analysis employing PITC or its homologues. The buffers we have chosen are simple to make, and hence consistant quality is easily achieved. The addition of sodium azide to a concentration of 100 µg per ml buffer will prevent the growth of microorganisms. As buffer B is an all organic mixture, pumping 75% B permits thorough rinsing of the column. The occasional rinsing of the column with 100% B is also recommended. The sample buffer, 0.2 M citrate buffer pH 2.2, is the normal sample buffer for classical amino acid analysis and its use has the advantage that samples can be analysed by both methods without any special preparation.

The use of an HPLC system for analysing amino acids has the advantage that in a laboratory where the number of samples to be processed is not sufficient to warrant a dedicated amino acid analyser, the system can be applied at different times to other problems, for example to peptide or protein separation with or without changing the column and buffer system. This can be an important consideration when working on a limited budget.

The presence of a thiol-group is necessary for the OPA derivatisation. We used the usual mercaptoethanol, but mercaptopropionic acid may also be used. In the latter case more hydrophilic derivatives are formed, which require gradients with lower concentrations of the organic eluent when separating them, by reversed phase HPLC. This reagent also has the advantage of being less volatile than mercaptoethanol, and therefore is odour free (17).

Disadvantages of the OPA and the Stop Flow System

The stop flow system of automatic derivatisation used here, which requires the pumps to be stopped during the injection of reagent and sample, has the disadvantage that the column is not continuously operating under constant flow and pressure conditions. This leads to a reduced column life. The serial injection of reagent and sample means that they lie contiguously in the capillary, and this results in poor mixing and derivatisation. Hence the sample to reagent volume ratio has to be kept constant throughout a series of analyses, e.g. 1:2 (see above). HPLC columns become easily blocked, although the use of a guard column which can be easily changed alleviates this problem.

A serious drawback with OPA derivatisation is the failure to detect the secondary amino acids, proline and hydroxyproline. There are several possible solutions to this problem, one is the addition of hypochlorite to the solution (18), which opens the proline-ring thereby generating a free amino-group. This procedure damages the HPLC columns, as the hypochlorite is not removed from the sample, and the labile amino acids will also be attacked. The second procedure (19) relies on enzymatic modification of proline or hydroxy-proline, with the respective dehydrogenases which we have not tested. This modification has the further disadvantage that proline is converted to glutamic acid and hydroxy-proline to glycine. A third solution to the proline problem would be the use of a different derivatising reagent that reacts with all amino acids (11,12) or the recently proposed FMOC-reagent (6). However, with these reagents the reaction takes more time or the coupling has to be followed by repeated extractions of the excess of reagent which substantially changes the composition of a polypeptide's hydrolysate due to extraction losses of some of the amino acids.

Concluding Remarks

The results presented here show that despite the practical problems discussed above, pre-column derivatisation with OPA followed by separation of the derivatives by reversed phase HPLC is a method that can be applied to routine amino acid analysis. The results demonstrate that the technique fulfils the requirements of an advanced analytical system: it is specific; the resolution is good; the detection response is linear; it is reproducible, accurate and sensitive. Further development of HPLC methodology, such as the use of microbore columns (20), and the availability of other derivatising reagents may well make amino acid at femtomole concentrations possible. But at such concentrations one has to draw attention to the fact that the more sensitive the methods become, the greater will be the difficulties encountered. The preparation of the samples and the general cleanliness of the equipment will be of paramount importance, as free amino acids and other contaminants are everywhere.

Acknowledgement

We thank Dr. I. Molnar for kindly supplying HPLC-columns.

References

1. Spackman, D.H., Stein, W.H., Moore, S.: Anal. Chem. 30, 1190-1206 (1958)
2. Wittmann-Liebold, B., Ashman, K. (in this volume)
3. Stein, S., Böhlen, P., Stone, J., Dairman, W., Udenfriend, S.: Arch. of Biochem. and Biophys. 155, 202-212 (1973)
4. Benson, J.R., Hare, P.E.: Proc. Natl. Acad. Sci. USA 72, No. 2, 619-622 (1975)
5. Barbarash, G.R., Quarles, R.H.: Anal. Biochem. 119, 177-184 (1982)
6. Einarsson, S., Josefsson, B., Lagerkvist, S.: J. of Chrom. 282, 609-618 (1983)
7. Heinrikson, R.L., Meredith, S.C.: Anal. Biochem. 136, 65-74 (1984)
8. Chang, J.-Y., Knecht, R., Braun, D.G.: Methods Enzymol. 91, 41-48 (1983)
9. Tapuki, Y., Schmidt, D.E., Linder, W., Karger, B.L.: Anal. Biochem. 115, 123-129 (1981)
10. De Jong, C., Hughes, G.J., van Wieringen, E., Wilson, K.J.: J. of Chrom. 241, 345-359 (1982)
11. Watanabe, Y., Imai, K.: J. of Chrom. 239, 723-732 (1982)
12. Ghosh, P.B., Whitehouse, M.W.: Biochem. J. 108, 155-156 (1968)

13. Roth, M.: Anal. Chem. 43, No. 7, 880-882 (1971)

14. Fleury, M.O., Ashley, D.V.: Anal. Biochem. 133, 330-335 (1983)

15. Ashman, K., Bosserhoff, A.: Königsteiner Chromatographie-Tage, H. Aigner, ed., Waters GmbH, Eschborn, pp. 198-207 (1984)

16. Simons, S.S., Jr., Johnson, D.F.: J. Org. Chem. 43, 2886 (1978)

17. Godel, H., Graser, T., Földi, P., Pfaender, P., Fürst, P.: J. of Chrom. 297, 49-61 (1984)

18. Nakazawa, K., Tanaka, H., Arima, M.: J. of Chrom. 233, 313-316 (1982)

19. Meiler, L., Leisinger, T.: Eur. J. Biochem. 129, 67-75 (1982)

20. Jones, B.N., Gilligan, J.P.: J. Chrom. 266, 471-482 (1983).

POTENTIALS OF HIGH-PERFORMANCE ION-EXCHANGE CHROMATOGRAPHY IN PROTEIN SEPARATIONS -

Synthesis and Characterization of Aprotinin Derivatives and Homologues

Herbert R. Wenzel, Jürgen Beckmann, Armin Mehlich, Jürgen Siekmann, Harald Tschesche
Universität Bielefeld, Fakultät für Chemie, D-4800 Bielefeld 1

Hermann Schutt
Bayer AG, Institut für Biochemie, D-5600 Wuppertal 1

Introduction

During the past decade, high-performance liquid chromatography has undoubtedly revolutionized the isolation and characterization of peptides and proteins. Resolution, speed and sensitivity of analysis have reached new dimensions. The broad scope and the rapid development of HPLC are well documented by several symposia and books entirely devoted to the various aspects of the new technologies [1 - 4].

So far, reversed phase systems, sometimes in combination with ion-pairing techniques, have predominantly been used to separate peptides and proteins. However, HPLC versions of gel-permeation and ion-exchange chromatography are being given increasing attention. In fact, high-performance ion-ex-

---

Abbreviations: TKI, basic trypsin-kallikrein inhibitor (Kunitz) from bovine organs or aprotinin (registered trade name Trasylol®). Numbers in parenthesis following "*", "**", or "reduced" denote the position of one split peptide bond, two split peptide bonds, or a reductively split disulfide bond respectively, according to the numbering of the TKI sequence shown in Fig. 1.

Modern Methods in Protein Chemistry, Volume 2

change chromatography has emerged as a universal high speed/high resolution method for the separation of biologically active macromolecules [5]. This holds especially true with the introduction of hydrophilic polymer based matrices, for example Bio-Gel TSK SP-5PW cation-exchange column and Bio-Gel TSK DEAE-5PW anion-exchange column from BIO-RAD, Richmond USA, or Ultropac TSK SP-5PW cation-exchange column and Ultropac TSK DEAE-5PW anion-exchange column from LKB, Bromma Sweden. Unlike traditional ion-exchange HPLC columns with silica matrix, these new columns with their resin based support permit a pH operating range of 2 - 12 and hence provide a wider application potential. Moreover, they have a capacity of several milligrams of sample and can separate proteins with molecular weights as high as 10^7.

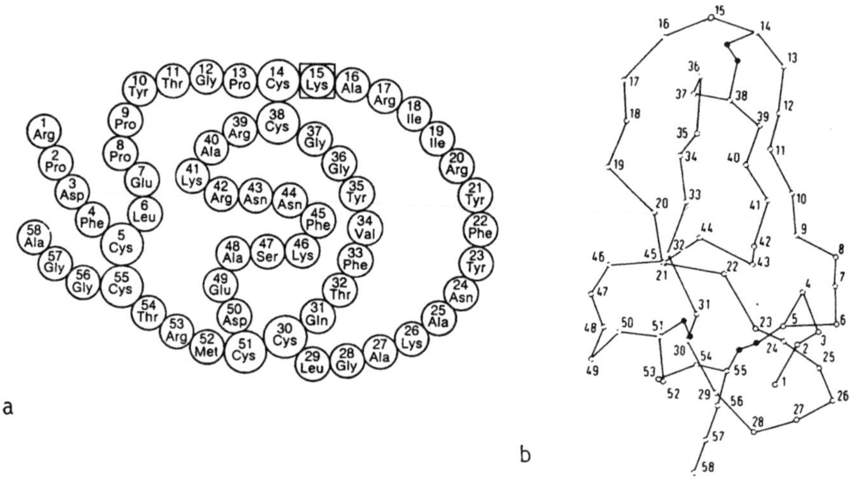

Fig. 1: Covalent structure (a) [9, 10] and projection of the three-dimensional structural model of aprotinin (b) [11], O = α-carbon atoms of the main chain, ● = sulfur atoms of the disulfide bridges.

The purpose of this paper is to present a high-performance cation-exchange chromatography system which has been exploited in our laboratory to characterize a variety of structurally related proteins: aprotinin, the trypsin-kallikrein inhibitor (Kunitz) from bovine organs (TKI) [6 - 8], naturally occurring isoinhibitors, chemically and enzymatically modified derivatives, and semisynthetic inhibitor homologues with single amino acid replacements.

These proteins, which differ from one another by more or less pronounced charge differences, provide an ideal model system to evaluate the potentials of ion-exchange HPLC.
Fig. 1 shows the primary structure [9, 10] and tertiary structure [11] of aprotinin. The single polypeptide chain consists of 58 amino acid residues which are cross-linked by three disulfide bridges. It is folded in such a way that a pear-shaped molecular structure results. The reactive-site peptide bond, Lys^{15}- Ala^{16}, which specifically interacts with the active site of the enzyme in a proteinase/inhibitor complex [6], is located at the top of the pear-shaped molecule.

Experimental Conditions

Liquid chromatograph	:	Hewlett-Packard 1084A equipped with LC terminal 79850A, variable wavelength UV detector 1030B and automatic sample transport system 79842A.
Column	:	Ultropac TSK SP-5PW, 7.5 x 75 mm, strong cation-exchanger, LKB, Sweden.
Oven temperature	:	30 °C.
Detection	:	UV absorbance at 220 nm.
Samples	:	About 10 μg of each compound are injected, protein concentration 1 mg/ml water.
Elution	:	10 % B for 5 min, then (insert "%B" in Fig. 2 - 7) linear gradient from 10 % B to 100 % B over 45 min. Solvent A: 0.01 M phosphate buffer, pH 7.0 Solvent B: 1 M NaCl in 0.01 M phosphate buffer, pH 7.0 The column is equilibrated with 10 % B for 10 min between two runs. Compounds which pass the column unretarded, e.g. succinic acid, reach the detector 2.5 min after injection.
Flow rate	:	1 ml/min.

Results and Discussion

The retention times t_R determined for a selection of bovine Kunitz inhibitors and derivatives thereof are summarized in the table. The times represent mean values from about seven runs for each compound with a standard deviation of ± 0.1 min for the same column.
Short outlines of synthetic routes to the individual compounds or references to published procedures will be given in the following sections together with representative HPLC chromatograms.

TKI and derivatives with split or closed peptide bonds (Fig. 2)

Incubation of aprotinin with catalytic amounts of certain serine-proteinases leads to a highly specific hydrolysis of the Lys^{15}-Ala^{16} reactive-site peptide bond [12, 13]. This reaction does not proceed to completion; an equilibrium between native or "virgin" TKI and "modified" TKI* (15, 16), characterized by a pH-dependent equilibrium constant K_{hyd}, becomes established. The cation-exchange HPLC system described appears very useful in measuring these conversions and determining values for K_{hyd}, because TKI* (15, 16) with its additional carboxyl end group is eluted prior to TKI with baseline-separation.
As bovine trypsin splits the reactive-site peptide bond of TKI extremely slowly [12], pure TKI* (15, 16) as reference substance is prepared by selective reduction of the disulfide bridge Cys^{14}- Cys^{38}, tryptic hydrolysis and air reoxidation of the half-cystine residues [14]. During the second step, the peptide bond Arg^{39}- Ala^{40} is also partly split leading to TKI** (15, 16; 39, 40) which elutes from a preparative CM-Sephadex C-25 column (0.01 M borate, pH 8.6, NaCl-gradient) in front of TKI* (15, 16). Surprisingly this sequence is reversed in our HPLC standard system at pH 7.0: TKI** (15, 16; 39, 40) is eluted between TKI* (15, 16) and TKI.
Recently a novel circular form of TKI was described with the amino and carboxyl termini linked together via a peptide bond by the action of a water-soluble carbodiimide [15, 16]. This circular TKI elutes shortly before TKI from the HPLC column, again contrary to the preparative run on a Whatman CM52 carboxymethyl-cellulose column at pH 9.0.

TKI derivative	t_R (min)
S-carboxymethylated TKI reduced (14, 38)	3.92
des-Lys^{15} des-Arg^{39} TKI** (14, 16; 38, 40)	11.31
des-Lys^{15} iso-TKI II* (14, 16)	11.87
des-Ala^{16} des-Arg^{17} TKI* (15, 18)	13.24
des-Lys^{15} TKI* (14, 16)	13.72
guanidinated des-Ala^{16} des-Arg^{17} TKI* (15, 18)	15.05
des-Arg^{39} Arg^{15}-TKI* (38, 40)	15.09
iso-TKI I	15.13
Val^{15}- TKI	15.62
Phe^{15}- TKI	16.26
iso-TKI II* (15, 16)	17.32
des-Ala^{16} TKI* (15, 17)	17.96
TKI* (15, 16)	18.55
S-carboxamidomethylated TKI reduced (14, 38)	19.18
S-methylated TKI reduced (14, 38)	19.59
TKI** (15, 16; 39, 40)	20.14
iso-TKI II	20.70
guanidinated TKI* (15, 16)	20.72
transaminated TKI	21.02
guanidinated des-Ala^{16} TKI* (15, 17)	21.04
circular TKI	21.67
TKI	22.05
Arg^{15}- TKI	23.92
guanidinated TKI	25.70
S-aminoethylated TKI reduced (14, 38)	34.08
TKI* (15, 16) hexamethyl ester	39.98
TKI pentamethyl ester	40.20
guanidinated TKI* (15, 16) hexamethyl ester	46.08

Table: Retention times of bovine Kunitz inhibitor derivatives on a high-performance cation-exchange chromatography column. (For details see experimental conditions.)

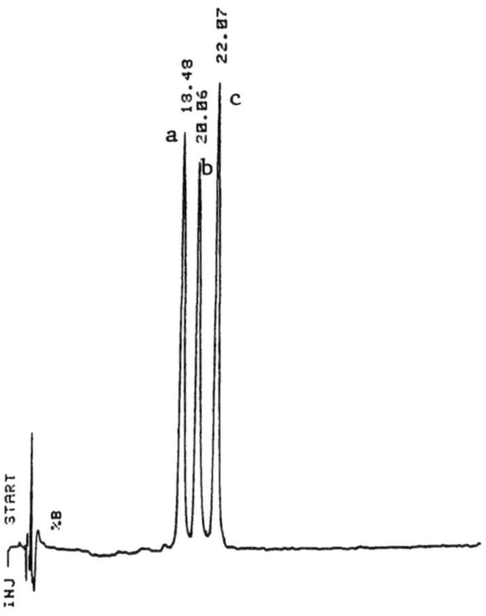

Fig. 2: a) TKI* (15, 16)
b) TKI** (15, 16; 39, 40)
c) TKI

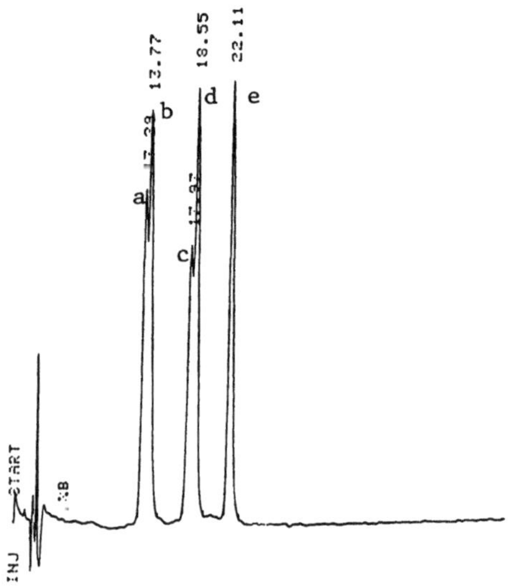

Fig. 3: a) des-Ala^{16}des-Arg^{17} TKI* (15, 18)
b) des-Lys^{15} TKI* (14, 16)
c) des-Ala^{16} TKI* (15, 17)
d) TKI* (15, 16)
e) TKI

TKI derivatives with missing amino acid residues (Fig. 3)

Native TKI is virtually resistant to the attack of exopeptidases. The newly formed amino and carboxyl termini of TKI* (15, 16), however, are susceptible to enzymatic hydrolysis. HPLC is a convenient tool to follow these reactions and to separate the products which sometimes have only marginal charge differences.
The low-molecular-weight aminopeptidase K splits off Ala^{16} and Arg^{17}successively, leaving des-Ala^{16} TKI* (15, 17) and des-Ala^{16} des-Arg^{17}TKI* (15, 18) [14, 17] with small and large decreases in retention time relative to TKI* (15, 16).

Carboxypeptidase B removes Lys^{15} yielding des-Lys^{15} TKI* (14, 16)[14, 18], eluting shortly after des-Ala^{16} des-Arg^{17} TKI* (15, 18) as expected. Removal of two basic amino acid residues from TKI** (15, 16; 39, 40) by carboxypeptidase B gives des-Lys^{15} des-Arg^{39} TKI** (14, 16; 38, 40)[19] with still a further decreased retention time.
Des-Arg^{39} Arg^{15}- TKI* (38, 40) was identified as a side product arising during the semisynthesis of Arg^{15}- TKI [19, 20].

TKI isoinhibitors and derivatives (Fig. 4)

Depending on the breed, bovine organs contain, besides TKI, several Kunitz isoinhibitors [21], which can be analysed by electrophoretic methods or by cation-exchange HPLC.
The sequences of two of them have recently been elucidated [22]:

The less basic inhibitor, iso-TKI I, shows seven amino acid replacements when compared with TKI: $Arg^{17} \rightarrow$ Lys, $Ile^{18} \rightarrow$ Met, $Leu^{29} \rightarrow$ Phe, $Gln^{31} \rightarrow$ Glu, $Arg^{39} \rightarrow$ Lys, $Arg^{42} \rightarrow$ Ser, $Lys^{46} \rightarrow$ Arg.

Iso-TKI II differs from TKI by an additional N-terminal pyroglutamic acid residue only. The loss of one amino group shifts the HPLC peak towards shorter retention times, comparable to the effect of chemical transamination yielding transaminated TKI [23]. Consequently, iso-TKI II* (15, 16) and des-Lys^{15} iso-TKI II* (14, 16) also elute earlier than the corresponding TKI derivatives.

Semisynthetic TKI homologues (Fig. 5)

Aprotinin is a powerful inhibitor of trypsin and kallikrein mainly because of its Lys^{15} reactive-site residue, the side-chain of which is accommodated in the specificity pocket of the enzymes.
Starting from TKI* (15, 16) it was possible to replace Lys^{15} by Phe [24] or Arg [19, 20] by a series of enzymatic reactions.

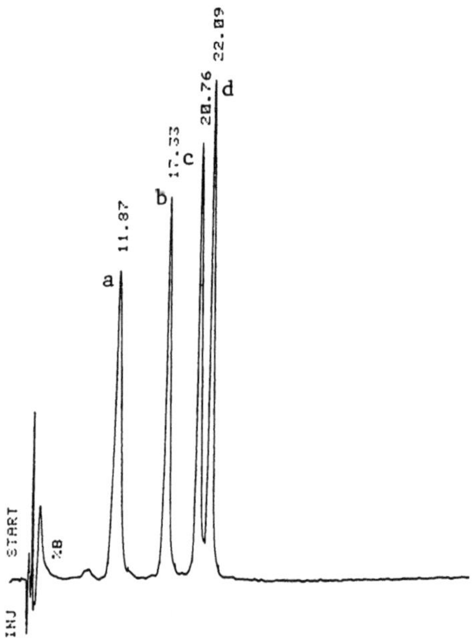

Fig. 4: a) des-Lys^{15} iso-TKI II* (14, 16)
b) iso-TKI II* (15, 16)
c) iso-TKI II
d) TKI

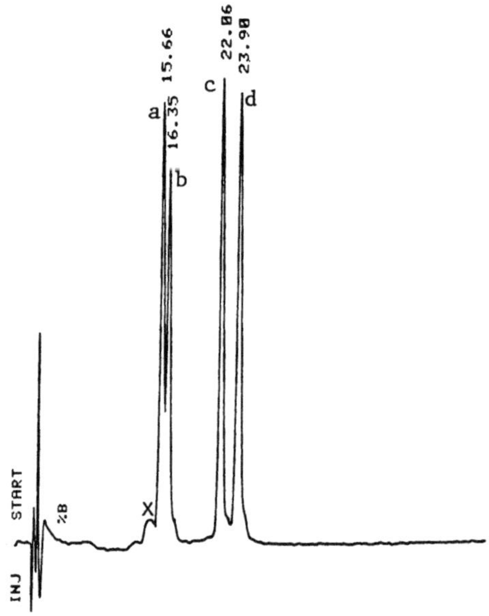

Fig. 5: a) Val^{15}- TKI
b) Phe^{15}- TKI
c) TKI
d) Arg^{15}- TKI
x) impurity of Phe^{15}- TKI sample

Carbodiimide-mediated coupling of amino acid esters to des-Lys^{15} TKI* (14, 16) was developed to introduce virtually any amino acid [25, 26]. In our earlier work coupling also took place to some extent at the five carboxyl groups already present in the native inhibitor. This difficulty has been overcome, however, by making use of protecting groups for the five carboxyls. Briefly, the reaction sequence consists of the following steps [27]:

a) Esterification of TKI* (15, 16) at its six carboxyl groups with 0.1 M HCl in methanol.

b) Specific saponification of Lys^{15} methyl ester with trypsin or endoproteinase Lys-C (Boehringer) in dioxane/0.1 M acetic acid, 1 : 1, pH 6.2.

c) Removal of Lys^{15} by carboxypeptidase B.

d) Coupling of an amino acid (e.g. valine) t-butyl ester to Cys^{14} with water-soluble carbodiimides at pH 4.75.

e) Hydrolysis of the t-butyl ester group with trifluoroacetic acid.

f) Reactive-site peptide bond closure with water-soluble carbodiimides at pH 4.0.

g) Saponification of all methyl ester groups at pH 10.5.

The behaviour of the TKI homologues during HPLC is as expected. Arg^{15}- TKI is more basic, Val^{15}- TKI and Phe^{15}- TKI are less basic than TKI.

Guanidinated TKI derivatives (Fig. 6)

Chemical modification has been an indispensable tool to study structure and function of proteins [28 - 31]. The potentials of ion-exchange HPLC to analyse the reaction products are exemplified by some selected TKI derivatives in the following.

Treatment of TKI or its derivatives with O-methylisourea allows one to selectively replace all ε-amino groups with even more basic guanidino groups [32, 33]. Consequently, guanidinated TKI, guanidinated des-Ala^{16} TKI* (15, 17), guanidinated TKI* (15, 16) and guanidinated des-Ala^{16} des-Arg^{17} TKI* (15, 18) are all eluted appreciably later than the corresponding starting compounds.

TKI derivatives modified at $cystine^{14,38}$ (Fig. 7)

Partial reduction of TKI with sodium borohydride results in the selective cleavage of the disulfide bond linking the half-cystine residues 14 and 38 [34]. Reaction of this product with various sulfhydryl group reagents leads to derivatives which can easily be analysed by cation-exchange HPLC. Iodoacetate yields the S-carboxymethylated derivative [35] which is only slightly retarded on the column and elutes before application of the NaCl-gradient. Iodoacetamide [35] and methyl-4-nitrobenzenesulfonate [36] both introduce neutral moieties and lead to the S-carboxamidomethylated and S-methylated derivative respectively, eluting prior to TKI. The S-aminoethylated derivative with its two additional basic groups produced by ethyleneimine [35] is eluted rather late as expected.

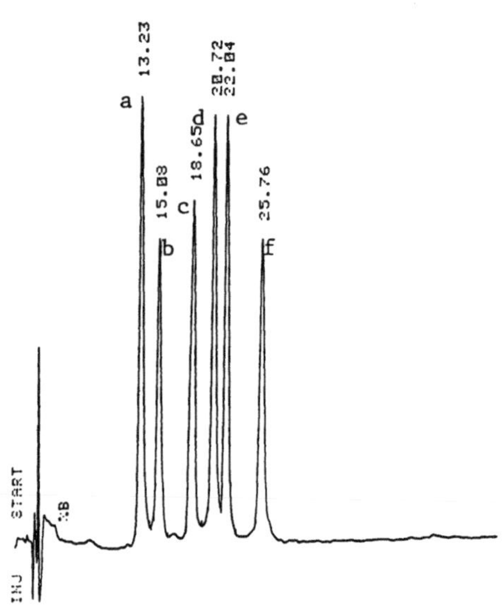

Fig. 6: a) des-Ala^{16} des-Arg^{17} TKI* (15, 18)
b) guanidinated des-Ala^{16} des-Arg^{17} TKI* (15, 18)
c) TKI* (15, 16)
d) guanidinated TKI* (15, 16)
e) TKI
f) guanidinated TKI

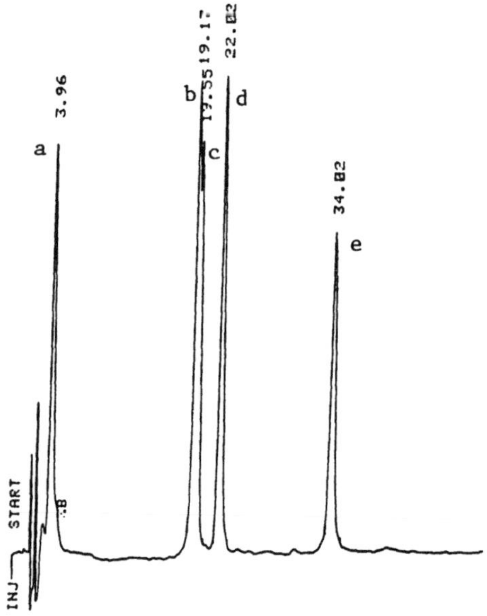

Fig. 7: a) S-carboxymethylated TKI reduced (14, 38)
b) S-carboxamidomethylated TKI reduced (14, 38)
c) S-methylated TKI reduced (14, 38)
d) TKI
e) S-aminoethylated TKI reduced (14, 38)

Esterified TKI derivatives

Treatment of TKI and derivatives with acidified methanol converts the negatively charged carboxyl groups into the neutral methyl esters [37, 38]. Our HPLC system is ideally suited to follow these rather slow reactions, the ester groups are sufficiently stable at pH 7.0. Very high NaCl concentrations are necessary to elute the esterified products from the cation-exchange column.

Financial support by the Deutsche Forschungsgemeinschaft and the Fonds der Chemischen Industrie is gratefully acknowledged.

References

1. Lottspeich, F., Henschen, A., Hupe, K.-P., eds.: High Performance Liquid Chromatography in Protein and Peptide Chemistry, Walter de Gruyter, Berlin 1981.
2. Hearn, M.T.W., Regnier, F.E., Wehr, C.T., eds.: High-Performance Liquid Chromatography of Proteins and Peptides, Academic Press, Orlando 1983.
3. Horváth, C., ed.: High-Performance Liquid Chromatography, Advances and Perspectives, Vol. 3, Academic Press, New York 1983.
4. Hancock, W.S., ed.: CRC Handbook of HPLC for the Separation of Amino Acids, Peptides and Proteins, Vol. I and II, CRC Press, Boca Raton 1984.
5. Regnier, F.E.: Meth. Enzymol. 104, 170 - 189 (1984).
6. Laskowski, Jr., M., Sealock, R.W.: in The Enzymes, Vol. III, Boyer, P.D., ed., Academic Press, New York 1971, pp. 375 - 473.
7. Tschesche, H.: Angew. Chem. 86, 21 - 40 (1974).
8. Fritz, H., Wunderer, G.: Arzneim.-Forsch./Drug Res. 33, 479 - 494 (1983).
9. Kassell, B., Laskowski, M.: Biochem. Biophys. Res. Commun. 20, 463 - 468 (1965).
10. Anderer, F.A., Hörnle, S.: J. Biol. Chem. 241, 1568 - 1572 (1966).
11. Huber, R., Kukla, D., Rühlmann, A., Epp, O., Formanek, H.: Naturwissenschaften 57, 389 - 392 (1970).
12. Tschesche, H., Kupfer, S.: Hoppe-Seyler's Z. Physiol. Chem. 357, 769 - 776 (1976).
13. Estell, D.A., Wilson, K.A., Laskowski, Jr., M.: Biochemistry 19, 131 - 137 (1980).

14. Jering, H., Tschesche, H.: Eur. J. Biochem. 61, 443 - 452 (1976).
15. Goldenberg, D.P., Creighton, T.E.: J. Mol. Biol. 165, 407 - 413 (1983).
16. Goldenberg, D.P., Creighton, T.E.: J. Mol. Biol. 179, 527 - 545 (1984).
17. Jering, H., Tschesche, H.: Angew. Chem. 86, 705 (1974).
18. Jering, H., Tschesche, H.: Angew. Chem. 86, 702 - 703 (1974).
19. Wenzel, H.R., Tschesche, H.: unpublished results.
20. Bode, W., Walter, J., Huber, R., Wenzel, H.R., Tschesche, H.: Eur. J. Biochem. 144, 185 - 190 (1984).
21. Arens, A., Rauenbusch, E., Schutt, H., Siekmann, J., Truscheit, E., Tschesche, H., Wenzel, H.R.: unpublished results.
22. Schröder, W., Siekmann, J., Wenzel, H.R., Tschesche, H.: unpublished results.
23. Brown, L.R., De Marco, A., Richarz, R., Wagner, G., Wüthrich, K.: Eur. J. Biochem. 88, 87 - 95 (1978).
24. Jering, H., Tschesche, H.: Eur. J. Biochem. 61, 453 - 463 (1976).
25. Wenzel, H.R., Tschesche, H.: Hoppe-Seyler's Z. Physiol. Chem. 361 345 (1980).
26. Wenzel, H.R., Tschesche, H.: Angew. Chem. 93, 292 - 293 (1981).
27. Wenzel, H.R., Mehlich, A., Beckmann, J., Schmuck, R., Schnabel, E., Truscheit, E., Tschesche, H.: unpublished results.
28. Means, G.E., Feeney, R.E.: Chemical Modification of Proteins, Holden-Day, San Francisco 1971.
29. Glazer, A.N., De Lange, R.J., Sigman, D.S.: Chemical Modification of Proteins Selected Methods and Analytical Procedures, North-Holland Publishing Company, Amsterdam 1975.
30. Lundblad, R.L., Noyes, C.M.: Chemical Reagents for Protein Modification, Vol. I and II, CRC Press, Boca Raton 1984.
31. Pfleiderer, G.: Chemical Modification of Proteins, this volume.
32. Chauvet, J., Acher, R.: Biochem. Biophys. Res. Commun. 27, 230 - 235 (1967).
33. Wenzel, H.R., Tschesche, H., von Goldammer, E., Netzelmann, U.: FEBS Lett. 140, 53 - 57 (1982).
34. Kress, L.F., Laskowski, Sr., M.: J. Biol. Chem. 242, 4925 - 4929 (1967).
35. Kress, L.F., Wilson, K.A., Laskowski, Sr., M.: J. Biol. Chem. 243, 1758 - 1762 (1968).
36. Heinrikson, R.L.: J. Biol. Chem. 246, 4090 - 4096 (1971).
37. Avineri-Goldman, R., Snir, I., Blauer, G., Rigbi, M.: Arch. Biochem. Biophys. 121, 107 - 116 (1967).
38. Quast, U., Engel, J., Steffen, E., Tschesche, H., Kupfer, S.: Biochemistry 17, 1675 - 1682 (1978).

SEPARATION OF CYANOGEN BROMIDE PEPTIDES OBTAINED FROM THE CATALYTIC DOMAIN OF THE NADPH-CYTOCHROME P-450 REDUCTASE (PIG LIVER) BY HPLC ON WIDE PORE COLUMN

Frank Vogel and Ludwig Lumper
Biochemisches Institut (Fachbereich Humanmedizin) der Universität Gießen
D-6300 Gießen

Introduction

The location of cysteine residues in the sequence of NADPH-cytochrome P-450 reductase (EC 1.6.2.4) and their accessibility to alkylating reagents in the native state are of great interest toward better understanding of the mechanism of this FAD and FMN containing enzyme. We have therefore isolated cysteine containing peptides from the catalytic domain (= trypsin-solubilized NADPH-cytochrome P-450 reductase) by cyanogen bromide (= CNBr) cleavage. This paper describes the purification of CNBr peptides by reversed phase hplc and the use of the fluorescent bimane-label to detect cysteine residues.

Results

1) Isolation of the CNBr peptides. Cleavage of the catalytic domain of NADPH-cytochrome P-450 reductase into large fragments was performed by the highly specific CNBr procedure (1). Preliminary experiments showed that cleavage of the enzyme at tryptophanyl bonds by iodosobenzoate or the thiocyanylation method produced unsatisfactory yields (< 30 %). As reported previously, gel filtration on Sephadex G-75 in 10 % (v/v) acetic acid separates the CNBr peptides derived from the S-carboxymethylated catalytic domain into seven size classes with peptide bands showing mobilities on sodium dodecyl sulfate polyacrylamide gel electrophoresis corresponding

Modern Methods in Protein Chemistry, Volume 2

to molecular weight ranges of ≧ 15 (pool I), ∿ 10 (pool II+III), 5 - 6 (pool IV), 3 - 4.5 (pool V), 2 - 3 (pool VI) and < 2 (pool VII, pool VIII) kDa (1) (Fig. 1).

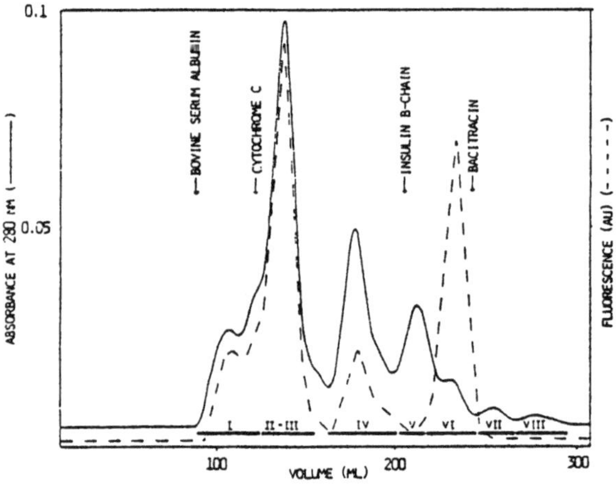

Fig.1: Separation of the peptides obtained by CNBr cleavage of the bimane-labelled catalytic domain (3.2 bimane-groups/mol enzyme) of the NADPH-cytochrome P-450 reductase on Sephadex G-75 column (1.7 cm x 130 cm; flow rate 13.8 ml/h) in 10 % (v/v) acetic acid (absorbance at 280 nm (——); fluorescence (λ_{ex}=395 nm, λ_{em}=475 nm) (- - -) of the eluate).

Sephadex G-75 pool I was found, on the basis of its methionine content, to contain incompletely cleaved protein and was discarded. Preliminary experiments to purify the S-carboxymethylated CNBr fragments of pools II - VIII by fractionated solvent extraction with 0.05 M NH_4HCO_3 pH = 8.5 and 5 % (v/v) acetic acid demonstrated that the peptides present in the pools I - IV were insoluble at pH > 3.0. Pools II - VII were subjected to reversed phase hplc with C18 Wide Pore material (Bakerbond (330 Å), Vydac (330 Å) A)) using 0.1 % (v/v) trifluoroacetic acid as mobile phase and acetonitrile as organic modifier. Separation of the individual Sephadex G-75 pools by gradient elution at room temperature led to the isolation of the CNBr fragments in a state of purity which allowed sequencing (Fig. 2A - 2E). Acetonitrile gradients (slope: 0.33 - 0.42 % / min in the eluent) with initial concentrations of 25 - 30 % (v/v) were utilized to elute the large and medium-sized CNBr peptides (Fig. 2A - 2C). Under the conditions applied, the large hydrophobic CNBr fragments CB1 and CB2 with nearly identical molecular weights of about 10 kDa could be separated and were recovered with a yield of 80 % in a single hplc step (Fig. 2A). Although the CB1 peak has a shoulder in its descending part, samples taken from various fractions within the peak were found to have an identical amino acid com-

position. In addition, automated sequencing by Edman degradation of CB1 and CB2 resulted in unequivocal N-terminal sequences of 37 and 40 amino acids, respectively (2). An analytical column (0.46 cm x 25 cm) filled with C18 Bakerbond Wide Pore material displayed high resolving power for the large CNBr fragments up to a load of 100 nmoles / peptide.
On C18 Wide Pore materials (330 Å) the peptides of Sephadex G-75 pools II+III were eluted in the order CB1 and CB2, whereas on cyanopropyl bonded silanol phases (Bakerbond (330 Å)) an inversion of this order and insufficient separation has been observed. With C18 n-alkyl bonded silanol phases of narrow pore size (∿ 100 Å) separation and yields of the peptides are essentially poorer. Smaller CNBr fragments could also be easily purified using Bakerbond C18 Wide Pore column by the solvent gradients as shown in Fig. 2B - 2E. Minor differences in elution profile of pool VI are due to the pooling of the Sephadex G-75 fractions, because some peptides (e.g. CB7 and CB14) are present in more than one Sephadex G-75 pool. The number of CNBr peptides isolated correlates well to the methionine content determined for the catalytic domain of pig liver NADPH-cytochrome P-450 reductase (2.9 ± 0.2 mol % methionine ∿ 14 - 16 methionine residues per mol reductase (68 kDa)). Loss of CNBr peptides due to incomplete cleavage of methionyl bonds or during the purification procedure cannot, however, be totally ruled out, because overlapping methionine peptides have not yet been characterized.

2) Modification of CNBr peptides by monobromobimane. Bromomethyl derivatives of syn-9,10-dioxabimane (= 1,5-diazabicyclo[3.3.0]octadienedione) are specific reagents for thiol groups in proteins as suggested by Fahey et al. (3) and demonstrated during our studies on the chemical modification of the NADPH-cytochrome P-450 reductase (1). The fluorescent (excitation at 395 nm; emission at 475 nm) bimane-group (3,4-methylene-3,6,7-trimethyl-9,10-dioxabimane) introduced at the cysteine residue is stable under the conditions of protein digestion by proteases and CNBr cleavage of methionyl bonds (1). The bimane-label is therefore an useful tool to monitor cysteine peptides in column eluates by fluorescence and in sequencer studies during Edman degradation (2). The bimane group increases the retention of CNBr peptides on all reversed phase columns used. Bimane-labelled CB10 co-

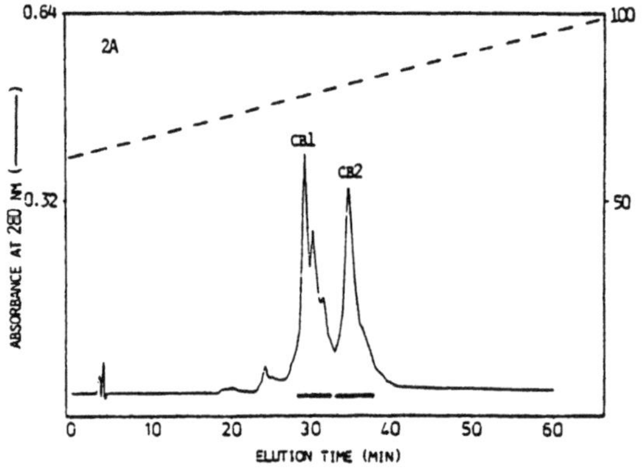

Fig.2A: Hplc pool II + III
Fig.2B: Hplc pool IV
Fig.2C: Hplc pool V
Fig.2D: Hplc pool VI
Fig.2E: Hplc pool VII

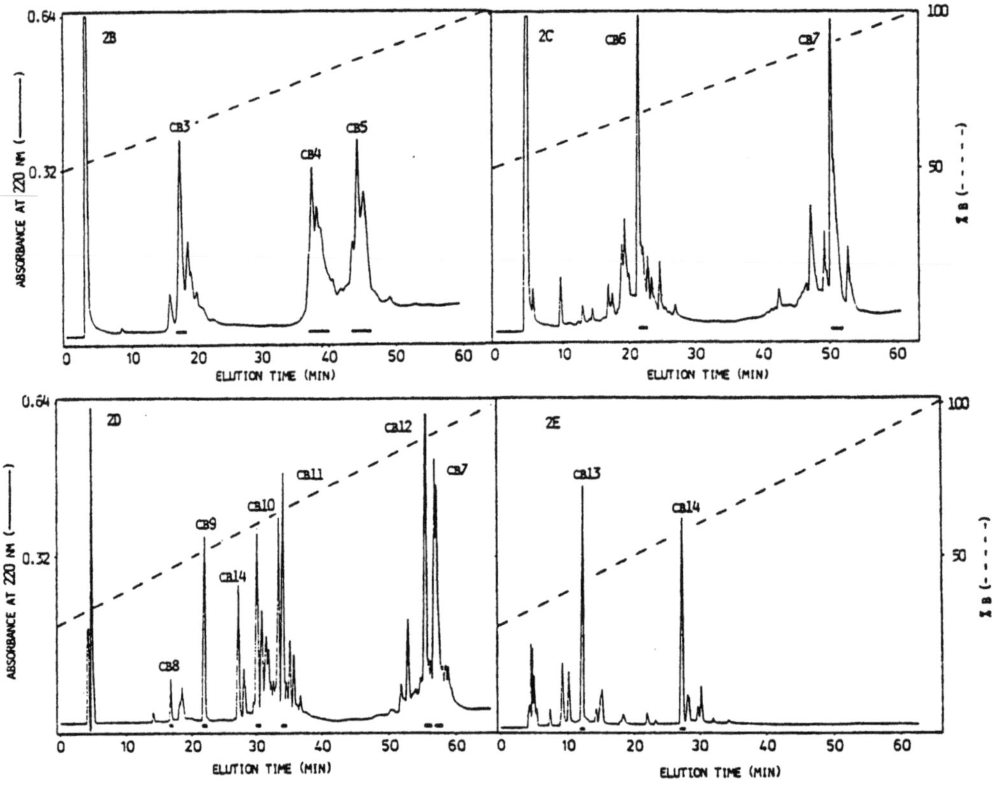

Fig.2A - 2E: Hplc of the Sephadex G-75 pools II - VII on Bakerbond Wide Pore column (330 Å; 0.46 cm x 25 cm; flow rate 0.7 ml/h) were developed with linear acetonitrile gradients (solvent A: 0.1 % (v/v) trifluoroacetic acid, solvent B: 0.1 % (v/v) trifluoroacetic acid in 50 % (v/v) acetonitrile).

elutes together with CB11 from Bakerbond C18 Wide Pore column (Fig. 3A), although the S-carboxymethylated form of CB10 is completely resolved (Fig. 2D). Separation of labelled CB10 and CB11 could only be achieved by rechromatography on Shandon C18 3μ material (Fig. 3B). On the other hand the separation of CB7 and CB12 was improved after labelling and the resolution of the remaining cysteine containing peptides was not negatively influenced. According to amino acid analysis, the catalytic domain of the NADPH-cytochrome P-450 reductase from pig liver contains eight cysteine residues (1). The distribution on the CNBr fragments could be determined using the CNBr digest of the exhaustively alkylated catalytic domain by monobromobimane (= 4-bromomethyl-3,6,7-trimethyl-9,10-dioxabimane) in the presence of 6 M guanidinium chloride. By the hplc procedure described, a set of six cysteine containing CNBr peptides could be identified (CB1, CB2, CB4, CB5, CB10 and CB12) (2).

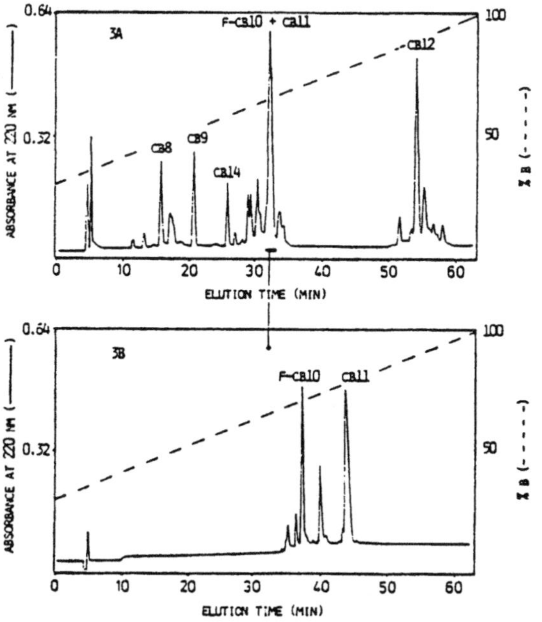

Fig.3A - 3B: Co-elution of the fluorescent bimane-labelled CNBr peptide CB10 (=F-CB10) and CB11 on Bakerbond Wide Pore column (330 Å, 0.46 cm x 25 cm)(Fig.3A). Rechromatography of the mixture of both peptides on Shandon C18 3μ material (0.46 cm x 25 cm) (Fig.3B). Solvent system as described on Fig.2.

The method established for the isolation of the CNBr fragments obtained from unmodified and bimane-labelled trypsin-solubilized NADPH-cytochrome P-450 reductase allows the location of the accessible, cosubstrate (NADPH)-protected and buried cysteines in this enzyme (2).

Acknowledgement. The authors thank Mrs. M. Wagner and Mr. F. Busch for their excellent technical assistance. They are grateful to Mrs. M. Linder for reading the manuscript.

References

1. Vogel, F., Lumper, L.: Biochem. J. 215, 159 - 166 (1983)
2. Vogel, F., Kaiser, C., Witt, I., Lumper, L.: in preparation
3. Fahey, R.C., Newton, G.L., Dorian, R., Kosower, E.M.: Anal. Biochem. 111, 357 - 365 (1981).

IMMUNOASSAYS OF PEPTIDES AND PROTEINS

M. Gemeiner

Institut für Medizinische Chemie, Veterinärmedizinische Universität Wien, A-1030 Vienna, Austria

Introduction

Immunoassays made an explosive impact upon endocrinology and other areas of medicine in which the accurate measurement of small concentrations of biologically active compounds is of crucial importance. The use of antigen/antibody reactions for the measurement of even trace amounts of a specific antigen has been elaborated from the initial discovery by Yalow and Berson (1) into the wide range of immunoassay systems in use today.

In radioimmunoassay (RIA) - the first well established of the immunoassays -, radioactive tracer competes with identical nonlabelled antigen providing a combination of specificity, precision and practicality for the direct determination of specific peptides or proteins even in a complex biological matrix. As radioactively labelled compounds produce a potential environmental hazard, many attempts have been made to devise nonradioactive systems showing the same performances.

A major development of the last decade has also been the extension of immunoassays into the field of hapten-specific assays, where small, biologically active peptides, which are not immunogenic themselves, are first conjugated to a "carrier" macromolecule.

There are some basic requirements for the setting up of a satisfactory immunoassay. It will depend on:

- the analyte (antigen)
- the antibody

Modern Methods in Protein Chemistry, Volume 2

- the "tracer"-molecule
- the separation system
- the immunoassay system itself.

In the following, an overview on problems related to antigen and tracer preparation and antibody production will be given, advantages and limitations of immunoassay systems will be discussed.

Antigen

Purity requirements

A purified preparation should be used for raising antiserum, a highly purified analyte is required for tracer preparation. Provided that the assay developed is specific, the standards used throughout the assay do not necessarily have to be of the same degree of purity, but they must be stable and perform identically to the pure analyte in the assay.

Structural implications

The chemical nature of an antigenic determinant is determined by the particular functional groups involved and by their spatial arrangement; the same is also true for the biologically active sites (2). For the clinical relevance it is not only important to have a method for the specific detection of a molecule but to measure the biologically active form or biologically active metabolites. Therefore it would be helpful to have an idea about the topographical relationship between biologically active sites and antigenic determinants. Proteins, as immunogens, can directly be used to raise antibodies. If the native protein is not available in adequate quantities or if immunochemical mapping approaches to correlate structural domains with functional domains are undertaken, then synthetic peptide immunogens containing at least one known immunoreactive site can be used to raise specific antibodies (3). Rather small peptides or polypeptide fragments have also be considered for the production of antibodies recognizing small biologically active peptides (gastrointestinal hormones, neuropeptides..).

Conjugation of haptens to carrier molecules

Since low molecular weight peptides are - with rare exceptions - not immunogenic, different procedures for establishing covalent linkages between peptide and carrier proteins (4) or synthetic polypeptides (5) have been employed. Based on the presence of different reactive groups on the peptide and the carrier to be conjugated, all kinds of chemical modifications on both molecules are conceivable (6,7). In fact, the most commonly used coupling reagents such as glutaraldehyde, bis(imido esters), toluene diisocyanate and carbodiimides (8-11) can lead to extensive inter- and intramolecular self-coupling of both components of the conjugate (12). These random reactions of linking functional groups on peptide and carrier result in a variety of products where the conjugation-sites remain undefined. Varying ratios of polymerized peptide, polymerized carrier and peptide-carrier conjugate are produced from preparation to preparation. Immunization with this kind of complex antigens leads to heterogeneous antibody populations. As a consequence it is desirable to obtain peptide-protein conjugates of uniform size, with high level of peptide attachment and fully exposed and unmodified antigenic determinants in the peptide moiety. Several approaches have been undertaken (3, 13-14), the ones based on work of Keller and Rudinger (15) directed towards the use of maleimide function as N-terminal "multipurpose anchor" group of the peptide molecule for a mild and selective conjugation to thiol carriers being very promising as shown by the synthesis of the N-terminal maleoyl-ß-alanyl derivative of human gastrin-[2-17] (16). This derivative enables coupling of gastrin to a macromolecular carrier bearing at least one thiol function (12).

Antibody Production

In vivo immunization

The two main advantages of immunoassay procedures are their sensitivity and specificity, both of which depend predominantly on the use of antibodies that possess a high avidity and selectivity for the material to be assayed. It is often said that raising of such antisera belongs more to art than to science and that success is often determined by chance. Nevertheless, immunization schedules can be chosen to give the maximum probability of success.
Factors influencing the success of an immunization are:

- adjuvants
- antigen
- route of administration
- frequencies of injection

Adjuvants: The antibody response to injected immunogen is increased by the inclusion of adjuvants such as aluminium hydroxide, Freund's complete or incomplete adjuvant (17). The favorable effect of adjuvants is associated with greater persistence of the antigen in vivo. Although the "depot" effect is very important, it is also recognized that adjuvants mediate complicated cell interactions (18). Interesting developments are to be expected from the use of synthetic adjuvants where muramyl dipeptide and its derivatives are best known (19).

Antigen: For the immunogen, the most important factors are the ability of the substance to provoke an antibody response and its availability in adequate amounts. Problems related to the structure and the purity of the antigen under concern have been discussed earlier. The mentioned possibility to produce chemically well defined hapten-carrier-conjugates offers an interesting approach to the sequence-specific recognition of antigens (20). Antisera obtained with cytochrome C-gastrin-conjugates or with ribonuclease-gastrin-conjugates did bind the C-terminal tetrapeptide amide (H-Trp-Met-Asp-Phe-NH_2) only to negligible extent, but this sequence is the most immunogenic portion of the gastrin molecule, as demonstrated by using conventionally prepared albumin-gastrin-conjugates. The antisera raised in rabbits against the gastrin molecule recognize the hormone molecule to sequence lengths which approach those of native fully active gastrin forms. One may assume that in chemically "unchanged" gastrin molecules the dominant immunodeterminant may possibly also be conformationally conditioned by the adjacent N-terminal elongation.

Route of administration: Only few critical studies have been performed dealing with the effects associated with the route of administration on the immune response to poor immunogens. Although the immune response is a systemic reaction, the site at which the immunogen is administered may influence the predominant site of antibody production. With subcutaneous and intramuscular injections, the highest concentration of immunogen is found in local granulomata and in regional lymph nodes; after intraperitoneal injections, in the mesenteric nodes and following intravenous administration, in the spleen and lungs. Evidence suggests that the sites ranked in decreasing order of effectiveness are as follows: intradermal, intramuscular, intraperitoneal, subcutaneous and intravenous (21).

Frequencies of injection: As to frequencies of antigen administration, very few theoretical studies exist in which the efficiencies of different methods have been compared directly and adequately; protocols in use are often based on good personal experiences (22-24).

"In vitro" immunization

Antisera raised by conventional immunization procedures contain the products of different antibody-secreting clones.

Besides the desired clonal products, these antisera may contain antibodies directed against impurities or molecules closely related to the immunogen. A new way has been opened by Köhler and Milstein (25), when they showed that fusing individual clones of normal antibody-secreting cells give place to immortalised, specific antibody-secreting hybridomas (26-28). This new methodology has seen a wide range of applications (29-32). Immunogens are often characterized by several epitopes; as selected monoclonal antibodies are specific for one epitope, it has been suggested to mix a series of well defined monoclonal antibodies and so obtain a very selective tool for identification of specific antigen structures.
Monoclonal antibodies may be produced following immunization with crude antigen preparations and each selected clonal antibody may be used for individual components.
A drawback of this new method is the need for intensive screening of numerous supernatants to eliminate those without the desired specifications and rather high overall costs to maintain the cell cultures and the test systems.

Crossreactivity

In case of heterogeneous antigen mixtures, as it usually occurs on conjugation of peptides with carrier proteins by means of nonspecific covalent linkages, a multitude of antibodies with different specificities **is** obtained. In view of these considerations, two concepts of antigen crossreactivity can be deduced:

- two antigens which share a small but precise detail on their topology and bind to the same binding site of an antibody with different affinities
- two antigens having dissimilar structures but having epitopes interacting with the same antibody molecule.

With heterogeneous antibodies both kinds of crossreactivities may be observed; with monoclonal antibodies only the former type (33).

Attention has been drawn to some of the differences between molecular recognition by conventional and monoclonal antibodies (34). The authors refer to examples of monoclonal antibodies which recognize crossreacting sites on molecules in which conventional obtained antibodies had failed to detect homology. This means that reaction with monoclonal antibody cannot always be interpreted as proving molecular identity.

Tracer

Immunoassays rely on adequate labelled molecules. These tracers are radioisotopes, enzymes, fluorescent-, chemiluminescent- or bioluminescent labels. These tracers have to satisfy a certain number of demands, i.e. stability, easy detectability with high sensitivity, chemically defined structure, identical behaviour as the antigen to be measured in the assay system (for competitive systems).

Radioligands

Radioimmunoassay is still one of the most widely applied of all immunoassay procedures. Many of the problems encountered with RIAs are caused by difficulties associated with the radioligand. On the other hand, there are advantages associated with the use of radioactively labelled tracer ligand, i.e. ease, reliability, sensitivity and precision of their quantitation. Radioligands can replace a non-radioactive isotope of the same element which is present in the native molecule (^{14}C, ^{3}H), or a foreign radioactive atom is used (^{125}I). ^{125}I is the preferred radioisotope used for the labelling of tracer proteins or peptides; the most frequently considered alternative isotopes to radioiodine for tracer preparation are ^{3}H and ^{14}C. It is often difficult however, to obtain a degree of incorporation consistent with the required sensitivity of the assay.

Tritium: Substitution at amino groups of lysyl residues leads to tritium-labelled polypeptides and proteins with different specific activities. The lysyl residues may be reductively methylated, amidinated, guanidinated, carbamoylated, acetylated and acylated (35). Some of these chemical modifications may alter the conformational stability and influence therefore the binding of the tracer molecule. Direct tritium-labelling produces tracers without modification of physical or chemical properties, unfortunately suitable unsaturated or halogenated precursors for catalytic reduction can only be obtained in a few cases (36).

Iodine: Iodination occurs by direct chemical or enzymatic substitution of hydrogen in tyrosine using chloramine-T or lactoperoxidase (37-39), in other procedures N-Bromo-succinimide (40-41) or Iodo-Gen (42) are used. Radioiodination of proteins or peptides gives rise to products chemically different from the immunogen. Reduction in affinity may be caused by structural alterations as a consequence of substitution of iodine into the protein or of chemical alterations of the protein caused by some reagents of the reaction mixture. Chemical alteration of amino acids by iodination reagents (oxidizing agents) has been demonstrated for tryptophan (39, 43), cysteine and methionine (44). Tyrosin is the major amino acid iodinated, although histidine is also substituted to a lesser extent (45). Besides these direct iodination procedures, there are methods in which a radioactive moiety is conjugated to a protein without direct exposure of the protein being labelled to oxidizing or reducing agents. The most used conjugation-labelling method is by Bolton-Hunter reagent (46-47).
The former mentioned maleoyl-peptides are well suited derivatives for a smooth and clean preparation of radio-labelled tracers, i.e. via reaction with $[^{125}I]$-desamino-tyrosyl-cysteine, $[^{3}H]$-acetylcysteine or $[^{35}S]$-cysteine (48). Radioactive tracers have limited shelf-life, peptides or

proteins can be damaged by radiolabelling and changes in binding affinity may occur therefore. However, radiolabelled antigens and antibodies will continue to be useful in the future.

Enzyme Labels

Immunoassays based on the use of enzymes as labels for antigens or antibodies rely on the determination of an enzyme by appropriate enzymatic techniques. The sensitivity, specificity and reproducibility of the various enzyme immunoassays depend upon the insoluble phase used to immobilize the antigen or the antibody, the enzyme used as the label and the effectiveness of the enzyme conjugate employed. Criteria for choice of an enzyme label are:
- availability of purified enzyme
- free functional groups on the enzyme through which it can be conjugated without loss of activity
- stable enzyme conjugates
- enzyme not present in the biological samples to be assayed
- simple assay method, but sensitive and precise.

Enzymes widely used for these immunoassays are: horseradish peroxidase, alkaline phosphatase, ß-D-galactosidase and glucose oxidase.
The methods used for the preparation of enzyme-antigen-(antibody)-conjugates underlie the same criteria as mentioned for hapten-carrier conjugates (49-52) with - in addition - the need for activity conservation of the enzyme moiety.
The sensitivity of enzyme-immunoassays can be improved by using fluorescent, radioactive or highly chromogenic substrates (53-55).
Possible advantages of enzyme labels are the longer shelf-life, the rather simple and inexpensive method, possibility for automating, no radiation hazard.
Possible disadvantages reside in the fact that substances in the sample to be measured may affect enzyme activity, the

assays are rather insensitive, the preparation of good enzyme labels is not so easy.

Fluorescent Labels

The efficiency of the conversion of incident photons by a fluorescent substance is influenced by many factors, i.e. bleaching, quenching. Moreover the natural fluorescence of many biological samples leads to high and changing "blanks". The magnitude of the background fluorescence lowers the detection limits of the assay (56). Fluorescence lifetimes of some of the common fluorophores are in the order of magnitude the background also has. The fluorescent lanthanide chelates form a unique group of fluorophores with much extended fluorescence decay times and a large Stokes shift. Successful exploitation of the lanthanide chelates in fluoroimmunoassay is governed by the choice of a bifunctional reagent which strongly chelates the metal ion, but also attaches to the molecule to be labelled; ß-diketone structures and europium salts represent such a system (57,58). Fluoroimmunoassays will play an increasingly important role, as improvements in methodology and instrumentation take place.

Chemiluminescent Labels

Chemiluminescence is the phenomenon observed when the vibronically excited product of an exoergic chemical reaction reverts to its ground state with photonic emission. All the reactions studied to date are of oxidative nature. The best known chemiluminescent reactions are those of luminol and its derivatives. Microperoxidase has been found to be the most efficient catalyst for these reactions. Recently, derivatives of isoluminol have gained popularity since they are less affected by structural alterations than is luminol itself (59). Of the other chemiluminescent compounds so far studied, only acridinium esters have attracted attention as probes in immunoassays (60). Little work has been reported so far concerning chemiluminescence immunoassays for proteins or peptides.

Separation Systems

Separation of antibody-bound ligand from non-antibody-bound ligand is done by very different methods. In radioimmunoassays dextran-coated charcoal (61) has been widely used for the separation of small molecules and the double antibody method has become a standard (62). Of the solid phase methods, covalent binding of antibody or antigen to finely divided particles such as cellulose (63), latex (64), or magnetized particles (65) has proved reliable. The choice of method from the many that have been described, will only depend on the kind of application required.

Immunoassay Systems

Immunoassays are performed following two basic principles: techniques relying on observation of the distribution of:
- labelled antigen following its reaction with an antibody
- labelled antibody following its reaction with an antigen.

The former system belongs to the "competitive" assay systems, the latter to the "immunometric" assays. Both techniques have been applied in a number of modifications using a large variety of different labels (64-72). In addition, end-point and kinetic nephelometry have become important immunological assay techniques for the measurement of specific proteins (73).

Parameters influencing immunoassays of peptides and proteins have been reviewed and the importance of preparing good, chemically defined immunoreagents has been underlined.

References

1. Yalow, R.S., S.A. Berson. 1960. J.Clin.Invest. 39, 1157.

2. Karush, F.. 1978. In: Comprehensive Immunology (G.W. Litman and R.A. Good, eds.). Plenum Press, p. 85.

3. Van Eldik, L.J., K.F. Fok, B.W. Erickson, D.M. Watterson. 1983. Proc.Natl.Acad.Sci.USA 80, 6775.

4. Jaffe, B.M., W.T. Newton, J.E. McGuigan. 1970. Immunochemistry 7, 715.

5. Sela, M., S. Fuchs. 1978. In: Handbook of Experimental Immunology, Vol. 1 (D.M.Weir, ed.). Blackwell Scientific Publications, p. 1.1..

6. Atassi, M.Z.. 1977. In: Immunochemistry of Proteins, Vol. 1 (M.Z. Atassi, ed.). Plenum Press, p. 1.

7. Kiefer, H..1979. In: Immunological Methods, Vol. 1 (I. Lefkovits and B. Pernis, eds.). Academic Press, p. 137.

8. Lübke, K., B. Nieuweboer. 1978. In: Immunologische Teste für niedermolekulare Wirkstoffe (K. Lübke and B. Nieuweboer, eds.). Georg Thieme Verlag, p. 11.

9. Erlanger, B.F..1980. In: Methods in Enzymology, Vol. 70 (H. Van Vunakis and J.J. Langone, eds.). Academic Press, p. 85.

10. Bauminger, S., M. Wilchek. 1980. In: Methods in Enzymology, Vol. 70 (H. Van Vunakis and J.J. Langone, eds.). Academic Press, p. 151.

11. Reichlin, M..1980. In: Methods in Enzymology, Vol. 70 (H. Van Vunakis and J.J. Langone, eds.). Academic Press, p. 159.

12. Wünsch, E..1985. In: Peptides 1984 (U. Ragnarsson, ed.). Almquist and Wiksell internat., p. 5.

13. Atassi, M.Z., A.L. Kazim, S. Sakata. 1981. Biochim. Biophys.Acta 670, 300.

14. Lee, A.C.J., J.E. Powell, G.W. Tregear, H.D. Niall, V.C. Stevens. 1980. Molecular Immunology 17, 749.

15. Keller, O., J. Rudinger. 1975. Helv.Chim.Acta 58, 531.

16. Wünsch, E., L. Moroder, R. Nyfeler, H. Kalbacher, M. Gemeiner. 1985. Biol.Chem.Hoppe-Seyler 366, 53.

17. Herbert, W.J.. 1978. In: Handbook of Experimental Immunology, Vol. 3 (D.M. Weir, ed.). Blackwell Scientific Publications, p. A3.1..

18. Osebold, J.W.. 1982. JAVMA 181, 983.

19. Chedid, L.A., M.A. Parant, F.M. Audibert, G.J. Riveau, F.J. Parant, E. Lederer, J.P. Choay, P.L. Lefrancier. 1982. Infect.Immun. 35, 417.

20. Geiger, R., L. Moroder, E. Wünsch. 1985. In: Peptides 1984 (U. Ragnarsson, ed.). Almquist and Wiksell internat., p. 451.

21. Hurn, B.A.L., J. Landon. 1971. In: Radioimmunoassay Methods (K.E. Kirkham and W.M. Hunter, eds.). Churchill Livingstone, p. 121.

22. Chase, M.W.. 1967. In: Methods in Immunology and Immunochemistry, Vol. 1 (C.A. Williams and M.W. Chase, eds.). Academic Press, p. 209.

23. Hurn, B.A.L., S.M. Chantler. 1980. In: Methods in Enzymology, Vol. 70 (H. Van Vunakis and J.J. Langone, eds.). Academic Press, p. 104.

24. Vaitukaitis, J.L.. 1981. In: Methods in Enzymology, Vol. 73 (J.J. Langone and H. Van Vunakis, eds.). Academic Press, p. 46.

25. Köhler, G., C. Milstein. 1975. Nature 256, 495.

26. Galfrè, G., C. Milstein. 1981. In: Methods in Enzymology, Vol. 73 (J.J. Langone and H. Van Vunakis, eds.). Academic Press, p. 3.

27. Goding, J.W.. 1980. J.Immunol.Methods 39, 285.

28. Oi, V.T., L.A. Herzenberg. 1980. In: Selected Methods in Cellular Immunology (B.B. Mishell and S.M. Shiigi, eds.). W.H. Freeman and Company, p. 351.

29. Zurawski, Jr., V.R., P.H. Black, E. Haber. 1980. In: Monoclonal Antibodies; Hybridomas: A New Dimension in Biological Analyses (R.H. Kennett, T.J. McKearn and K.B. Bechtol). Plenum Press, p. 19.

30. Antczak, D.F.. 1982. JAVMA 181, 1005.

31. Secher, D., D.C. Burke. 1980. Nature 285, 446.

32. Wang, L., C.S. Hexter, M. Inbar. 1981. In: Monoclonal Antibodies and T-Cell Hybridomas (G.J. Hammerling,

U. Hammerling and J.F. Kearney, eds.). Elsevier/North-Holland Biomedical Press, p. 357.

33. Berzofsky, J.A., A.N. Schechter. 1981. Molecular Immunol. 18, 751.

34. Lane, D., H. Koprowski. 1982. Nature 296, 200.

35. Tack, B.F., R.L. Wilder. 1981. In: Methods in Enzymology, Vol. 73 (J.J. Langone and H. Van Vunakis, eds.). Academic Press, p. 138.

36. Morgat, J.L., J.P. Girma, P. Fromageot. 1977. In: First Intern.Symp.on Hormonal Receptors in Digestive Tract, Inserm Symposium No.3 (S. Bonfils, P. Fromageot, G. Rosselin, eds.). Elsevier/North Holland, p. 43.

37. McConahey, P.J., F.J. Dixon. 1980. In: Methods in Enzymology, Vol. 70 (H. Van Vunakis and J.J. Langone, eds.). Academic Press, p. 210.

38. Morrison, M.. 1980. In: Methods in Enzymology, Vol. 70 (H. Van Vunakis and J.J. Langone, eds.). Academic Press, p. 214.

39. Alexander, N.M.. 1974. J.Biol.Chem. 249, 1946.

40. Reay, P.. 1982. Ann.Clin.Biochem. 19, 129.

41. Butt, W.R., S.S. Lynch, P. Reay, W. Robinson, J. Sage. 1983. In: Immunoassays for Clinical Chemistry: A Workshop Meeting (W.M. Hunter and J.E.T. Corrie, eds.). Churchill Livingstone, p. 286.

42. Ferguson, K.M., M.M. Hayes, S.L. Jeffcoate. 1983. In: Immunoassays for Clinical Chemistry: A Workshop Meeting (W.M. Hunter and J.E.T. Corrie, eds.). Churchill Livingstone, p. 289.

43. Mourier, G., L. Moroder, A. Previero. 1984. Z.Naturforsch. 39b, 101.

44. Schechter, Y., Y. Burstein, A. Patchornik. 1975. Biochemistry (USA) 14, 4497.

45. Krohn, K.A., M.J. Welch. 1974. Int.J.Appl.Radiat.Isot. 25, 315.

46. Bolton, A.E. W.M. Hunter. 1973. Biochem.J. 133, 529.

47. Langone, J.J.. 1981. In: Methods in Enzymology, Vol. 73 (J.J. Langone and H. Van Vunakis, eds.). Academic Press, p. 112.

48. Moroder, L., M. Gemeiner, H. Kalbacher, R. Nyfeler, E. Wünsch. 1982. In: Peptides 1982 (K. Blaha and P. Malon, eds.). Walter de Gruyter, p. 667.

49. O'Sullivan, M.J., V. Marks. 1981. In: Methods in Enzymology, Vol. 73 (J.J. Langone and H. Van Vunakis, eds.). Academic Press, p. 147.

50. Kabakoff, D.S.. 1979. In: Enzyme-Immunoassay (E.T. Maggio, ed.). CRC Press, Inc., p. 71.

51. Ishikawa, E., Y. Yamada, Y. Hamaguchi, S. Yoshitake, K. Shiomi, T. Ota, Y. Yamamoto, K. Tanaka. 1978. In: Enzyme Labelled Immunoassay of Hormones and Drugs (S.B. Pal, ed.). Walter de Gruyter, p. 43.

52. Krüger, D., A. Maelicke. 1978. In: Enzyme Labelled Immunoassay of Hormones and Drugs (S.B. Pal, ed.). Walter de Gruyter, p. 82.

53. Harris, C.C., R.H. Yolken, H. Krokan, I.C. Hsu. 1979. Proc.Natl.Acad.Sci.USA 76, 5336.

54. Puget, K., A.M. Michelson, S. Avrameas. 1977. Anal.Biochem. 79, 447.

55. Moroder, L., R. Nyfeler, M. Gemeiner, H. Kalbacher, E. Wünsch. 1983. Biopolymers 22, 481.

56. Soini, E., I. Hemmila. 1979. Clin.Chem. 25, 353.

57. Siitari, H., I. Hemmila, E. Soini, T. Lovgren, V. Koistinem. 1983. Nature 301, 258.

58. Dakubu, S., R. Ekins, T. Jackson, N.J. Marshall. 1984. In: Practical Immunoassay (W.R. Butt, ed.). Marcel Dekker, Inc., p. 71.

59. Simpson, J.S.A., A.K. Campbell, M.E.T. Ryall, J.S. Woodhead. 1979. Nature 279, 646.

60. Weeks, I., F. McCapra, A.K. Campbell, J.S. Woodhead. 1983. In: Immunoassays for Clinical Chemistry: A Workshop Meeting (W.M. Hunter and J.E.T. Corrie, eds.). Churchill Livingstone, p. 525.

61. Odell, W.D.. 1980. In: Methods in Enzymology, Vol. 70 (H. Van Vunakis and J.J. Langone, eds.). Academic Press, p. 274.

62. Midgley, A.R., M.R. Hepburn. 1980. In: Methods in Enzymology, Vol. 70 (H. Van Vunakis and J.J. Langone, eds.). Academic Press, p. 266.

63. Hales, C.N., J.S. Woodhead. 1980. In: Methods in Enzymology, Vol. 70 (H. Van Vunakis and J.J. Langone, eds.). Academic Press, p. 334.

64. Magnusson, C.G.M., D.L. Delacroix, J.P. Vaerman, P.L. Masson. 1984. J.Immunolog.Methods 69, 229.

65. Forrest, G.C., S.J. Rattle. 1983. In: Immunoassays for Clinical Chemistry (W.M. Hunter and J.E.T. Corrie, eds.). Churchill Livingstone, p. 147.

66. Platts-Mills, T.A.E., M.D. Chapman, E. Tovey. 1981. In: Immunoassays for the 80s (A. Voller, A. Bartlett, D. Bidwell, eds.). MTP Press Limited, p. 289.

67. Rubenstein, K.E.. 1981. In: Immunoassays for the 80s (A.Voller, A. Bartlett, D. Bidwell, eds.). MTP Press Limited, p. 127.

68. Clark, B.R., E. Engvall. 1980. In: Enzyme-Immunoassay (E.T. Maggio, ed.). CRC Press, Inc., p. 167.

69. Voller, A.. 1980. In: Enzyme-Immunoassay (E.T. Maggio, ed.). CRC Press, Inc., p. 181.

70. Mattiasson, B., C. Borrebaeck. 1980. In: Enzyme-Immunoassay (E.T. Maggio, ed.). CRC Press, Inc., p. 213.

71. Masson, P.L., D. Collet-Cassart, C.G. Magnusson, C.J.M. Sindic, C.L. Cambiaso. 1981. In: Immunoassays for the 80s (A. Voller, A. Bartlett, D. Bidwell, eds.). MTP Press Limited, p. 35.

72. Guesdon, J.L., T. Ternynck, S. Avrameas. 1979. J.Histochem. Cytochem. 27, 1131.

73. Whicher, J.T., D.E. Perry. 1984. In: Practical Immunoassay (W.R. Butt, ed.). Marcel Dekker, Inc., p. 117.

CHEMICAL MODIFICATION OF PROTEINS

G. Pfleiderer

Institut für Organische Chemie, Biochemie und Isotopenforschung der Universität Stuttgart, Pfaffenwaldring 55, 7000 Stuttgart 80

Introduction

The chemical modification of proteins is a useful method to change their physical or biochemical properties {1,2,3,4}. Originally one would expect that a distinct chemical reaction with a highly sensitive molecule like the protein is difficult to carry out, but under conditions, not too extreme (pH 5 - 9) and in aqueous solution most proteins can react like a normal organic molecule characterized by a high molecular weight. Not discussed in the following will be reactions known from the sequence analysis, e.g. the blocking of ε-amino lysine groups with dicarbonic acid anhydrids in order to hinder the tryptic attack to get peptide fragments with arginine residues on the carboxyl end or bromocyane splitting of a protein at the carboxylic end of methionine. These methods are normally applied under conditions leading to denaturation. They are helpful for the determination of the primary structure of proteins and can be looked up in all handbooks dealing with sequence analysis.

What is the purpose of a chemical modification? The chemical modification is, above all, the important method to recognize essential side chains in the active center. Usually we understand by essential groups areas of an enzyme important for structure or function. In most cases the chemical modification of such side chains is followed by a decrease, but in some cases even by increase or change of the catalytic activity. After identification of the modified side chain one can calculate with the help of additional kinetic experiments in comparison to the native enzyme (for example pH-dependence of the enzyme reaction, need for essential part of substrate or coenzymes, etc.) the mechanism of the enzymic catalysis. The

Modern Methods in Protein Chemistry, Volume 2

same considerations apply to other aspects concerning the problem structure-function relationship (allosteric regulation, interconvertibility from inactive to active state or vice versa, subunit interaction, dissociation of oligomers, folding or enfolding process). As a rule functional amino-acid side chains with a high reactivity (so called 'superreactive groups') are preferentially localized in the active center of an enzyme. By 'active center' we understand those groups which are involved in the substrate binding and catalysis. The active center is often to be seen in tertiary structure as a cleft in the interior. We can distinguish between substrate binding sites and active sites. The former are mainly charged groups in the protein which form, together with opposite ionic groups of substrates and/or coenzyme, ionic interchanges. Otherwise hydrophobic areas could, by means of hydrophobic forces, attach to a corresponding part of the ligand. Active sites are called those groups which are directly involved in the catalytical step. Their high reactivity appears related to their catalytic function.

When the product of modification is colored, fluorescent or isotopically marked, one can easily recognize the localization in the primary structure and, finally in the tertiary structure. If, on the other hand, specific chemical modification has no effect, one can exclude the importance of such a functional group in the active center.

Otherwise the accessibility of a distinct amino-acid side chain by chemical modification reaction (also without inactivation) is a helpful method to distinguish between buried and exposed groups. It is possible, by this technique, to provide an idea about the topography of a protein without any knowledge of the tertiary or quaternary structure. For instance Perham et al. {5} modified all accessible ε-amino groups of lysines in aldolase under native conditions with unlabeled imido esters and, after urea denaturation, again with isotopically marked ones. The authors were able to distinguish between those groups localized on the surface of the enzyme and others lying in the interior of the protein. The same principle is used for a long time in numerous analogous experiments with tyrosine, histidine, SH-groups and others in order to differentiate between accessible and non-accessible side chains. The accessibility of functional groups in proteins determined by chemical modification often correlates with physico-chemical results.

The chemical modification of essential, but even non-essential groups can additionally be used to introduce reporter groups, containing chromophoric, fluorescent or radical substituents, helpful for physical measurements under conditions which produce conformation changes, dissociation, enfolding, etc., controlling the change of these signals.

The chemical modification of non-essential groups, mainly ε-amino groups of lysine plays an important role for the biotechnology; the technical use of enzymes, especially for the immobilization after covalent binding on insoluble matrices. Non-essential groups otherwise can be used to form cross-links between homologous or heterologous proteins. A practical use in the immunochemistry is the formation of cross-links between antibodies and so-called reporter enzymes. In this case it is necessary to preserve the antigen-binding capacity <u>and</u> the activity of the reporter enzyme. Very well suited for investigations of the quaternary structure of oligomeric enzymes or multi-enzyme systems is the introduction of cross-links between subunits.

All the above discussed problems can be investigated by chemical modifications and are discussed in the following, demonstrated on practicable examples chosen by the author, who is long-experienced in this field.

Principles for a careful chemical modification and correct explaining of the effects

If available, a reagent as specific as possible for a distinct functional group, should be used. Absolutely specific are only a few reagents, such as imidoesters for amino groups, disulfides for thiols or tetranitromethane for tyrosines. Other reagents are very unspecific although in some special cases required. Iodoacetic acid or iodoacetamide were originally known as SH-blockers forming the corresponding carboxymethylthioether of cysteine (1).

$$\textcircled{P} - SH + JCH_2 - C\genfrac{}{}{0pt}{}{\nearrow O}{\searrow OH} \rightarrow \textcircled{P} - SCH_2 - C\genfrac{}{}{0pt}{}{\nearrow O}{\searrow OH} + HJ \qquad (1)$$

It was found out later that RNAse which does not contain free SH-groups, reacts with ^{14}C-iodoacetate at pH 5.5, forming the N-carboxymethyl derivatives of the His_{119} under inactivation of the enzyme {6}. Under distinct conditions either His_{12} or His_{119} has been reacted. The two monoalkylated derivatives were isolated chromatographically; since both were inactive, His_{12} as well as His_{119} is most probably concerned with the enzymic activity {7}. We demonstrated the carboxymethylation of an essential methionine residue in malate dehydrogenase as cause of an inactivation {8}; finally also the carboxymethylation of amino groups was described. In such cases it is useful to control the inactivation kinetics under different pH-conditions. The reactivity of the thioether group of methionine is only slightly influenced by pH, but the amino and the imidazole groups are only reactive in the unprotonated form. By choice of pH it is possible to increase the specificity of the chemical attack. Variation in reactivity as a consequence of pH, therefore is a convenient way to control the course of some modification reactions. Unfortunately the p_K of various side chain groups are not only very similar, but those of members of the active center can even vary under the influence of the protein microenvironment (table 1).

Table 1: p_K Values of Side Chains in Proteins

	25° p_K
αCOOH group	3.0 - 3.2
βCOOH group of aspartic acid	3.0 - 4.7
γCOOH group of glutamatic acid	ca. 4.4
phenol group	9.8 - 10.4
SH group of cysteine	8.3 - 8.6
imidazole	5.6 - 7.0
αamino group	7.6 - 8.4
εamino group of lysine	9.4 - 10.6
guanidino group of arginine	11.6 - 12.6

Because of the relative specificity of modification reactions pure reagents should be only used. Impurities with high inhibitor effects are often observed. Some reagents are instable and must therefore be recrystallized from time to time. The excess of chemical reagents, in relation to the number of identical amino acids, should be as low as possible. Then the side reactions are lowered. All experiments must be controlled by a blank (especially the control of the enzyme activity), that means that all procedures during and after modification have to be carried out parallel without addition of the reagents. If then a biochemical effect is observed for the first time, a characterization of the modification product is required and carefuld analysis should be carried out. In some cases it is possible to follow the grade of modification directly by a spectrophotometric change (see below: Ellman reaction). Normally, parallel to the observation of the inactivation kinetics at distinct points the modification reaction was stopped by destruction of the residual reagents. Then the reaction mixture should either be dialysed or the modified protein must be separated from the excess of reagent by gel filtration (G-25).The latter method is sufficient if one wants to remove even radioactively labeled reagents completely. After concentration of the eluted peak, an exact protein value has to be estimated and all analytical data are to be correlated for this particular value (amino acid-analysis, spectrophotometric-, radioisotopic analysis). The chemical product must be identified in cases, if it is not clearly recognized. In some cases the product of the chemical modification is very instable and has, therefore, been destroyed after acidic or alkaline hydrolysis of the peptide chain. The mildest form to characterize the modified amino acid itself is the use of a peptidase cocktail which splits off all peptide bonds under mild conditions, preserving the modified side chain. We were very successful using a mixture of trypsin/chymotrypsin followed by a digestion through pronase and a partially purified aminopeptidase from kidney. For instance we obtained after inactivation of cytoplasmic malate dehydrogenase from pig heart with ^{14}C-iodoacetate and separating the excess of reagent with the help of peptidase cocktails carboxymethyl methionine identical with a synthetic model substance {8}.In another case Woenckhaus modified lactate dehydrogenase from pig heart with {^{14}C}-3-(1-hydroxyethyl)pyridine under inactivation.

The radioactively marked enzyme was treated by $NaBH_4$ and tryptically digested after urea denaturation and the isolated radioactive peptide incubated with carboxypeptidase B and A and finally with aminopeptidase M. The radioactive amino acid derivative was identical with synthetic N{2-hydroxy-2-(pyridyl)-(3)-ethyl}-histidine confirming our expectations of an alkylation of an essential histidine in the enzyme {9}.

The next step is the exact determination of the number of essential groups. Under optimal conditions (protein + modification reagent concentration, buffer, pH, temperature) exact kinetics of the decrease of enzyme activity and modification grade of a distinct side chain has to be measured with a control of the blank. The best is to follow the kinetics in an independent experiment. Then conditions can be chosen for not too fast a reaction period. In presence of a high excess reagent mostly first order kinetics are found, which indicates that semilogarithmic plot for loss of activity and loss of functional groups (SH-, NH_2, Tyr, His) is a straight line. When activity against the number of modified groups/subunit is a straight line and extrapolated to 100 %, inhibition results in the modification of a single amino acid, a high probability for an identification of one essential group exists. If several groups are modified, then the importance of each of them is not clear.

An essential side chain must not be the most reactive group. Nearly all mammalian lactate dehydrogenases of heart type contain one highly reactive thiol group, after the blocking of which the enzymic activity is lost. The corresponding essential cysteine peptide {10} is found in numerous species with identical sequence. Kaplan and coworkers discovered lactate dehydrogenase in other species with a large number of non-essential SH-groups, which react faster than the above-mentioned essential thiol group without decrease of enzymic activity. At last the essential group reacts {11}.

Another type of inactivation kinetics is shown in case of the modification of tyrosines in pig heart lactate dehydrogenase. Here we find two modification rates {12}. At first an essential tyrosine is nitrated with tetranitromethane and correlates with the decrease of enzyme activity. After one group per subunit had been nitrated, a slower nitration rate with a low further decrease of activity is observed (Fig. 1 and 2). When, after modification of one tyrosine group the reaction is stopped, only one specific tyrosine in the peptide chain is modified.

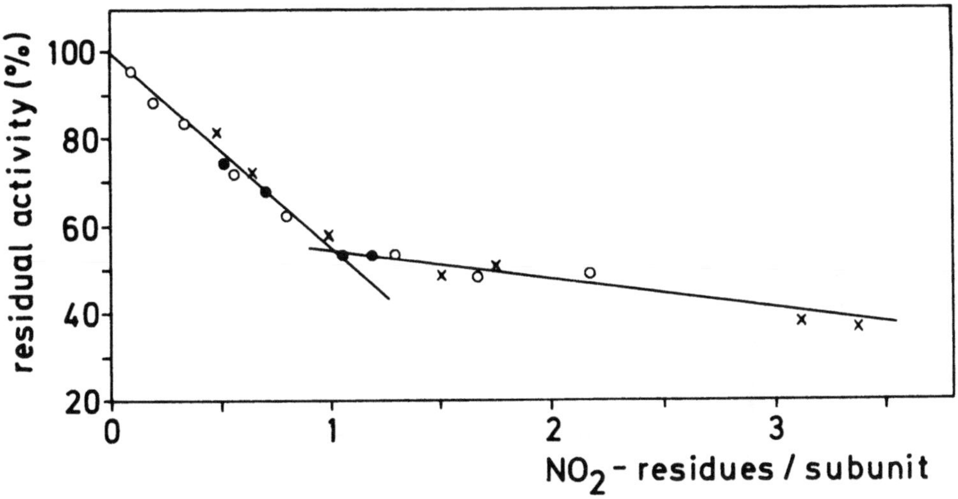

Fig. 1: Nitration of lactate dehydrogenase from pig heart by tetranitromethane

We obtained identical results by azo coupling of tyrosine residues in the same enzyme by diazonium salt of sulfanilic acid {13}. The preparation of lactate dehydrogenase containing one azo tyrosine group showed only one enzymic and protein-specific component in the membrane foil electrophoresis, migrating faster than the original enzyme (SO_3^--group attached). Generally it is optimal when the yield and purity of the product is characterized by physical methods.

Another important aspect must be mentioned. In many cases aliquots of the enzyme are denatured during the modification. It is useful for future kinetic experiments to purify the native part of the modified product. We, for instance, perform after amidination or nitration of lactate dehydrogenase a brief affinity chromatography with AMP-sepharose which binds only the native part. In a gel filtration on Sephadex G-200 a small part of enzyme appears sometimes in the exclusion volume as a sign of aggregation.

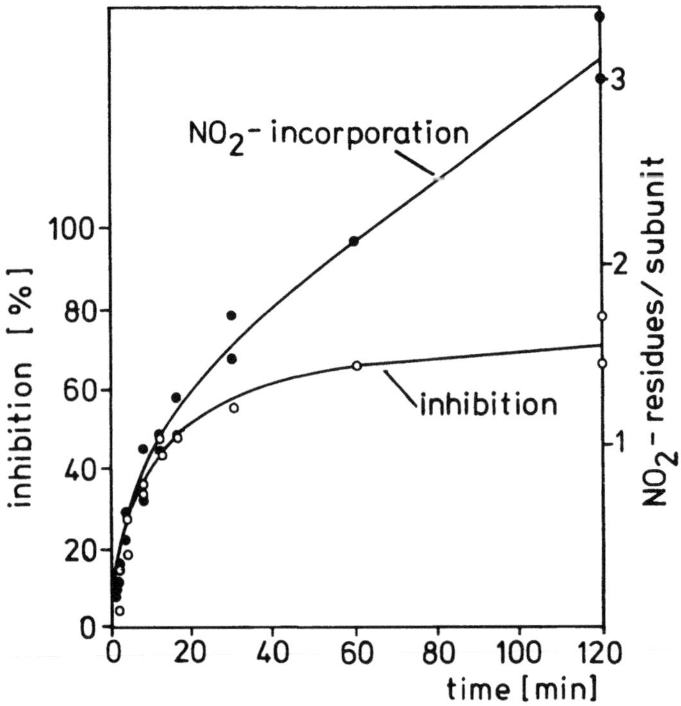

Fig. 2: Kinetics of inhibition and nitration of pig heart lactate dehydrogenase

The degree of native and modified protein enzymes also can quickly be estimated by gel electrophoresis.

The best way to find out that only one essential group had been modified is the fragmentation of the modified protein and the isolation of the peptide carrying the labeled amino acid.

In our biochemical laboratory we use an excellent example for demonstratio or learning: Lactate dehydrogenase of pig heart, which is easy to crystallize within a few days, reacts very rapidly with the Ellman reagent under decrease of the enzyme activity. After extrapolation to 100 % inactivation one thiol group has reacted. The inhibition is totally reversible. By use of radioisotopically labeled aromatic maleimide and after tryptic digestion in the fingerprint, only one peptide spot is radioactive, containing over 92 % of the incorporated radioactivity {14}.

The best information about the modification of amino acid side chains is the sequence analysis of the labeled peptide. Recently we have published an amidination reaction of fluorescent imidoesters with glucose dehydrogenase from bacillus megaterium {15}. After analysation of all fluorescent tryptic peptides only one lysine, residue no. 201 is completely modified. 50 % of all fluorescent lysines lie in this fraction at 50 % inhibition of the enzymic activity. Therefore we are sure that lysine no. 201 is essential.

The next step is to explain the cause of the change of enzymic activity. The best is to control in case of coenzyme-dependent enzymes the binding capacity of coenzymes after the modification. NADH binding is easy to measure by fluorescence titration {16}, equilibrium dialysis or by gel filtration in presence of excess of NADH {17}.

Investigations on dehydrogenases also allow to follow the formation of ternary substrate, coenzyme and apoenzyme complex, e.g. dehydrogenases, which were inhibited by $SO_3^{=}$ can form an NAD-SO_3^{-}-complex with an absorption maximum in the region of 320 nm. By a simple titration the residual binding capacity can be estimated within a few minutes {18}. Other titration techniques follow the increase of fluoresence in the ternary complex or changes in the CD spectrum.

The decrease of enzymic activity under standard conditions could also be effected by a change of pH optimum or K_m values. An example for that is the change of biochemical properties of lactate dehydrogenase H_4 after nitration of one tyrosine residue (Fig. 3, {12}). Therefore the modified enzyme must be tested by careful kinetic experiments.

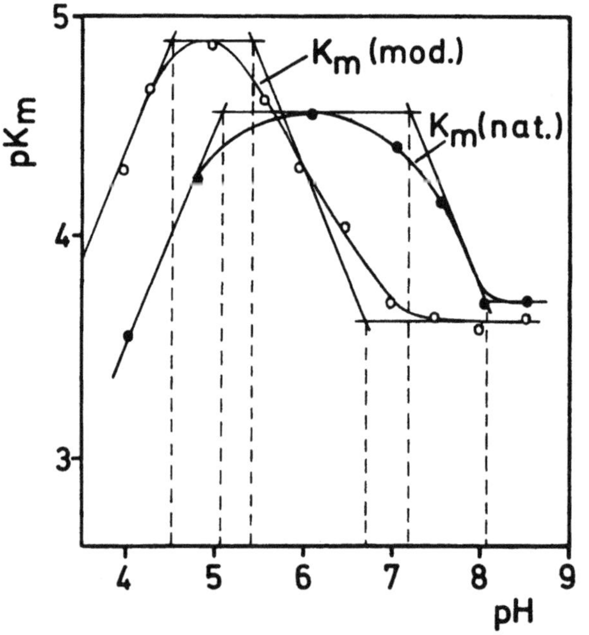

Fig. 3: Change in biochemical properties of lactate dehydrogenase from pig heart after nitration of one tyrosine residue

The active center of an enzyme mostly contains various essential groups which all together effect the enzyme reaction. When it is seen that a modification reagent reacts with two or more different functional groups, then it is necessary to protect reversibly one group covalently against the chemical attack of the reagent. After modification of the other essential group, the first one is converted into its native form. If then an irreversible inactivation is observed, the second modified group is essential. The reversible marking of the first group under complete reactivation must be controlled in an independent experiment {9}.

In some cases it is necessary to make a differential labeling. One modification is carried out in presence of substrate, coenzyme or inhibitor, which protect the active center, the second modification in absence of the ligands. If enzymic activity is retained in the presence of ligand, but is lost in its absence, one can assume that the ligand has protected the active center. In such a case at first unlabeled reagents were used and after removal of excess of reagents and ligand radioactive reagents in order to label the active center.

Another important aspect must be proved for the explanation of a modification effect. When the secondary, tertiary and quaternary structure is changed, then this can be the origin of an inactivation. Therefore it is necessary to compare the native molecular weight (ultra centrifugation, gel filtration), the circular dichroism spectra of the backbone (222, 210, 194 nm) or other characteristics of the native structure with those of the modified molecule. So far the migration in the gel electrophoresis or in the gel filtration can give confirmation of the native structure of the modified protein. We often use also quantitative immunotechniques to compare the native and the modified protein. Only small changes in the structure can produce loss of the antigenity by changes in the determinants. In a spectacular experiment we find a strict correlation between change in the enzymic activity, ORD values and antigenity {19}.

Chemical modification of specific amino acid side chains

Thiol groups

The most applied routine method for quantitative analysis of SH groups in proteins was developed by Ellman {20}, who used 5,5'(dithiobis(2-nitrobenzoic acid) (DTNB) as reagent. It reacts with free SH groups (2) libe-

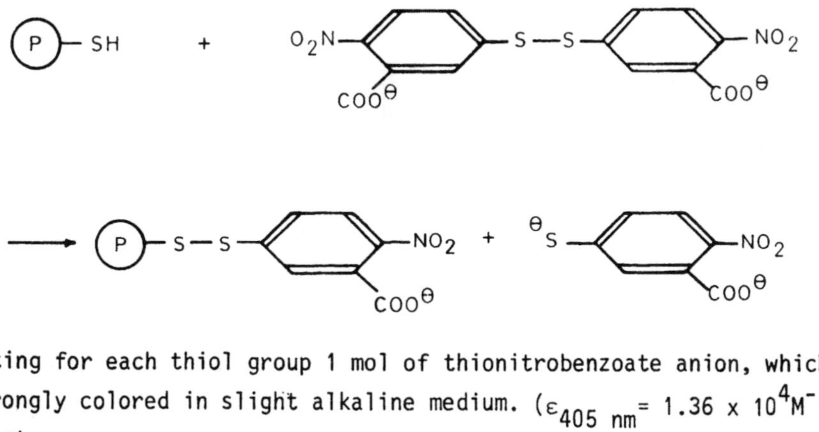

(2)

rating for each thiol group 1 mol of thionitrobenzoate anion, which is strongly colored in slight alkaline medium. ($\varepsilon_{405\ nm} = 1.36 \times 10^4 M^{-1} cm^{-1}$ pH 8).

Thus one can follow spectrophotometrically the course of the reaction, which is reversible upon addition of an excess of thiols. In the case of essential SH groups it is often possible by such a way to protect reversibly this group against the attack of reagents used for chemical modification of other functional amino acid side chains in the active center. Theoretical SH values were obtained between pH 7,8 and 10.4, but below pH 7 no reaction occurs.

A disadvantage is the low reactivity obtained with many enzymes. For instance with lactate dehydrogenase of pig heart we find only four reactive from a total of 20 SH groups in the native enzyme. One must usually at first enfold the proteins in urea to reach all thiol groups. This means that most thiol groups in a protein are buried. A reagent similar to DTNB is 2,2'-dithiopyridine (2-PDS). It reacts with SH groups to also form mixed disulfide and the corresponding 2-thiopyridones. The disulfides of five-membered heterocycles have a higher oxidation potential than dipyridyl disulfides and a smaller steric hindrance. We were successful using 2,2'-dithiodithiazole, 2,2'-dithiobis(4-methylthiazole) and 5,5'-dithiobis(3-methylisothiazole) to block in the native state all 20 thiol groups of lactate dehydrogenase H_4 (Fig. 4) located partially in the interior of the enzyme. In addition these reagents are available in the pH range ≤7 {21}. Extinction coefficient of 2-mercaptothiazole pH 5 at 310 nm 1.48 x 10^4 M^{-1}/cm^{-1} (3).

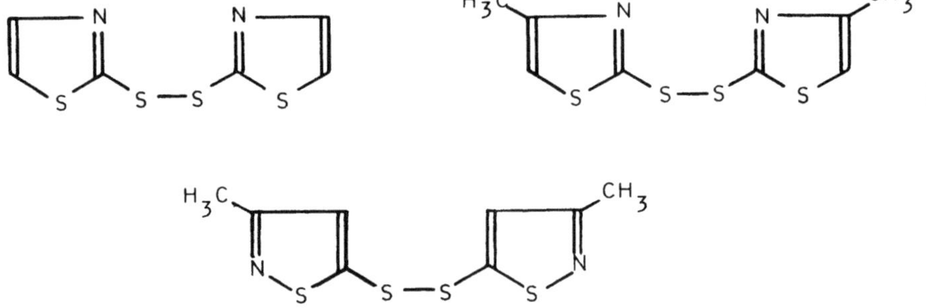

(3)

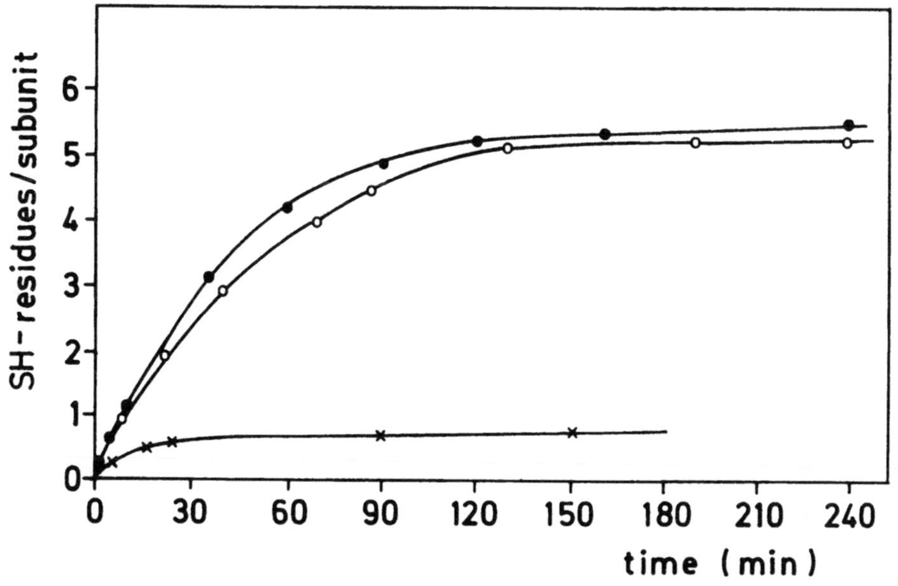

Fig. 4: Time dependence of the modification of lactate dehydrogenase (LDH) with 2,2'-dithiodithiazole (•), 5,5'-dithiobis(3-methylisothiazole) (o) and 2-PDS (x) in 67 mM PP, pH 7.2

<u>Maleimides</u>

N-ethylmaleimide (NEM) is a useful sulfhydryl reagent (4), R = C_2H_5

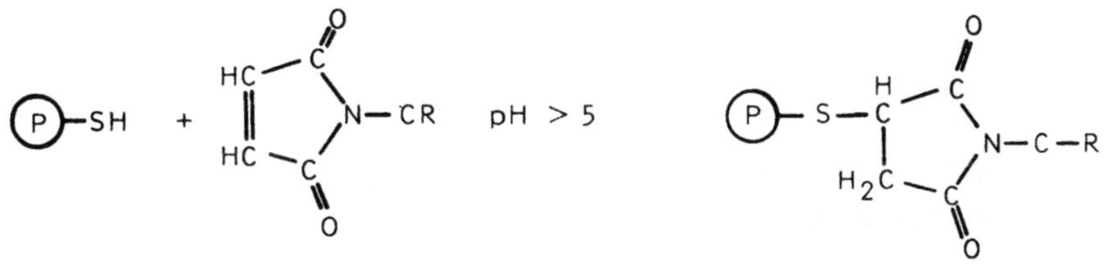

(4)

NEM has an absorption band near 300 nm (ε = 620 $M^{-1}cm^{-1}$). The S-alkylation could be controlled by the decrease of absorption at 300 nm for quantitative thiol estimation. The above reaction is to a certain extent specific for protein sulfhydryl groups below pH 7. At higher pH values reaction with amino groups becomes more and more significant. Highly reactive ε-amino groups have been observed to react faster with maleimides than thiols.

A further disadvantage is the relative instability of the five-membered ring system. During peptidase digestion or separation of the split peptides a partial hydrolysis to half amide occurs. In the case of isotopically marked maleimides we find often during peptide separation two isotope marked peptides with identical sequences: one with closed succinimid ring, the other with the open half amide form. For the preparative chemist it is simple to synthesize maleimides with radioactive, chromophoric, fluorescent or radical substituents in order to introduce reporter groups in thiol proteins. In the case of aromatically substituted maleimides we were able to test the deepness of the active center preparing maleimide with phenyl-, benzyl or phenethyl residues, etc. While the phenylmaleimid is ineffective in some thiol enzymes because the steric hindrance of the benzene residue phenethyl maleimide was reactive {22}. Long chain alkyl maleimides are sometimes more effective than NEM for reaction with sulfhydryl groups in a polar medium.

In a few cases it is practical to use radioactive iodoacetic acid or iodoacetamide for thiol blocking. But, as it is mentioned above, the identification of the carboxymethyl cysteine product is important.

Mercurials react rapidly and specifically with thiol groups of proteins (5).

$$\text{Ⓟ} - S^- + {}^+Hg - R \rightarrow \text{Ⓟ} - SHgR \qquad (5)$$

Often used are organic mercurials like p-chloro-mercuribenzoate. The increase of absorbancy in UV during mercaptide formation with thiols allowed the quantitative estimation of thiol groups in proteins. Because mercurial and protein absorb strongly in the UV, blanks had to be measured exactly. According to Boyer {23}, the difference of $\varepsilon_{250\ nm}$ is 7.6×10^3. The mercaptid formation can be split by addition of excess thiol reagents.

Tyrosine

The iodination of proteins is a very simple method to introduce radio isotopes using ^{124}I or ^{131}I. At mild alkaline conditions preferentially mono- and diiodotyrosine derivatives are formed, but also histidine modification and oxidation of cysteine and methionine side chains are observed. The iodination reaction is mostly dubious, other methods for tyrosine modification are preferred. The O-methylation of tyrosine hydroxyl is easily performable by treatment with N-acetyl-imidazole. Like acetic acid anhydride, N-acetylimidazole reacts with both amino- and tyrosyl groups of proteins, but it is more selective for tyrosine. Under distinct conditions one can selectively acetylate tyrosyl residues without affecting amino groups. With this method it is further possible to distinguish between free and buried tyrosyl residues. The O-acyl derivatives could easily be hydrolyzed by hydroxylamine in contrast to N-acyl groups which are stable. The strong change of UV absorption spectrum while tyrosine ionization (red shift and increase of the absorbance) allows to differentiate between free, buried and modified tyrosine residues {24}.

$$\Delta\varepsilon_{278} = 1160\ M^{-1}\ cm^{-1}$$

The widely used modification reagent for tyrosine is tetranitromethane (TNM) introduced by Sokolovsky {25}. According to the reaction scheme tyrosyl radicals are formed. Therefore intra- and intermolecular cross-linking of proteins is available. During our investigations we have never found such cross-links and were very successful using the TNM technique. Another advantage is the easily performed reduction of the nitro group into an amino group. This aromatic amino group has a p_K of 4.8; this means that a special alkylation of this group in neutral or slight acidic solution is possible without alkylation of α- or ε-amino groups of the protein. It is notable to perform the nitration with as small as possible an excess of the reagent related to all tyrosine residues per subunit. The strong chromophoric characteristics of nitro tyrosine allow a simple quantitative follow up of the modification reaction ($\varepsilon_{428} = 4.4 \times 10^3\ Mol^{-1}cm^{-1}$ at pH >9). Also after proteolytic digestion the nitro tyrosine peptides could be easily detected during the isolation. The tetranitromethane may oxidize cysteinyl, methionyl and tryptophanyl residues. The absorption spectrum of 3-nitrotyrosyl residues is highly sensitive to the polarity of their environment. Hence, they may be used as 'reporter group' by itself.

Modification of the phenolic ring of tyrosine results in characteristic changes in phenolic p_K or absorption spectrum. For example the p_K of 3-nitrotyrosine hydroxyl is 7,2 (25°C) against 10.5 for the unsubstituted phenolic ring. λ_{max} at high pH is 428 nm instead of 293 nm.

Azo coupling of histidine or tyrosine residues

In spite of heterogenous nature of the reaction products, reaction of proteins with aromatic diazonium reagents has been widely used. The colored products of histidine or tyrosine azo derivatives are the basis of some useful colorimetric determinations of such modification reactions. Usually different products can be formed (mono-, bis- azo derivatives, triazene). Also guanido- and indol moieties show similar reactions. The substituents of diazonium compounds could be important for the specificity and reactivity with a functional group.

The best reagent is diazonium 1-H-tetrazole (DHT). Its application allows an accurate colorimetric procedure for quantitative determination of reacted histidine or tyrosine (6). No colored hydrolysis products are formed.

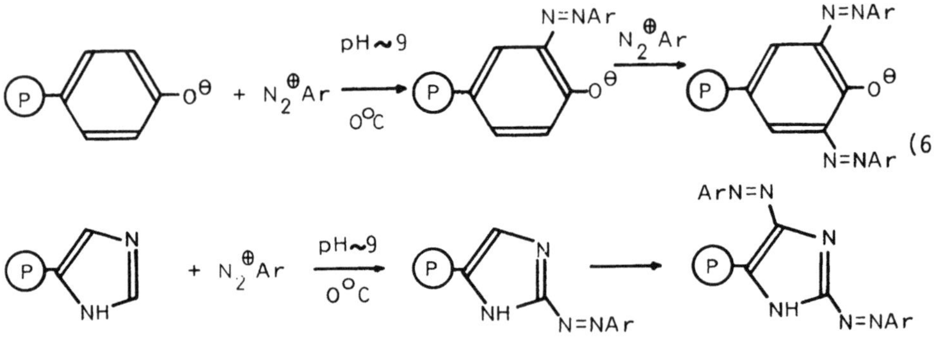

(6)

Under distinct pH-conditions, ~7 histidine reacts sometimes faster than tyrosine {26} (table 2).

Table 2: Spectral Properties of Azo Derivatives of Histidine and Tyrosine (pH 8.8)

	λ_{max}	$\varepsilon \times 10^{-3}$
Azotetrazole of N-acetylhistidine	360	11.5
Bisazotetrazole of N-acetylhistidine	480	21.0
Azotetrazole of N-acetyltyrosine	478	5.2
Bisazotetrazole of N-acetyltyrosine	548	14.1

We were successful to form, after coupling of diazotized sulfanilic acid with lactate dehydrogenase native monoazo-LDH with decreased specific activity in order to identify an essential tyrosine residue in the active center {13}. In another experiment we could demonstrate under careful spectral control (mono-, bis-azo derivatives) a good correlation between histidine coupling and decrease in the enzymic activity of alkaline phosphatase from calf intestine. Exactly 2 histidine residues/subunit have been coupled for 100 % inactivation, synchronic with the loss of 1 Zn atom. In some cases it is necessary at first to amidinate all amino groups on the surface to prevent the formation of triazene (Fig. 5) {27}. A further control is the amino acid analysis of unmodified residual histidine residues. The azo bridge can be used furthermore as reporter group, when localized in the active center.

Modification of histidyl residues by diethylpyrocarbonate {28}

Diethylpyrocarbonate has been shown to react specifically or stoichiometrically with a single histidyl residue in certain proteins (yeast alcohol-, lactate and glutamate dehydrogenase). Only the unprotonated form of the imidazole residue is nucleophile to attack the pyrocarbonate (7).

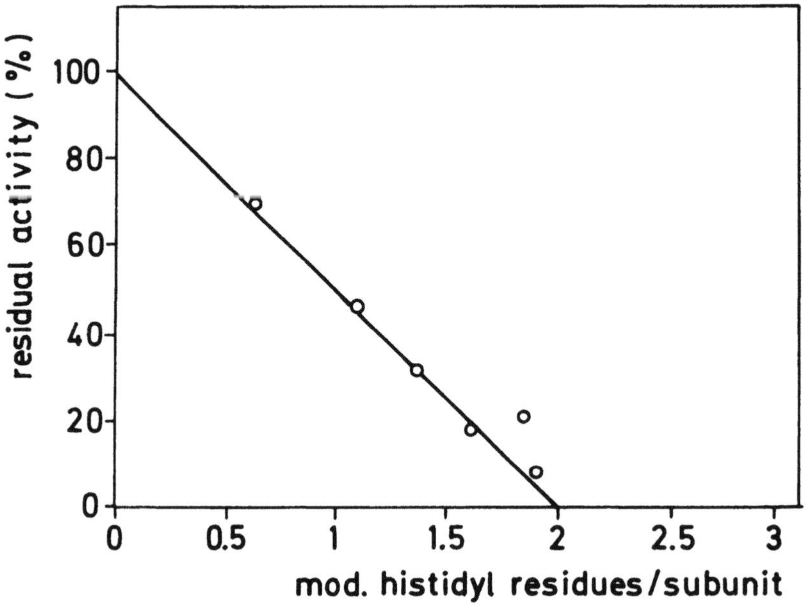

Fig. 5: Azo coupling of acetamidinated alkaline phosphatase from calf intestine

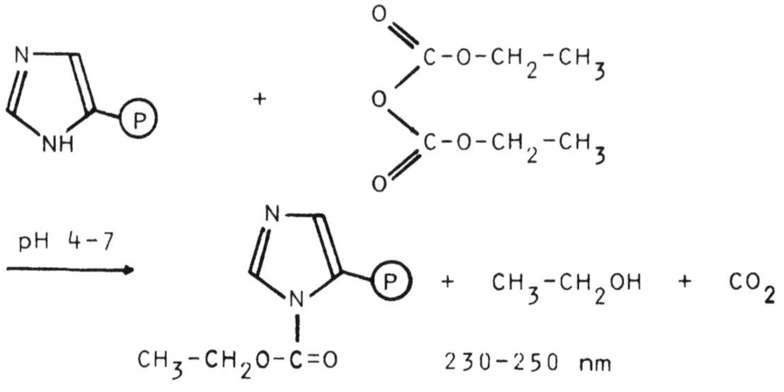

(7)

Therefore the reaction should be carried out in a pH range of 6 - 8 (K-phosphate buffer). The number of modified histidyl residues is calculated from the molar absorption differences for N-carbethoxy histidine at 240 nm 3200 $M^{-1}cm^{-1}$. Inactivation of an enzyme by diethylpyrocarbonate should be correlated with the modification of a histidyl residue if hydroxylamine reactivates the enzyme.

Diethylpyrocarbonate can also react with other nucleophiles like -SH-, arginyl- and tyrosyl-residues. Also the active site serine in chymotrypsin has been modified by this reagent, but the side reaction has only been reported.

Photo-oxidation

Although dye sensitized photo-oxidation of proteins by irridiation with visible light was introduced by Weil et al. in 1951, this method has been applied rarely. Many authors score for this technique because only under distinct conditions the selective modification of one type of side chain or a limited number of amino acid residues is effected. In addition the characterization of photo-oxidation products is uncertain. There is now a general agreement, where the mechanism of the process is concerned. The dye-sensitized photo-oxidation proceeds by two major pathways involving either $^1O_2$ or free radicals. By energy transfer from the triplet state of the sensitizer singlet oxygen was generated. The electrophilic singlet oxygen may attack electron-rich chains of cysteine, methionine, histidine, tryptophan and tyrosine.

By choice of different pH conditions and appropriate sensitizer (rose bengal, methylene blue, thiopyronine, riboflavine, proflavine) it is possible to modify selectively distinct side chains without breaking the backbone. The strong pH-dependence of photo-oxidation of histidine derivatives mediated by methylene blue demonstrates the conclusion that only the free imidazole base is oxidized {29}. Suitable chromophores bound non-covalently at the active center of enzymes have been successfully applied as photo-oxidizers. The observation that many dyes interact preferentially with coenzyme binding sites (see dyes for affinity chromatography) could explain the unexpected high selectivity of photo-oxidation. Also naturally occurring sensitizers are successful. This is the case in the presence of pyridoxalphosphate at the catalytic site of the β_2 subunit of tryptophan synthetase of E. coli and of glutamate decarboxylase, where these were used for the production of the specific photo-oxidation of one essential histidine {30,31}.

In many cases the number and localization of attacked amino acid can indicate hindrance to the susceptibility in the tertiary or quaternary structure. Jovi and Galiazzo were able to differentiate between 5 tryptophyl residues in papain, depending on the surrounding medium, which was chosen for the proflavine sensitized photo-oxidation. They concluded from their result that some areas of the papain molecule may be different from that in the crystalline state as deduced from X-ray diffraction studies {32}.

Carboxyl groups

The participation of aspartyl residues in the 'charge relay' systems of numerous hydrolytic enzymes has directed increasing interest in the importance roles for carboxylic groups in active centers of enzymes. On the other side the knowledge of internal salt bridges involved in carboxylic groups is important for structural constellations. The oldest modification of carboxylic groups was the esterification with alcohol + HCl. It is only applicable with extremely stable proteins like lysozyme, RNAse or bovine serum albumin. Esterification by diazomethane, a highly reactive compound, is not specific. But diazoacetate esters and amides are more stable and react more specifically with carboxyl groups of proteins (8).

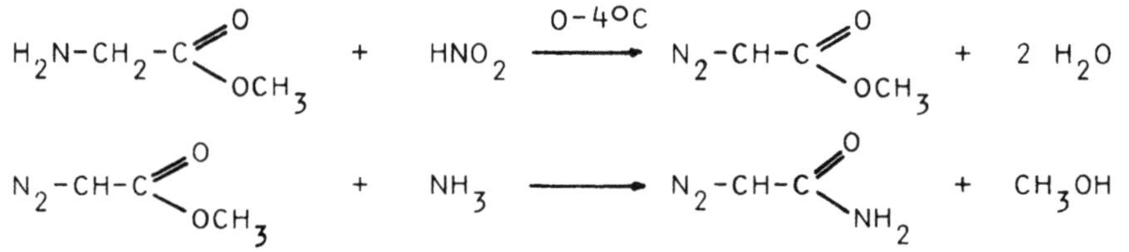

(8)

Because of the rapid destruction by water, only most accessible or reactive carboxylic groups are attacked. The mildest and most suitable modification reaction is performed with carbodiimide in two steps. Under mild conditions accessible or reactive carboxylic groups are converted by water-soluble carbodiimides in an intermediate O-axyl isourea product (9).

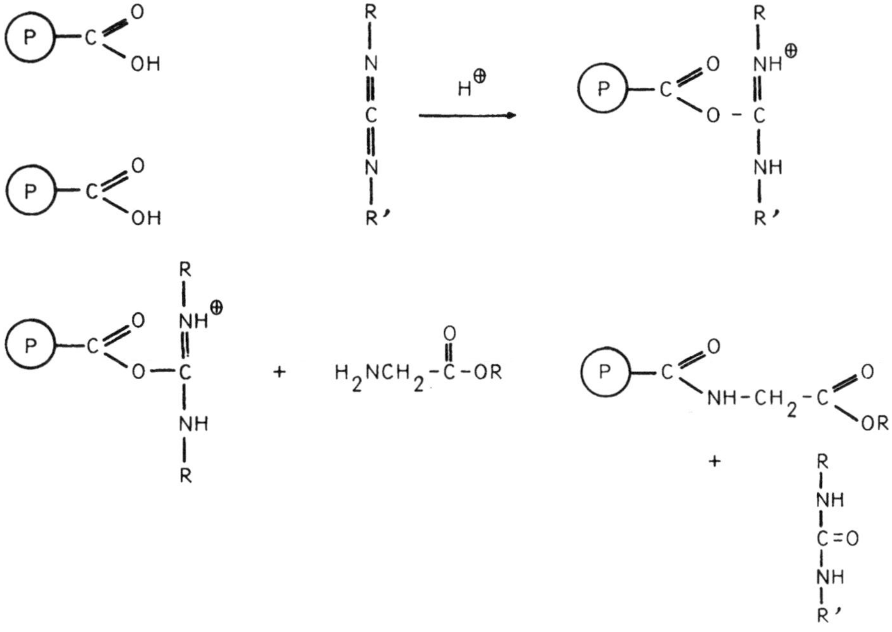

(9)

In presence of nucleophiles an acyl nucleophile originates. The application of different substituted nucleophiles allows a high variability of the products.

The modification of carboxylic groups in proteins with isoxazolium salts represents a similar approach to the carbodiimide-nucleophilic reaction. Woodward had demonstrated that N-alkyl-5-phenylisoxazolium salts were very reactive toward carboxylic groups, activating them for peptide synthesis {33}. Intermediate arises an enolester, which is shown to be an efficient acylating agent (10). Woodward's reagent K is commercially available and suited in some cases to activate COOH groups in enzymes. The absorption of the reagents can be used for quantitative estimation (change of OD_{280}/OD_{250}). Displacement of the protein enols with $\{^{14}C\}$nucleophiles results in proteins with modified carboxylic groups {34} , but also a base catalyzed rearrangement to the relative inert imide was reported.

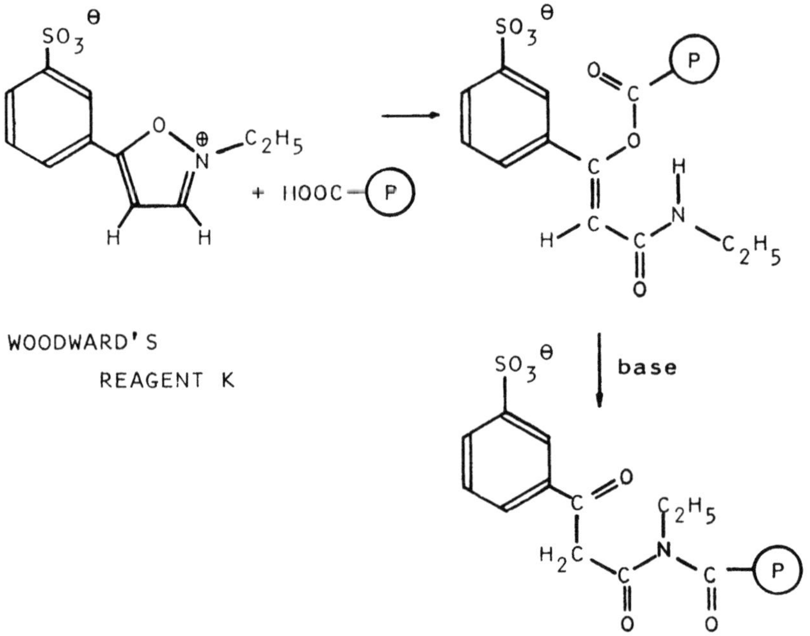

(10)

Tryptophan

The importance of tryptphan in the mechanism of enzyme reaction or for conservation of the native conformation in proteins is poorly understood. Beside the photo-oxidation which is sometimes very complicated, no ideal modification reaction for the indole part of tryptophan is known. The best is the use of Koshland's reagent I dimethyl-(2-hydroxy-5-nitrobenzylsulfonium bromide (HNB) {35}. In acidic solution the reagent is stabile and reacts specifically with the indole moiety of tryptophan and with thiol groups in proteins, but the acid pH leads to enzyme denaturation (11).

In neutral or alkaline medium the reagent is rapidly hydrolyzed, therefore a high excess must be added in portions. The HNB incorporation can be calculated at 410 nm with

$$\varepsilon = 1.85 \times 10^{4}\ M^{-1}cm^{-1} \quad pH \quad >10$$

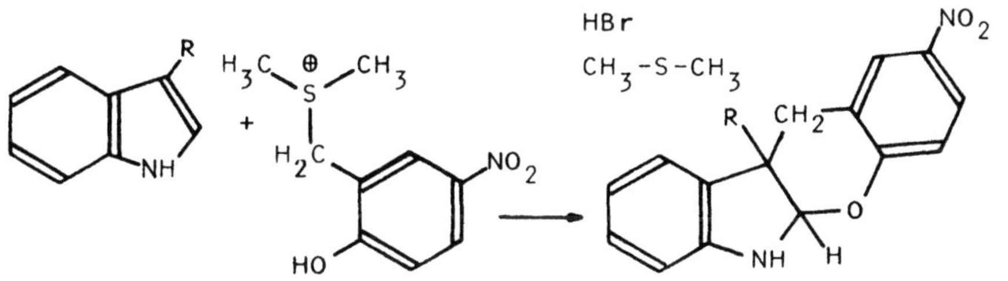

(11)

We were able, after reversible protection of the reactive cysteine residue of glyceraldehyde phosphate dehydrogenase from yeast to specifically modify tryptophan no. 193 under loss of the enzymic activity. In order to remove adsorbed non-covalently bound reagent, we unfolded the protein in urea solution, before gel filtration and dialyzation {36}.

Arginine

Now it is well-known that the guanido group of arginine in proteins is an anion recognition site for ligands. So far arginine binds in carboxypeptide the free COOH group in peptide substrates or the aspartate side chain in the case of glutamate oxalacetate amino transferase. Further arginine residues participate in the binding of phosphateester-substrates of alkaline phosphatase or some oxidoreductases form ionic interchanges between arginine and the pyrophosphate group of NAD/NADH. Nearly all phosphoesters of the glycolytic intermediates are bound by arginine residues, etc. Therefore the chemical modification of guanido groups in enzymes could be very important for detecting binding sites of ligands. Procedures of this kind are based upon the reaction with aldehydes and with 1,2 or 1,3-dicarbonyl compounds. The aldehyde adducts are labile, but condensation with dicarbonyl group leads to the formation of relatively stable heterocyclic rings. Patthy and Smith {37} reported a successful application of selective modification of arginyl residues with 1,2-cyclohexanedione in sodium borate buffer between pH 8 and 9. Only a single condensation product: N^7, N^8-(1,2-dihydroxycyclohex-1,2-ylene)-L-arginine is obtained under this condition. The product is stable in both: borate buffer (pH 8 - 9) and in acidic solution (12).

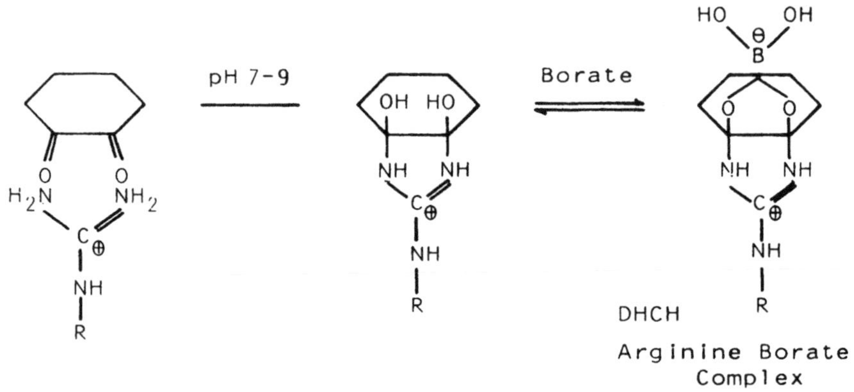

(12)

Treatment of proteins with this reagent at neutral pH results in the primary modification of guanido groups. The product is destroyed during acid hydrolysis. Reaction products with phenylglyoxal,the structures of which are not known, are stable at acidic pH, but arginine is slowly regenerated in the absence of excess reagent in neutral or alkaline medium. Therefore phenylglyoxal is useful for reversible modification of accessible arginine residues in enzymes. In the presence of high excess of reagent also ε-amino groups of lysine may react. Very simple to observe is the adduct formation of the guanido group with 4-hydroxy-3-nitrophenylglyoxal {38}. The selectivity for arginine is biggest at pH 9, when, after only 15 minutes arginine is at 73 % modified and all other amino acids were less than 7 % reacted. The adduct is also labile. Absorption coefficient of the adduct with a maximum at 405 nm is 4750 $M^{-1}cm^{-1}$. Creatine kinase f.e. is rapidly inactivated by the reagent. Loss of 85 % of the activity coincided with the incorporation of 1.1 mole reagent/subunit.

Amino groups

The chemical modification of amino groups of proteins is in three aspects important for the enzymologist:

1) For identifaction - as in other side chain reactions - of single essential amino groups in the active center and the explanation of their function.

2) For studies of structural properties of proteins. The charged groups, preferentially ε-amino and carboxylic groups located mostly on the surface of globular proteins, can form interionic forces for stabilization of tertiary or quaternary structure. Change of the charge by modification often results in a change of the conformation solubility or state of the aggregation.

3) Amino groups, especially non-essential ones, are very well suited for the introduction of specific reagents as indicator molecules for UV and visible absorption spectroscopy, fluorimetry or ESR spectroscopy, for cross-linking or preparation of immobilized enzymes.

Experiments for 1) are carried out in a similar way as described for other amino acid side chains. In the case of essential amino groups one observes during the chemical modification a strong loss of enzyme activity correlated with the chemical reaction. On the other hand Plapp found that amidination of a single amino group in horse liver alcohol deyhdrogenase enhanced significantly the enzymic activity. By use of {^{14}C}acetimidate during the differential labeling procedure, only lysine 228 was modified {39}.

Acetylation

Fortunately we know various amino group reagents with different influence on the charges. Succinylation or reaction with other dicarbonic acid-anhydrids produce instead of the positive charge of the amino group (neutral pH) a negatively charged group (+1 → -1). Some acyl groups of this kind can reversibly split off under mild conditions. For instance the tetrahydrophthalic acid part is hydrolyzed while stored at pH 5. This reversibility is important to exclude secondary effects. Succinylation or maleoylation often cause a dissociation of oligomeric protein into subunits. After intensive succinylation a total enfolding occurs as in the case of urea-enfolding.

Simple acylation, the best with acetic acid anhydride, N-acetylimidazole or $F_3C\text{-}\overset{O}{\overset{\|}{C}}\text{-}S\text{-}C_2H_5$ form N-acyl groups with loss of the positive charge. The trifluoroacyl residues can be removed in 1 M piperidine at 0°C.

Amidination {40}

A group of most specific modification reagents are the imidoesters. Under distinct conditions only α- and ε-amino groups in peptides or proteins react in aqueous solution forming amidino groups which preserve the positive charge of the original $NH_3^{\oplus}$ group at physiological pH. Therefore the mobility of the modified proteins in the usual buffers (pH 8.6) is identical with the native one (13).

The p_K of the ε-amino groups of lysine (10.5) is after the amidination reaction changed to 12.5 (the guanidino group of arginine has an identical p_K-value). Using different R-substituents of imido ester, it is possible to introduce in the protein structure, especially in the surface, chromophore, fluorophore, radicals and others without change of the net charge of the proteins, as signals for quantitative estimation of the modification yield and for conformational change. Furthermore the amidino groups are protected against tryptic hydrolysis. For instance the complete amidination of all 14 ε-amino groups of ß-trypsin by methylacetimidate hydrochloride in the presence of competitive inhibitors effects no loss of enzymic activity and hinders the well-known autolysis in the absence of Ca^{++} {41}.

A similar product was formed with O-methylisourea. It reacts with ε-amino groups of proteins converting lysine residues into homoarginine residues {42}. In all modification experiments it is important to test a possible change in the secondary, tertiary or quaternary structure.

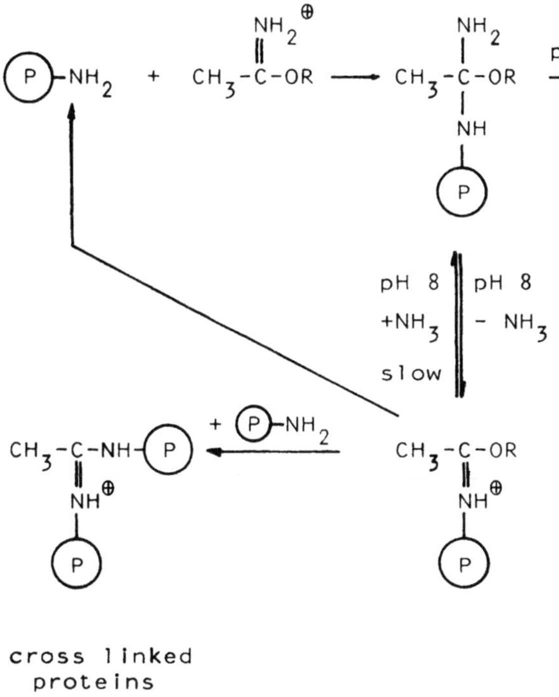

(13)

In the past the importance of electrostatic forces in a protein molecule was rarely discussed. Some effects, like higher temperature stability, can be caused by simple additional electrostatic forces, as it can be seen from a few changes in the primary structure of corresponding enzymes, e.g. proteins from thermophylic and mesophilic species. As we have shown, the amidination of lactate dehydrogenase from pig heart causes a strong increase of stability against heat, urea or alkali. Physical investigations confirmed the increase of molecula stability {43}. Müller could explain on the basis of this chemical experiment the extremely high stability of chicken heart lactate dehydrogenase {44}. In chicken lactate dehydrogenase only one lysine residue is substituted by arginine in position 241 and can form a strong ionic bridge with aspartic acid 57 in the next subunit. This is not a common effect. Of course, it depends on the localization of ionic groups in the contact region of oligomeric subunits.

Basical for this kind of experiments and their explanation are good analyses. Modification of non-essential amino groups on the surface of a globular protein occurs mostly statistically. One can calculate only the average of reacting amino groups, but cannot give a correlation to distinct lysine residues in the primary structure. The protein species, of which the NH_2 groups are modified to different extents can be detected by electrophoretic techniques. After acylation or succinylation the mobility of the modified protein is changed. The amount of the components in electrophoresis is a hint to the distribution of modified groups. The usual analytical method is the determination of free amino groups before and after the modification process. The best reagent for this purpose is trinitrobenzenesulfonic acid. The amount of modified NH_2-groups can be calculated from the molar extinction coefficient at 342 nm $\varepsilon = 1.44 \times 10^4\ M^{-1}cm^{-1}$. Because some amino groups could be placed in the interior of the protein, enfolding in urea is required.

Reductive alkylation is a convenient method to convert amino groups in proteins into their alkylamino derivatives {45}. As a rule the reaction of proteins with formaldehyde and a following reduction with $NaCNBH_3$ is used. So far amino groups in proteins can be selectively and quantitatively converted into their dimethyl derivatives at neutral pH under mild reaction conditions. Only the NH_2 terminus and ε-amino groups are modified. The changes in the physical properties of the modified proteins are minimal and the charge of the dimethyl amino groups is only slightly changed. The application of ^{13}C formaldehyde provides a spectroscopic probe for detailed ^{13}C-NMR investigations of properties of amino groups in protein

$$6\ HCHO + 3\text{Ⓟ} - NH_2 + 2\ NaCNBH_3 \rightarrow 3\text{Ⓟ} - N(CH_3)_2 + 2\ HCN + 2\ NaH_2BO_3$$

Reaction of a carbonyl group with amino groups of proteins is one of the most common natural reactions. It is the key reaction in many enzymic processes. The carbonyl-amino reaction is also a valuable tool for the determination of modifying amino groups of proteins {46}.

Pyridoxal phosphate can beside its coenzyme function be used for chemical modification of amino and thiol groups in proteins. After incubation with this reagent some enzymes are inactivated forming Schiff bases with essential lysine residues, for instance rabbit muscle aldolase, glyceraldehyde-3-phosphate dehydrogenase, 6-phosphogluconic-dehydrogenase, glutamate dehydrogenase, ribonuclease and others. The reaction was made irreversible by reducing the Schiff bases with borohydride. With the use of radioactive reagent the attacked lysine in the primary structure can be detected (14).

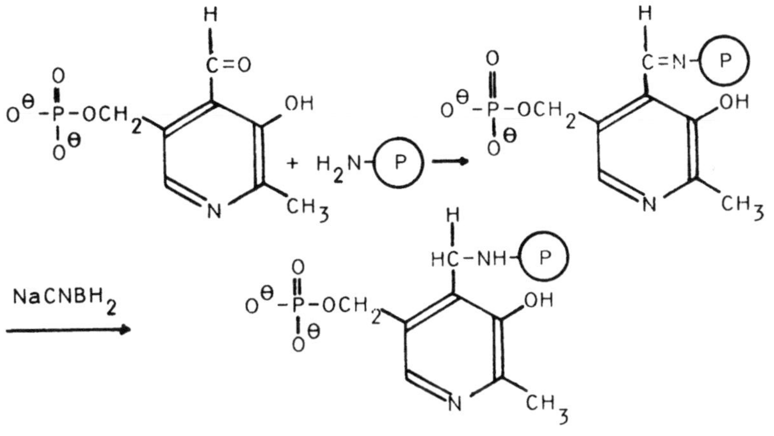

(14)

Aminoacylation of ε-amino groups in proteins

For some aspects the artificial incorporation of additional amino acid residues to form branches of the peptide chain by aminoacylation of ε-amino groups could be interesting. In a classical experiment Anfinsen, Sela and Cookre reacted ribonuclease with N-carboxi-α-amino acid anhydrides for instance D, L-alanyl derivatives. Instead of originally 10 ε-amino lysine residues only 2 - 3 are finally free and 70 alanyl-residues were incorporated. This artificial enzyme was fully active and could also be totally reactivated after urea and thiol treatment by air oxidation of the free SH groups. In distinct limits also introduction of tyrosine residues was able to form a nearly insoluble protein {47}.

Affinity labeling

The method of choice for the specific modification of functional groups in the active center of an enzyme is the affinity labeling or photo-affinity labeling technique. To achieve it, coenzyme-, substrate- or inhibitor-analoga, which contain a chemical reactive group have to be synthesized. When the enzyme possesses sufficient affinity for this substance it is adsorbed in the active or allosteric center and gets in supposition for the chemical reaction with an amino acid side chain, localized near the reactive group. A classical experiment was the high specific stoichiometry reaction of p-toluenesulfonyl-phenylalanine-chloromethylketone (15) (TPCK) with an imidazole residue of histidine in α-chymotrypsin {48} and the same specific inactivation of trypsin by p-toluenesulfonyllysine-chloromethyl-ketone (16) {49}. The first demonstrated affinity labeling was the use of diisopropylfluorophosphate (17) to modify essential serine residue in peptidases, esterases and phosphatases.

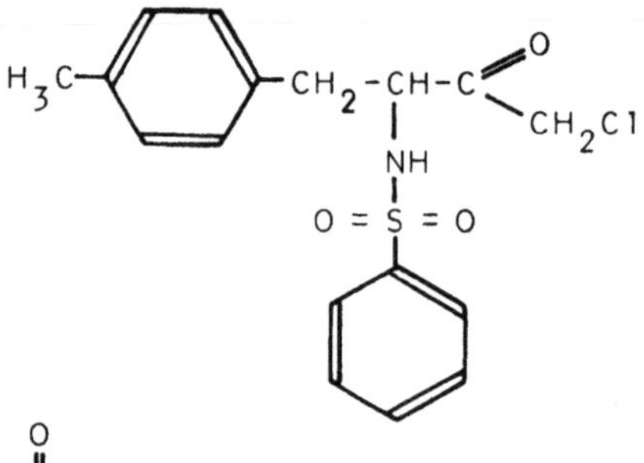

(15)

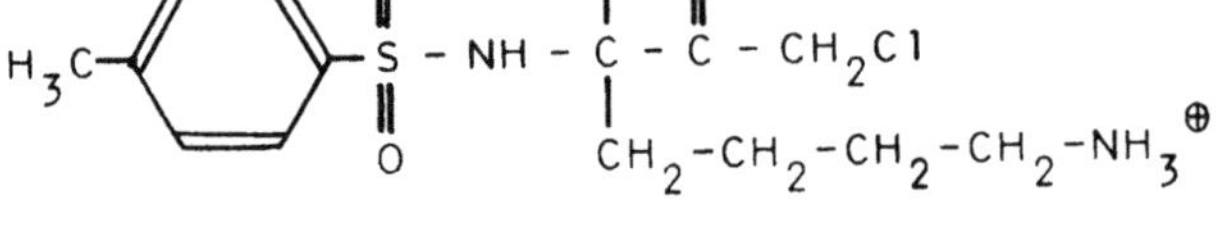

(16)

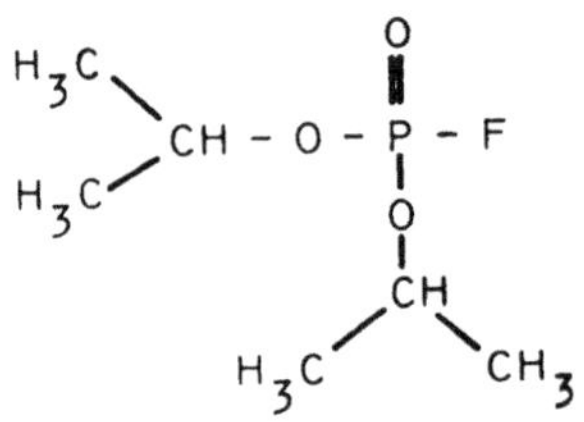

(17)

In some cases a 1:1 reaction between inhibitor and hydrolase was observed. Since then many affinity labels were synthesized and successfully applied. Only distinct samples should be discussed. In most cases such an affinity label effects the modification or inactivation of the active protein, which is only sometimes reversible (DIPF inhibition could be reverted by nucleophiles like NH_2OH and its derivatives). Interesting for the enzymologist is the covalent connection of coenzyme analogue with the apoenzyme to enzymatically active holo enzymes.

Excellent examples of affinity labeling of dehydrogenases were described by the team of Woenckhaus during the last years. They synthesized NAD analoga with bromoacetyl residues instead of the carbon amide or introduced reactive groups in the adenine part of this coenzyme or of adenine mononucleotide itself. In collaboration with the Jörnvall group the Woenckhaus team could demonstrate that in yeast alcohol dehydrogenase cysteine-43 was labeled after incubation with 3-(3-bromoacetylpyridinio)-propyl/adenosine-pyrophosphate (18, n = 3). The same group was attacked by nicotinamide-5-bromoacetyl-4-methyl-imidazole-dinucleotide (19).

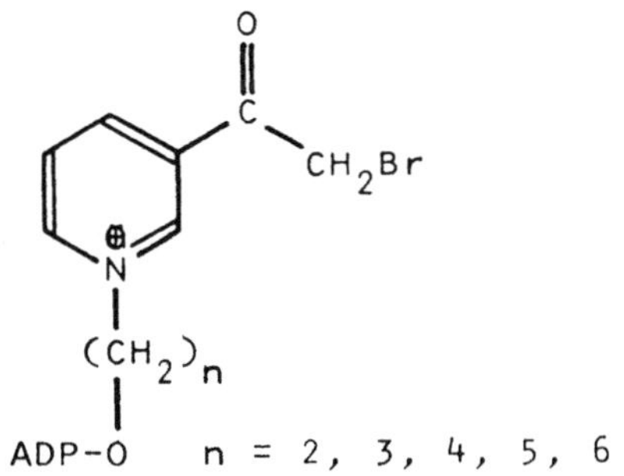

(18)

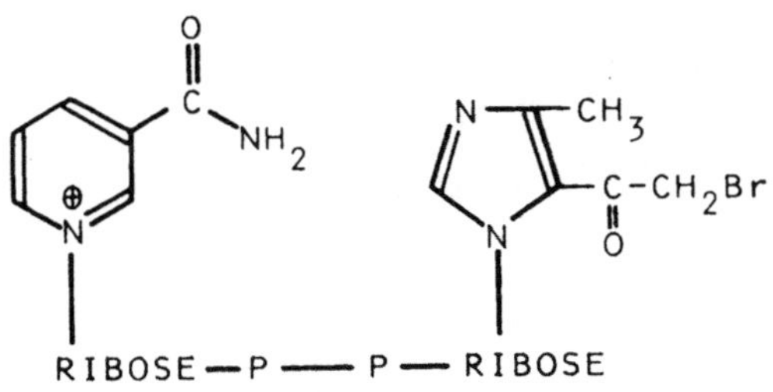

(19)

Different NAD-analogues produced small variation in modification and demonstrated a limitation in the possibility of mapping particular residues in the active center {50}.

The butyl-linked inactivator modifies in the liver enzyme mainly cysteine-46, two propyl-linked isomers either modify no residue or mainly cysteine-174. These observations show that cysteine-46 and 174 are closely associated at the active center in agreement with crystallographic data {51}.

Finally Woenckhaus et al. were successful to synthesize NAD analogues which contain in 6-position of the adenine part arylamines. The diazonium form of these derivatives (20) binds through azobridges to the active sites of alcohol and lactate dehydrogenase. Addition of the corresponding substrates causes an immediate reduction of the covalently bound analogues {52}.

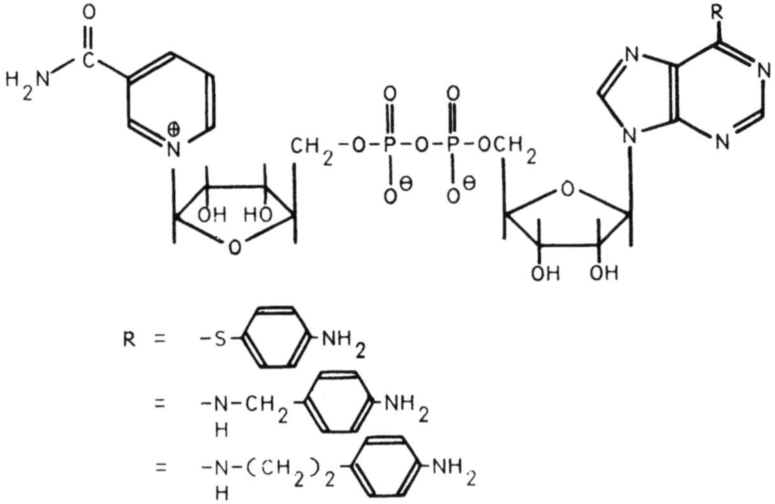

(20)

An affinity labeling of NADH inhibition site of bovine glutamate dehydrogenase was demonstrated by Batra and Coleman using 6-{(4-bromodioxybutyl) thio}-6-desaminoadenosine 5'-diphosphate (21).

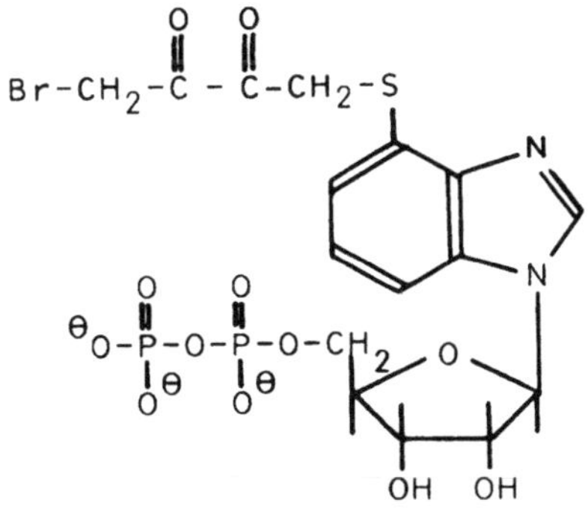

(21)

After incorporation of about 1 mol reagent/mol enzyme subunit a complete loss of its normal ability to be inhibited by high concentrations of NADH, which binds at a regulatory site was observed {53}.

Fasold, whose group has lots of experience in the synthesis and application of cross-linking reagents, was recently very successful with the affinity labeling of proteins, especially membrane proteins. His group synthesized various purine derivatives containing in 6-position thioether of dinitrophenyl- or 3-nitrobenzoic acid residues. By nucleophilic attack the dinitrothiophenyl moiety is split off and in 6-position a covalent binding, for instance with amino or thiol groups in proteins is formed (22) {54}.

The latter AMP analogue forms stabile thiol ether with an essential thiol group of rabbit skeletal muscle phosphorylase. The affinity label produced an activation of the enzyme. The activation showed stoichiometric correlation to the incorporated amount of nucleotide. In another investigation Fasold et al. synthesized the ATP analogue S-dinitrophenyl-mercaptopurine-ribose 5'-triphosphate. This product was covalently bound by rabbit muscle actine. The nucleotide was able to depolymerize N-ethyl maleimide-F-actine and became attached to the G-form at a ratio of 0.8 molecules per actine monomer. A single labeled peptide was found in the fingerprints of tryptic digestion.

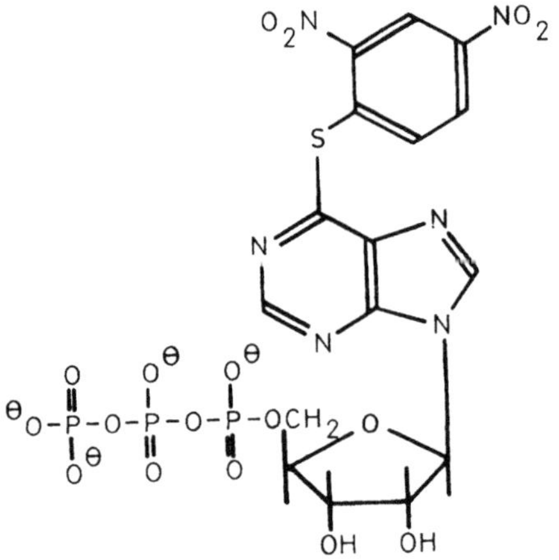

(22)

Similar experiments were carried out by Fasold et al. to characterize SH-groups, essential for polymerization of tubulin. Synthetic isotopic ^{32}P-labeled GTP analogue 2-amino-6-(S-dinitrophenyl)-thiopurineribosid-tri-phosphate was used to couple with depolymerized tubules (1 SH group/ 60 000 dalton protein). They could demonstrate that N-ethyl maleimide and the GTP affinity label compete for the same binding site in tubulin {55}.

The observation that isothiocyanate derivatives of stilbene disulfonic acid inhibit anion transport in red cells and react exclusively with the band 3 were used to study the molecular mechanism of anion transport. Most widely performed were experiments with diisothiocyanate derivatives of dihydrostilbene-2,2'-disulfonic acid. At first the reagent was rapidly reversibly bound, then a slow covalent fixation takes place. Under suitable conditions an intramolecular cross-link between two adjacent amino groups results. By kinetic experiments the cross-linking reaction could be divided into several steps and the mechanism was discussed {56}. Fasold and Frimmer have in this field also used brominated taurodehydrocholic acid to label anion transport protein in rat liver cells. The above-named diiso-thiocyanate derivative and the taurochol derivative lead to very similar labeling pattern suggesting that the anion transport system of liver cells cannot distinguish between bile acides and the anionic stilbene derivative. This and other experiments were very interesting because the strongly toxic phalloidine was transmembrally transported by the same system {57}.

Active-site-directed reagents of glycolytic enzymes {58}

The use of substrate analogues or competitive inhibitors with chemically highly reactive groups, which were bound by the corresponding enzyme in high concentration in the active site was also applied for modifying glycolytic enzymes. Most of the glycolytic enzymes are inactivated by incubation with pyridoxalphosphate. It appears likely that the phosphate groups of the glycolytic metabolites are attracted to the site normally occupied by the phosphate group and can form Schiff bases with ε-amino groups of enzymes. The authors have published a list of various site-specific reagents for glycolytic enzymes with the information of the modified amino acid residue. Some examples are given in this paper (23). As there can be seen, most reagents contain epoxy- or α-haloketo groups beside the substrate-like structure.

Photo affinity labeling {59}

This technique is a variant of affinity labeling. In the reagent an aliphatic diazo group or an arylazide as photochemical labeling reagent was introduced.

The diazo group can be photolytically converted by illumination $\lambda > 315$ nm to a reactive carbene, the arylazide upon irradiation $\lambda > 400$ nm to a nitrene. Both products react rapidly with the surrounding of the active center when the structure of the molecule has a high affinity to this essential part of an enzyme, as is necessary for affinity labeling. The products are mostly complex because the photo labile reagents are able to insert also in C-H bond, thus permitting the labeling of aliphatic residues (24).

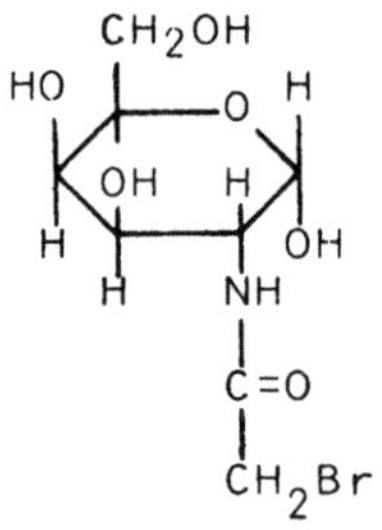

HEXOKINASE
(Cys)

N-Bromoacetyl-D-galacto-
pyranosylamin

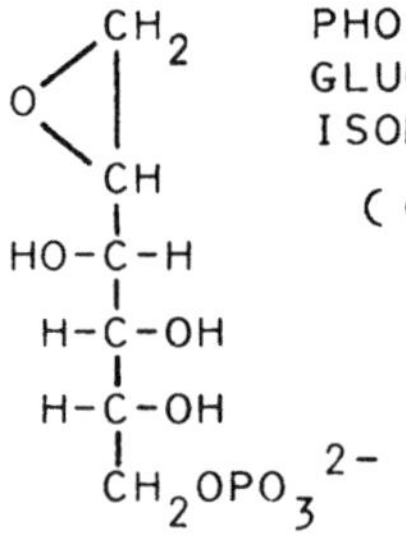

PHOSPHO-
GLUCOSE-
ISOMERASE
(Glu)

1,2-Anhydro-D-mannitol-
6-phosphate

(23)

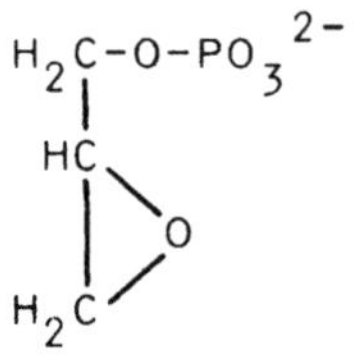

Glycidolphosphate

Triosephosphat-
isomerase
(Glu)

$$BrCH_2-\overset{O}{\overset{\|}{C}}-NH-CH_2-CH_2-OPO_3^{2-}$$

Bromoacetylethanolaminphosphate

Fructosebiphosphate-
aldolase (His or Lys)

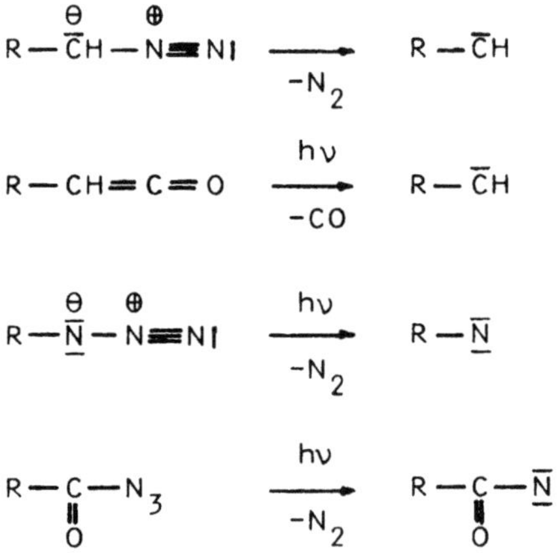

(24)

A classical experiment is the photo affinity labeling of chymotrypsin. Firstly the active serine in chymotrypsin was acylated by p-nitrophenyl-{2-^{14}C}-diazoacetate yielding an inactive enzyme. Photolysis of the ^{14}C-labeled diazoacetyl chymotrypsin with long wavelength UV light for 3-4 h led to 20-25 % irreversible incorporation of the radioactive label in the protein. Beside other radioactive products (also glycolic acid) O-carboxymethylserine is produced (25) {60}.

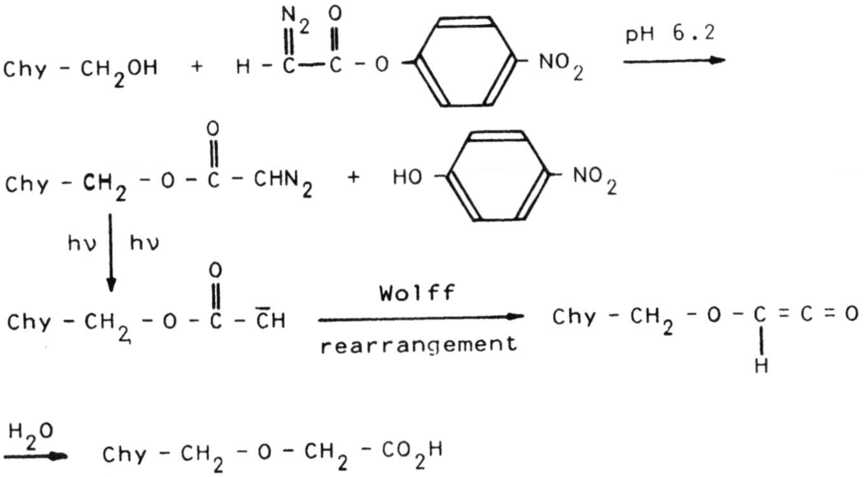

(25)

Two principal types of photolabile reagents are applied: one, in which the photolabile reagent is covalently bound to the protein at a specific site prior to photolysis, and the other in which the reagent is associated with the macromolecule as a reversible complex, but where photoactivation leads to irreversible incorporation of the bound ligand.

Numerous photo affinity labels are synthesized and in form of heterofunctional cross-links used for marking essential groups in a protein. A non-specific labeling results from the relatively long lifetime of the reaction species generated on photolysis of an aromatic azide. Even when the dissociation constant of the label is low, photolysis can lead to covalent labeling at points, removed from the active site.

One of the recent developments is the use of 8-azidopurine analogues which were very successful for investigation of the nucleotide binding sites in enzymes {61}. The analogues must be bound in a similar way as to the natural specific nucleotide. The azide portion of the bound analogue can then be activated to a nitrene via the appropriate wavelength of light. Then the activated nucleotide analogue is getting covalently attached to the protein. The concentration of the photo affinity reagent should be compatible with the binding affinity of the protein to be labeled. Too high concentrations lead to an increase of non-specific labeling. Enzymes with binding constante in the mM range are not effectively labeled with the above-mentioned substances. The optimal concentration of photo affinity analogues should previously be determined experimentally.

Cross-linking of proteins by bifunctional reagents {62,63}

Cross-linking reagents are usually reagents with two reactive groups, which can be used to introduce inter- or intramolecular bridges in proteins. By modification of two functional groups in proteins, which are in close neighborhood, covalent bridges are formed. Nearly all bifunctional reagents react mainly with amino- or thiol-groups, but also with histidine- and tyrosine side chains.

The first example was the cross-linking of cysteine-25 and histidine-106 in the active center of papain by dibromoacetone {64}. Different types of bifunctional reagents are applied: hetero- or homobifunctional reagents and both can be classified as non-cleavable and cleavable reagents (see below). In the latter compounds the cleavable bonds are disulfides, azobridges, sulfone, ester and vicinale glycol moieties.

The kind of functional group in the reagent is variable and depends on the expected attack to the amino acid side chains. Some bifunctional reagents contain photoactivable groups. They were at first covalently linked with one site through an originally reactive group.

Some non-cleavable and cleavable bifunctional reagents

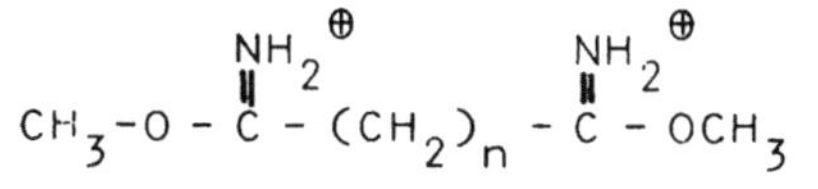

(26)

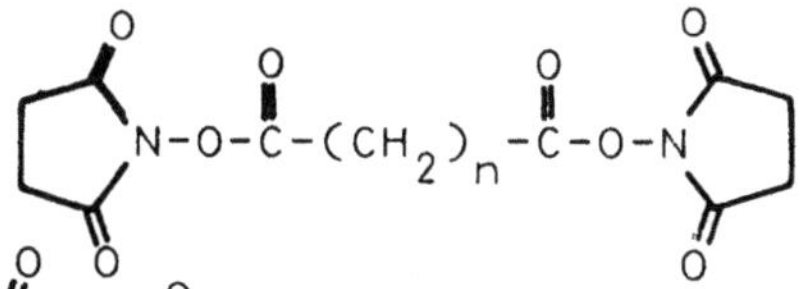

(27)

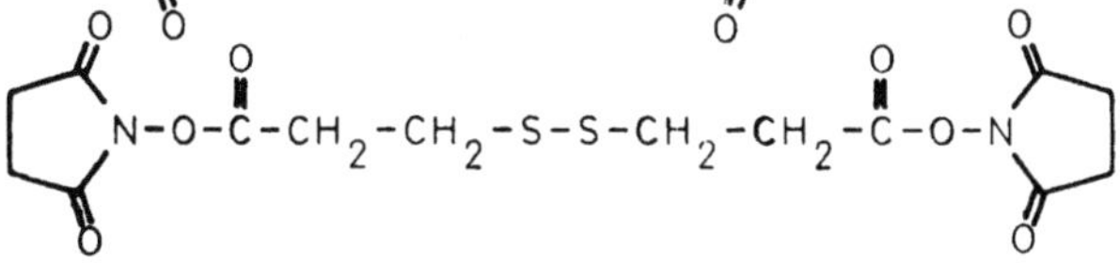

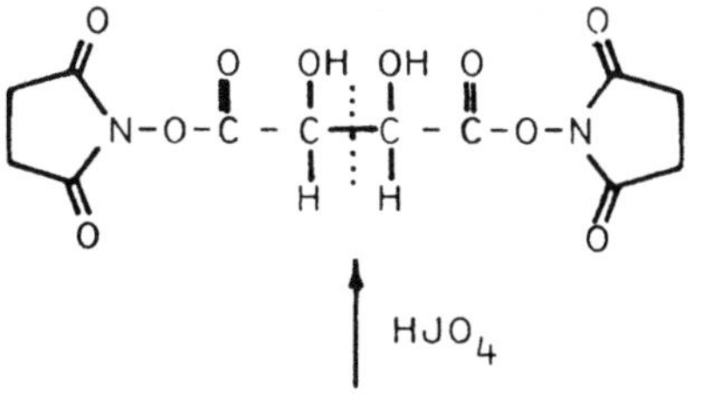

Then the second group was photo-activated to form high reactive groups (see photo affinity labeling). They react non-specifically with a wide variety of chemical bonds including C-H or O-H bonds.

The length of the spacer between the cross-linking reagent or its hydrophobicity influence the reactivity. Cross-linking is mainly used to investigate the subunit structure of oligomeric proteins. In not too diluted protein solution the subunits of an oligomeric protein were preferentially covalently linked intramolecular and minimal cross-linking was observed between different oligomers (intermolecular). Further by this technique the proximity of proteins in more complex assemblies, such as multienzyme-complexes, ribosomes or membranes could be investigated. In the immunochemistry antibodies and reporter enzymes are cross-linked by bifunctional reagents to chemical units. Fasold has determined distances in a monomer (myoglobin) making cross-links with different lengths between two amino acid side chains in the tertiary structure and compared his results with the X-ray data of Kendrew {65}. This is one of the most important ways to compare distances in the crystallic structure with those of soluted protein.

Intramolecular cross-linkages are nowadays used for stabilization experiments of proteins. At last the joining of enzymes to insoluble carriers is sometimes formed by bifunctional reagents.

The chemical products of protein cross-linking experiments are usually examined in the SDS electrophoresis in order to determine the smallest molecular weight under denaturating conditions. Also ultracentrifugation experiments under native and denaturating conditions are helpful. Mostly a mixture of different protein complexes were detected. The preparation of a distinct cross-linking product is very difficult and is sometimes successful after gel filtration with the recently developed gels. For a detailed analysis the cross-linked protein had to be split by endopeptidase and the bridge peptides to be separated in order to estimate the bridge foot points by sequence analysis.

One of the biochemically most popular reagents, because of the mild reaction conditions, are the bisimidoester dimethylsuberimidate (26, n = 4) and the bis-N-hydroxy-succinimideester of suberic acid (27, n = 6). The subermidate is reacting optimally between pH 9-10, the other one in the region of pH 6-9. In the former case the positive charge of amino groups is retained, in the second case it is lost.

Another simple homobifunctional reagent is the group of -bismaleimide. Their use allows the cross-linkage of proteins containing free thiol groups. Thiol groups can be artificially incorporated by use of 2-iminothiolan (4-mercapto-butyrimidate) (28).

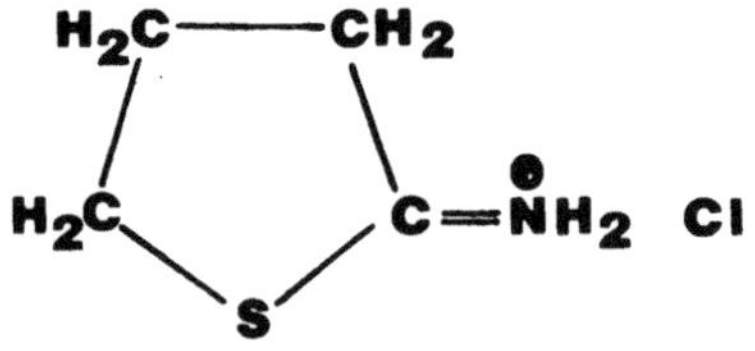

(28)

2-iminothiolan HCl

After amidination of amino groups of proteins with thiolan the modified proteins contain free thiol groups on their surface, which are very reactive with maleimides.

A very simple homobifunctional reagent is 1,5-difluorodinitrobenzene, firstly used bei Zahn {66} for cross-linking. The reagent is insoluble in water and must therefore be added generally as concentrated solution in acetone, dioxan or alcohol to an aqueous solution of protein. In all these and other experiments it must be examined before, if the addition of organic solvent does not effect a denaturation of an enzyme. Dayer et al. {67} could demonstrate that at pH 7 aspartate aminotransferase forms a single covalent bond with amino groups of lysine-258 in the active center. On shifting to pH 9, the second fluorine atom of the reagent is substituted by cysteine-390 or tyrosine-70. The cysteine-390 and tyrosine-70 are thus located at a distance of approximately 5 Å.

Other homobifunctional reagents useful for cross-linking are diisocyanates or thioisocyanates. Fasold has used and described numerous bifunctional reagents {68} with data for the distances of the linkage. The application of heterobifunctional reagents is principally identical and is carried out sometimes in two steps, especially with the use of photo-activable groups (29). Some reagents and examples will be described later. A very interesting heterobifunctional cross-linking reagent was synthesized by Friebel et al. (30) {69}. The advantage of this reagent is based on the selenolester function which serves a selective and reversible thiol reagent.
The first step is the splitting of the $-\overset{O}{\overset{\parallel}{C}}-Se$ bound by thiol groups forming S-acyl derivatives. Their azido groups react photochemically via nitrene generation under cross-linking. The cleavage of the first anchor point, the thiol ester, can be achieved by nucleophile-carrying labels.

The cross-linking of subunits in an oligomeric enzyme effects mostly a partial, sometimes strong stabilization. We were successful to introduce 2-3 cross-links per subunit of glycerol dehydrogenase from B. megaterium with suberimidate. The modified enzyme shows a strongly increased stability against alkali, urea or heat {70}. This is not a common effect, but depends on the topography of the bridge in a protein. Torchilin et al. {71} investigated the influence of the spacer length during cross-linking on the thermostability of proteins using carbodiimides and diamines with different chain lengths.

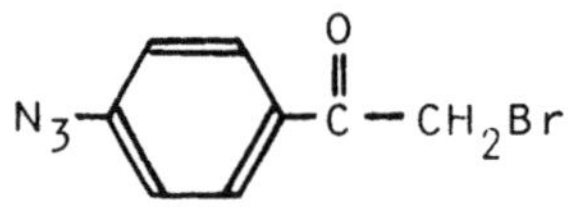

p-azidophenacyl bromide

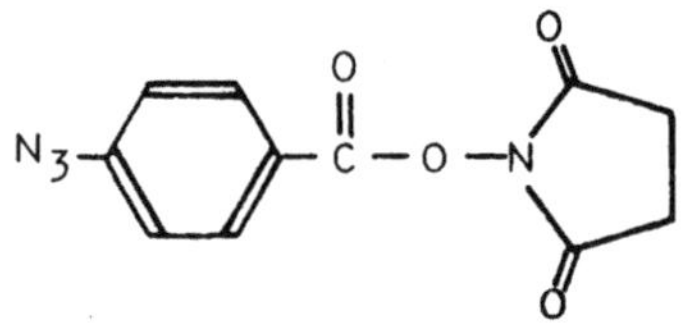

N-hydroxysuccinimidyl-4-azidobenzoate

(29)

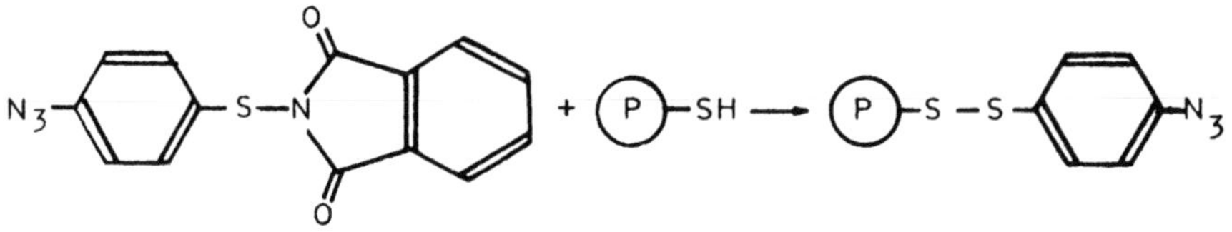

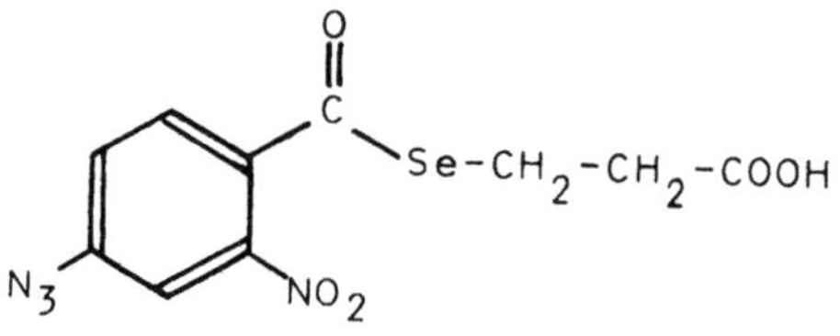

(30)

Enzyme labeling of antibodies {72}

One of the most applied chemical modification techniques is the covalent conjugation of antibodies or their fragments with reporter enzymes used for qualitative and quantitative immuno methods. The effects can easily be followed by photography or spectroscopy and do not require radio isotopes. Under favorable conditions enzyme-immunoassay may be even more sensitive than radio-immunoassay. Only the most applicable methods, especially the recently developed ones, are described here. The principle is the following. Enriched antibody preparations (after protein A-treatment, immunoadsorbêns, salt fractionation) or the papain fragments were cross-linked by bifunctional reagents with enzymes, the activity of which can be measured very easily and sensitively. This means that they should possess high specific activity and the product of their substrates is strongly colored. As reporter enzymes are preferably used: horseraddish peroxidase, alkaline phosphatase, glucose oxidase, ß-galactosidase. During the cross-linking neither the antibody binding capacity nor the specific enzyme activity should be decreased.

Techniques for histochemistry and quantitative immunoassays

One of the oldest methods and nowadays again available is the conjugation of enzymes with IgG by using glutaraldehyde. Glutaraldehyde was added to a mixture of reporter enzyme and IgG and then the reaction was allowed to proceed at pH 6.8 at room temperature. The conjugates produced were highly polymeric and heterogenous. Especially the IgG's were largely polymerized and a major loss of antibody activity was observed. Better yields are obtained in a two-step method. The enzymes are allowed to react for 1 h with glutaraldehyde and then the antibody is to be added. More useful is the perjodate method.

Perjodate method

The carbohydrate moiety of antibodies (Fc part) or of the reporter enzyme, if it is a glycoprotein (alkaline phosphatase, peroxidase), generates aldehyde groups after treatment with perjodate, which form with amino groups of the second protein Schiff's bases. After borohydride reduction by this way covalent conjugation between antibody and reporter enzyme originates. Usually by this procedure the biological activity decreases. Before the perjodate oxidation it is necessary to block chemically the amino groups of the macromolecule with imido esters or 1-fluoro-2,4-dinitrobenzene to minimize the self conjugation.

Maleimide methods

Maleimides react under mild conditions rapidly with easily accessible thiol groups and usually very slowly with other functional groups. With the help of di-maleimides it is possible to cross-link proteins containing free thiol groups (N,N-o-phenylendimaleimide). Antibodies contain various disulfide groups but no free thiols and must therefore be partially reduced by low molecular thiols in the hinge region. After treatment with excess of dimaleimides maleimide groups were introduced. Then the maleimide IgG molecules were allowed to react with the only thiol group in the native ßD-galactosidase. The dimaleimide method is practicable at pH 6.0 - 6.5. No loss of ßD-galactosidase activity was observed. The recovery of IgG activity is about 30 %. The cross-link is stable at 4°C, pH 6 - 7 for at least 1 year. Also in the case of thiol-free or disulfide-proteins the dimaleimide method is available after introduction of artificial thiol groups by thiolane (see cross-linking). Caution is required for two aspects: 1) The maleimides are instable in the region 7 or higher by opening the five-ring and splitting off the half-amide binding groups. 2) Thiol groups are autoxidable before they were cross-linked. Therefore the use of EDTA is necessary, which otherwise inactivates alkaline phosphatase, when it is used as reporter enzyme. Therefore one must work under strong anaerobic conditions up to the cross-linking reaction.

In many cases cross-linking experiments are simpler while using hetero-bifunctional reagents. The best reagent is now the succinimidylester of maleimides. The first reactive group is reacting with amino groups at pH 7 or lower and introduce by this way maleimide groups which in a second step can react with thiol proteins. N-succinimidyl 4-(N-maleimide-methyl)-cyclohexane-1-carboxylate is now commercially available (31). The maleimide

IgG molecule was allowed to react with only one thiol group in the native ßD-galacosidase {73}.

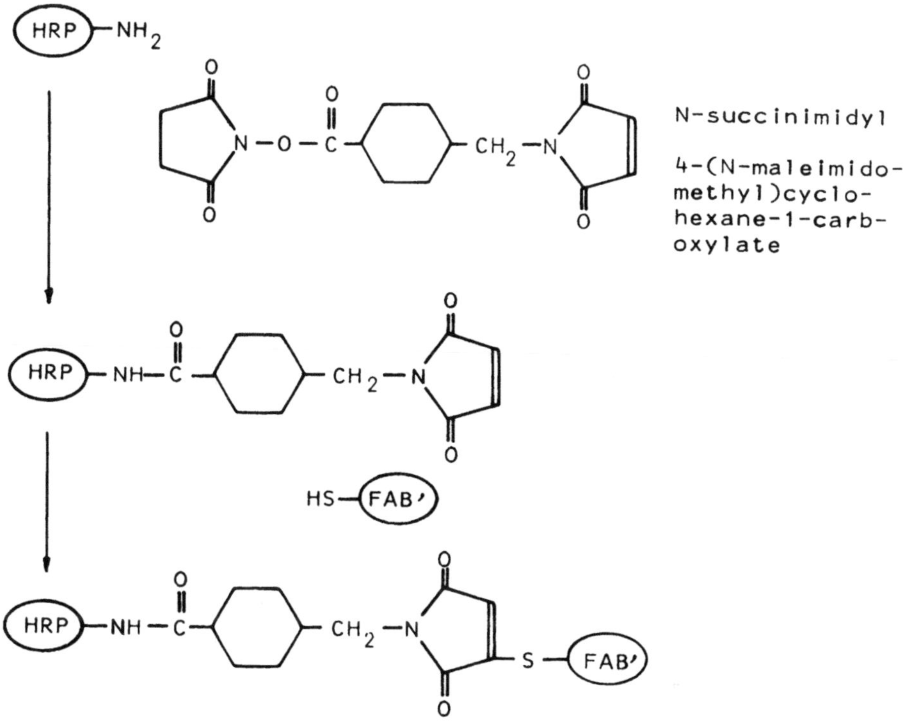

(31)

Immobilized enzymes {74}

Proteins can be immobilized in a number of different ways. It is impossible to discuss all techniques or applications in this article. Here should briefly be described only the most useful procedures for covalent coupling of proteins to a supporting material for application in the laboratory.

The most important techniques are:

a) Activation of the supporting material for reaction with functional groups of the proteins;

b) Use of coupling reagents to link proteins with a matrix.

The functional groups should be non-essential side chains, commonly ε-amino groups, which react in the deprotonated form. Therefore coupling reactions were carried out mostly at pH values of about 9.

<u>Cyanogen bromide activation of sepharose</u> is one of the most useful procedures for covalent coupling of proteins to insoluble matrix. The mechanism of the activation reaction of polysaccharide is the following (32){75}.

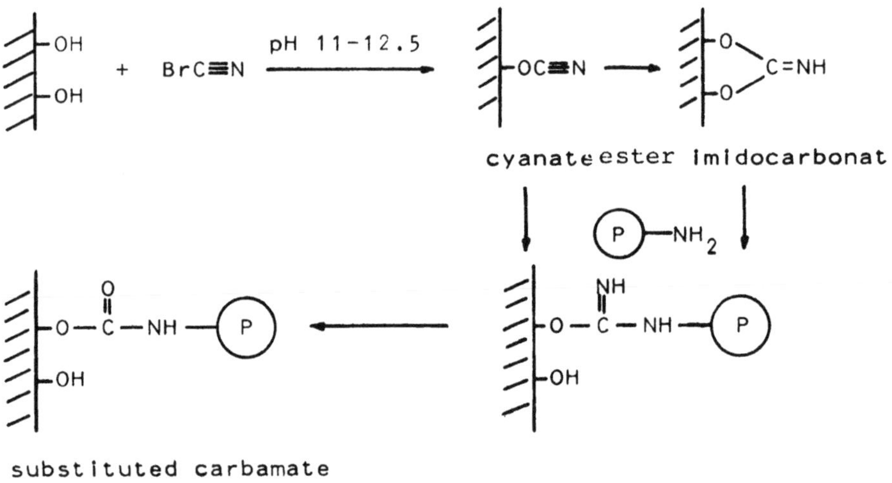

(32)

Coupling of the proteins is preferentially thought by forming subsituted isourea derivatives. Sepharose, sephadex or cellulose are predominantly responsible for coupling. Coupling capacity of activated residues toward small ligands can be predicted by determining the amount of cyanate esters and imidocarbonates. Cyanate esters are predominantly responsible for coupling reaction {75}.

Carbodiimide activation

As is described above (COOH modification) carboxylic copolymer (acrylic acid) could be activated by addition of carbodiimide (neutral pH, room temperature). By nucleophilic attack of the amino groups in proteins a covalent binding occurs with the formation of dicyclohexyl urea. Azo coupling could be carried out using polymers containing aromatic amines (33).

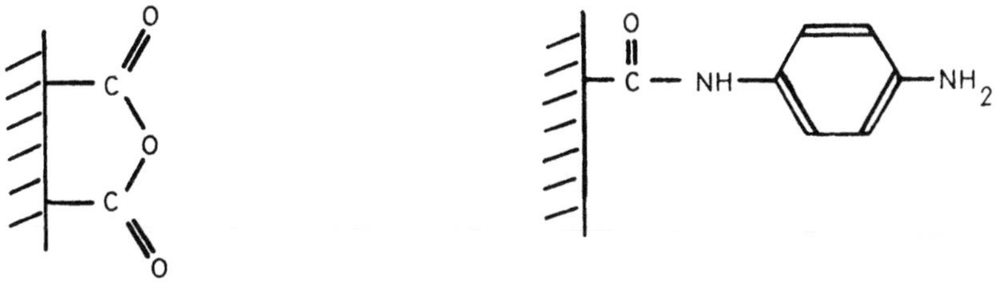

(33)

maleicacidanhydride

The diazonium components react extremely well with tyrosyl- and histidine side chains under formation of azo bridges. Also supports containing anhydride groups or azido groups can react with nucleophilic groups of proteins. Epoxides introduced in the support material are also capable to covalently bind enzymes under mild conditions (oxirane) (34). Proteins with free thiol groups can form with matrix containing disulfide bridges mixed disulfides. After use the protein can be split off again (35).

OXIRANE:

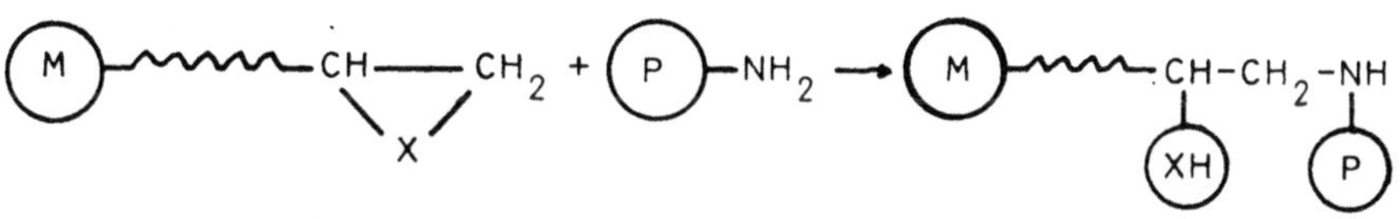

(34)

X = NH, O or S

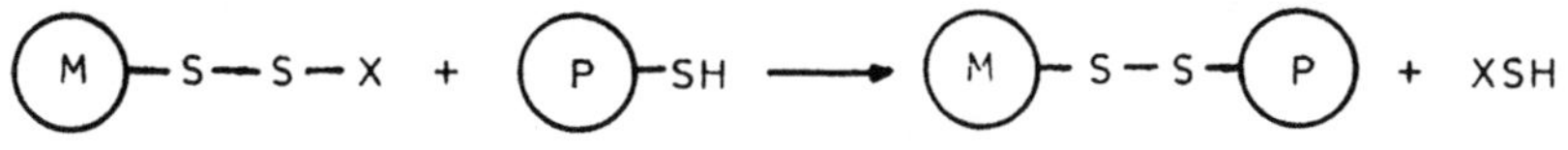

(35)

Very important is the control of the specific activity after covalent fixing of the enzyme to the matrix. When the enzymic activity is strongly decreased, sometimes the addition of coenzymes or subunits both protect a reaction with side chains in the active center. The major cause of the observed strong decrease of specific activity of immobilized enzymes is diffusional limitation. Most of the carriers for immobilization of enzymes bind the enzyme throughout the porous structure. Enzyme molecules within the inner parts of carrier particles have to get their substrates by diffusion from the surface and lose the products by the same way. If the activity enclosed in the particles is high enough, diffusion becomes rate-limiting. By model experiments Müller was able to demonstrate that decreased specific activity of an enzyme during immobilization can be reverted after resolubilization of enzymic digestion of the carrier {76}.

At last immobilization can be achieved by protein copolymerisation. Proteins are vinylated with acetylating or alkylating monomers and copolymerized with comonomers (36). Depending on the concentration of the alkylating or acetylating monomer soluble to insoluble protein polymers are obtained. In presence of cross-linkers insoluble immobilized enzymes are produced.

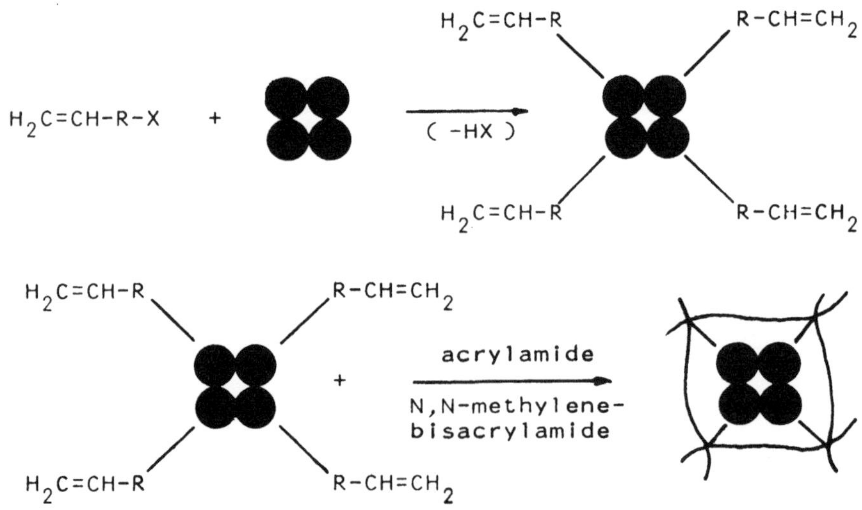

(36)

References

1. Means, G.E., Feeney, R.E.: Chemical Modification of Proteins, Holden Day Inc., San Francisco (1971).

2. Glazer, A.N., DeLange, R.J., Sigman, D.S.: Chemical Modification of Proteins, Natl. Holland Publ. Comp. Amsterdam (1975).

3. Glazer, A.N.: In: The Proteins, Third Edition, Vol. II, Acad. Press, New York (1976).

4. Special Annual Periodical Report: Amino Acids, Peptides and Proteins, The Royal Society of Chem., Burlington House, London W1VOBN, printed by J. Wright & Sons, Bristol.

5. Lambert, J.M., Perham, R.N., Coggins, I.R.: Bioch. J. 161, 63-71 (1977).

6. Gundlach, H.G., Stein, W.H., Moore, S.: J. Biol. Chem. 234, 1754-1764 (1959).

7. Crestfild, A.M., Stein, W.H., Moore, S.: J. Biol. Chem. 238, 2413-2420 (1963).

8. Leskovac, V., Pfleiderer, G.: Hoppe-Seyler's Z. Physiol. Chem. 350, 484-492 (1969).

9. Woenckhaus, C., Berghäuser, I., Pfleiderer, G.: Hoppe-Seyler's Z. Physiol. Chem. 350, 473-483 (1969).

10. Holbrook, J.J., Pfleiderer, G., Mella, K., Volz, M., Leskovac, V., Jeckel, D.: Eur. J. Bioche. 1, 476-481 (1967).

11. Fondy, J., Everse, T.P.J., Discoll, G.A., Castillo, F., Stolzenbach, F.E., Kaplan, N.O.: J. Biol. Chem. 240, 4219-4234 (1965).

12. Jeckel, D., Anders, R., Pfleiderer, G.: Hoppe-Seyler's Z. Physiol. Chem. 352, 769-779 (1971).

13. Pfleiderer, G., Jeckel, D.: Eur. J. Biochem. 2, 171-175 (1967).

14. Holbrook, J.J.: Bioch. Z. 344, 141-152 (1966).

15. Bozler, H., Jany, K.D., Pfleiderer, G.: Bioch. Biophys. Acta 749, 238-243 (1983).

16. Velick, S.F.: J. Biol. Chem. 233, 1455-1467 (1958).

17. Pfleiderer, G., Auricchio, F.: Bioch. Biophys. Res. Commun. 16, 53-59 (1964).

18. Pfleiderer, G., Jeckel, D., Wieland, Th.: Bioch. Z. 328, 187-194 (1956).

19. Boll, M., Falkenberg, F., Jeckel, D., Pfleiderer, G.: Hoppe-Seyler's Z. Physiol. Chem. 351, 1268-1272 (1970).

20. Ellman, G.L.: Arch. Bioch. Biophys. 82, 70-77 (1959).

21. Schmidt, U., Pfleiderer, G., Bartkowiak, F.: Analyt. Bioch. 138, 217-222 (1984).

22. Holbrook, J.J., Pfleiderer, G., Schnetger, J., Diemir, S.: Biochem. Z. 344, 1-14 (1966).

23. Boyer, P.D.: J. Am. Chem. Soc. 76, 4331-4337 (1954).

24. Riordan, J.F., Wacker, W.E.C., Vallee, B.L.: Bioch. 4, 1758-1765 (1965).

25. Sokolovsky, M., Riordan, J.F., Vallee, B.L.: Bioch. 5, 3582-3589 (1966).

26. Sokolovsky, M., Vallee, B.L.: Bioch. 5, 3574-3581 (1966).

27. Ensinger, H.A., Pauly, H.E., Pfleiderer, G., Stiefel, T.: Bioch. Biophys. Acta 327, 432-441 (1978).

28. Miles, E.W., in: Methods in Enzymology, Vol. XLVII, ed. C.H.W. Hirs, S.N. Timasheff, Acad. Press, N.Y., pp. 431-442 (1977).

29. Weil, L.: Arch. B. Bioph. 110, 57-68 (1965).

30. Miles, E.W., Kumagai, H.: J. Biol. Chem. 249, 2843-2851 (1974).

31. Cozzani, I., Santoni, C., Jori, G., Gennari, G., Tamburo, A.M.: Biochem. J. 141, 463-468 (1974).

32. Jori, G., Galiazzo, G.: Photochem. and Photobiol. 14, 607-619 (1971).

33. Woodward, R.B., Olofson, R.A.: J. Amer. Chem. Soc. 83, 1007-1009 (1961).

34. Brake, A.J., Weber, B.H.: J. Biol. Chem. 249, 5452-5457 (1974).

35. Horton, H.R., Tucker, W.P.: J. Biol. Chem. 245, 3397-3401 (1970).

36. Heilmann, H.D., Pfleiderer, G.: Bioch. Biophys. Acta 384, 331-341 (1975).

37. Patthy, L., Smith, E.L.: J. Biol. Chem. 250, 557-564 (1975).

38. Borders, C.L. Jr., Pearson, L.J., McLaughlin, A.E., Gustafson, M.E., Vasiloff, J., An, F.Y., Morgan, D.J.: Bioch. Biophys. Acta 568, 491-495 (1979).

39. Dworschak, R., Tarr, G., Plapp, B.V.: Bioch. 14, 200-203 (1975).

40. Imman, J.K., Perham, R.N., DuBois, G.C., Apella, E.: in: Meth. in Enzymol., ed. C.H.W. Hirs, Vol. 91, 559-569, Acad. Press, N.Y. (1983).

41. Nureddin, A., Inagami, T.: Bioch. Biophys. Res. Comm. 36, 999-1005 (1969).

42. Klee, W.A., Richards, F.M.: J. Biol. Chem. 229, 489-504 (1957).

43. Tuengler, P., Pfleiderer, G.: Bioch. Biophys. Acta 484, 1-8 (1977).

44. Müller, J.: Bioch. Biophys. Acta 669, 210-215 (1981).

45. Means, G.E., in: Meth. in Enzymol., ed. C.H.W. Hirs, Vol. XLVII, 469-478 (1977).

46. Feeney, R.E., Blankenhorn, G., Bixon, B.F., in: Adv. in Prot. Chem., 136-302, Acad. Press N.Y. (1975).

47. Anfinsen, C.B., Sela, M., Cooke, I.P.: J. Biol. Chem. 237, 1825-1831 (1962).

48. Schochmann, G., Shaw, E.: Bioch. 2, 252-255 (1963).

49. Shaw, E., MaresGuia, M., Cohen, W.: Bioch. 4, 2219-2224 (1965).

50. Jörnvall, H., Woenckhaus, C., Schättle, E., Jeck, R.: FEBS Lett. 54, 297-301 (1975).

51. Woenckhaus, C., Jeck, R., Jörnvall, H.: Eur. J. Bioch. 93, 65-69 (1979).

52. Woenckhaus, C., Korb, R., Burkhard, A., Schäfer, H.G.: Bioorg. Chem. 12, 45-57 (1983).

53. Betra, S.P., Colman, R.F.: Bioch. 23, 4949-4946 (1984).

54. Faust, U., Fasold, H., Ortanderl, Fr.: Eur. J. Bioch. 43, 273-279 (1974).

55. Mann, K. Giesel, M., Fasold, H., Haase, W.: FEBS Lett. 92, 45-48 (1978).

56. Ziegler, K., Frimmer, M., Möller, M., Fasold, H. : Arch. Pharm. 319, 249-253 (1982).

57. Kampmann, L., Lepke, S., Fasold, H., Fritsche, G., Passow, H.: J. Mem. Biol. 70, 199-219 (1980).

58. Hartmann, F.C., Norton, I.L., in: Meth. in Enzymol., Vol. XLVII, Acad. Press, N.Y., 479-498 (1977).

59. Chowdhry, V., Westheimer, F.W., in: Ann. Res. Bioch., Vol. 48, 293-325, Palo Alto Calif., USA (1979).

60. Shafer, J., Baronowsky, P., Lursen, R., Finn., F., Westheimer, F.H.: J. Biol. Chm. 241, 421-427 (1966).

61. Polter, R.L., Halay, B.E., in: Meth. in Enzymol., Vol. 91, p. 613-633, Acad. Press, N.Y. (1983).

62. Tae, H., Ji, in: Meth. in Enzymol., Vol. 91, 580-609, Acad. Press, N.Y. (1983).

63. Kia-Ki, H., Richard, C., Delacourte, A. : Int. J. Bioch. 16, 129-145 (1984).

64. Husain, S.S., Lowe, G.: Bioch. J. 108, 855-859 (1968).

65. Fasold, H.: Bioch. Z. 342, 288 and 294 (1965).

66. Zahn, H.: Angew. Chem. 67, 56 (1955).

67. Deyer, S.M., Afanasenko, G.A., Polynowsky, O.L.: Bioch. Biophys. Acta 534, 358-367 (1978).

68. Fasold, H., Kloppenberger, J., Mayer, E., Remolt, H.: Angew. Chem. 83, 875-882 (1971).

69. Friebel, K., Huth, H., Jany, K.-D., Trommer, W.E.: Hoppe-Seyler's Z. Physiol. Chem. 362, 421-428 (1981).

70. Ganzhorn, A., Scharschmidt, M., Pfleiderer, G.: Ztschr. Naturf. 39c, 575-583 (1984).

71. Torchilin, V.P., Maksimenko, A.V., Smirnov, V.N., Berezin, I.V., Klibonow, A.M., Martinek, K.: Bioch. Biophys. Acta 522, 277-283 (1978).

72. Istikawa, W., Imagawa, M., Hoshida, S., Yoshitake, S., Hamaguchi, Y., Veno, T.: J. of Imm. Chem. 4, 209-327 (1983).

73. Yashitake, S., Yamada, Y., Yshikawa, E., Masseyeff, R.: Eur. J. Bioch. 101, 395-399 (1979).

74. Meth. in Enzymol., ed. Mosbach, K., Vol. XLIV, Acad. Press, N.Y. (1976).

75. Kohen, J., Wilchek, M.: Analyt. Biochem. 115, 375-382 (1982).

76. Müller, J., Pfleiderer, G.: Hoppe-Seyler's Z. Physiol. Chem. 361, 675-681 (1980).

77. Jaworek, D., Botsch, H., Maier, I., in: Meth. in Enzymol., Vol. XLIV, 195-201, Acad. Press, N.Y. (1976).

CROSS-LINKING REAGENTS

Hugo Fasold, Hans Georg Bäumert, Christa Meyer
Institut für Biochemie der Johann Wolfgang Goethe-Universität Frankfurt am Main
D-6000 Frankfurt am Main 70

Bifunctional reagents, designed to establish two covalent bonds to aminoacid side chains, have been widely used in analytical biochemistry during the last two decades. In a few cases, they have even been proposed in industrial chemistry as stabilizing components (1,2).

Originally, cross-linkers were devised for the measurement of intramolecular distances in protein tertiary or quarternary structures (3,4,5). For this purpose, however, they have probably been of little avail, although several of such yardstick measurements were in very good agreement with the final resolution of the complete structure of the protein molecule by x-ray crystallography (5,6). It quickly became evident however, that even a rough resolution of tertiary structures would make a large number of exactly analyzed attachment points necessary, therefore, the occasional identification of a cross-link has mostly been used as additional evidence to enhance structural analyses by other methods (7).

Moreover, most reagents contain several highly flexible bonds at least, and may bind to longer side chains, e.g. to the amino group of a lysine. This leads to evaluations of distances within a certain range, usually between 6 and 10 Å only.

Modern Methods in Protein Chemistry, Volume 2

Very successful applications of several bifunctional reagents have been found in the elucidation of the array of physiological aggregates of proteins, and their functional interactions. A prominent example was furnished by the determination of the number of histone subunits within the nucleosome ooro (8). The functional properties of several allosteric proteins could be changed by cross-linking between their subunits (9), this kind of information may lead to more detailed insight into their intramolecular interactions (9).

Only during the recent years the technique has been introduced into the field of membrane protein chemistry. This aspect produced new experimental problems, as the position of the cross-link, once it has been successfully instituted and characterized in its attachment to a given protein, should be known in its orientation within the membrane.

Clearly, the application of reagents that cannot penetrate most of the biological membranes will be useful for the investigation of iuxtapositions on the outside, or, in the case of defined vesicles, on the inside of biological surfaces. Practically no selective reagents are available for the intrinsic regions of the membrane.

In a few propitious instances the partially hydrophobic, and frequently used cross-linkers, as suberimidate, were able to demonstrate protein neighbourhoods, e.g. on the internal region of the red cell membrane (10). In retrospective, these results were enhanced by the close contact of the covalently linked protein species, and also by the low yield of background cross-links due to the collision of proteins bearing half-side reacted cross-linker molecules on their surface (see below).

Usually, strong anionic or cationic groups of the reagent are supposed to confine them to one side of the membrane. These properties are however easily overridden by distinctly hydrophobic parts in the molecule. Thus, 2-nitro-4-azido-phenyltaurine (NAP-taurine) (11) does penetrate the red cell membrane slowly, and only molecules bearing two sulfonic acid groups, as 4,4'-diisothiocyanato-stilbene-disulfonic acid (DIDS) achieve a true label from the outside only. Usually it is applied in the form of its hydrogenated, and therefore highly labeled derivative (H_2DIDS) (12):

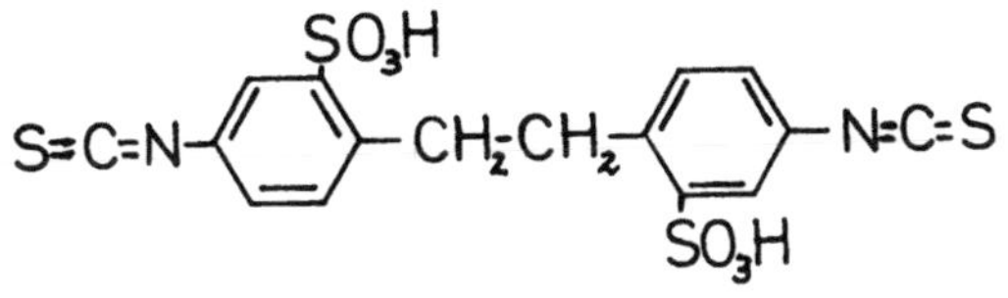

Its application to the cross-linking of two amino groups of different regions of the anion transport protein of the red cell membrane was enhanced by the high affinity of the reagent to the specific binding site of the molecule with a K_D value of 10^{-5} M. In such cases, the reagent, despite its strongly hydrophilic properties, is very effectively secluded from the aqueous milieu. Thus, the azo dye,

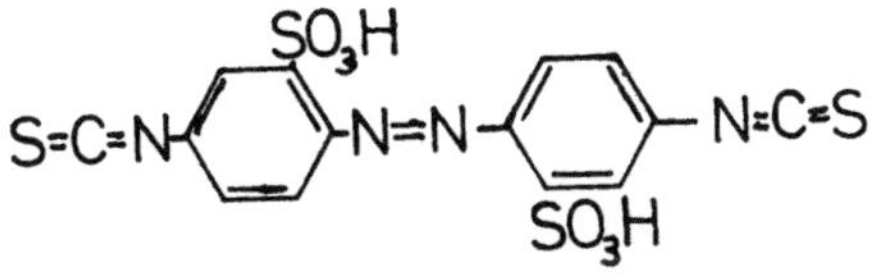

obtained by oxidation of p-acetamido-anilino-5-sulfonic acid with the aid of sodium perborate, hydrolysis of the acetyl groups, and subsequent derivatization to the isothiocyanate, binds to the anion carrier with the same affinity as DIDS. In aqueous medium, its azo bond is easily cleaved by low concentrations of dithionite ions during a few minutes at room temperature. Within the membrane, however, even prolonged reaction at temperatures up to $50^{o}C$ leaves the cross-link entirely intact.

Hydrophobicity and Length

The first cross-linkers devised for protein structure investigations used the rigid backbone of aromatic compounds to define the span of the reagent (3). When substances of greater length were introduced, this principle had to be abandoned, as compounds carrying four or more aromatic rings tend to denature most proteins even at low concentrations. A notable exception is the pyridoxal phosphate derivative:

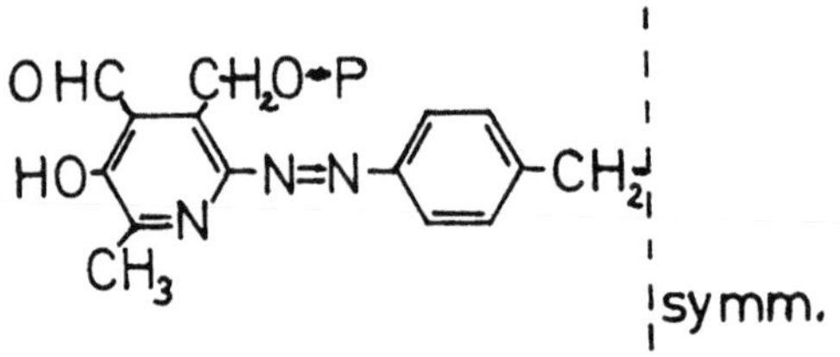

It appears to be highly selective for lysine side chains, to which it is attached by reduction with the aid of sodium boron hydride. As the azo bond is not cleaved during short reaction times of a few minutes at physiological pH, this offers a good possibilty of introducing a radioactive label during coupling. A cleavage of the reagent is easily effected by dithionite after isolation of cross-linked protein fragments.

The search for long reagent molecules, which were thought to be evidently more suitable for intermolecular cross-linking, sought to introduce rigidity in hydrophilic backbones. The inclusion of short pieces of polyproline forms one example (2). Indeed, a comparison of two reagents of approximately the same length, the strongly hydrophilic tartryl-bisglycinazide, and the slightly longer and partially hydrophobic azophenyl-bisureidoprolyl-azide, showed

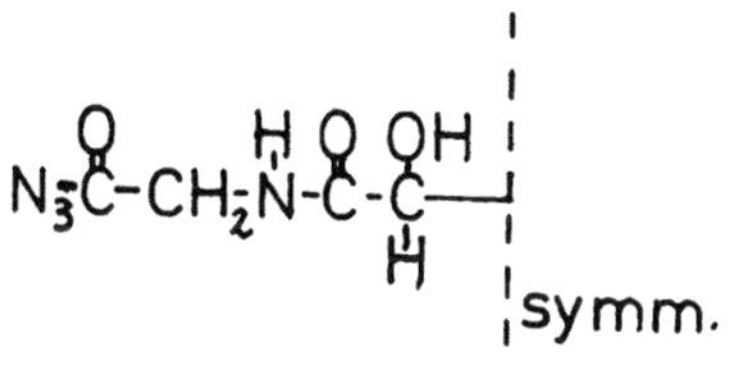

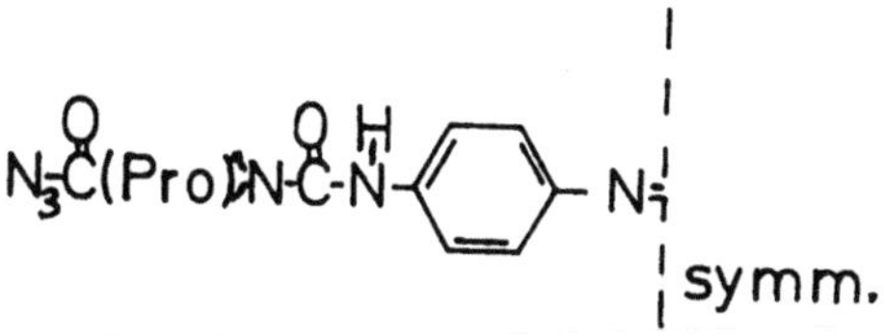

that the yield of cross-links between tertiary structures versus intramolecular links was much higher in the case of the former. For this evaluation, the cross-linking of hemoglobin (6) or F-actin furnish dependable systems, as the yield of intermolecular cross-links is easily determined after dissociation of the subunits by molecular sieve chromatography (13).

Two-step Reagents

Two different functional reactive groups are sometimes introduced into the linking reagent. This is of advantage when proteins, used as labels for various counterparts, as ferritin, avidin, or immunoglobulins, are to be coated by the reagent in the first step, and then, after the excess of reagent has been removed, activated to the linking reaction in the presence of the macromolecule to be labeled. The technique has been used for the binding of peroxidase or other enzymes to immunoglobulins for the development of various ELISA tests, and for the linkage of hormones to their receptor (14). Frequently, the functional group for the second step is an aromatic azide or diazoketone group, as in p-azidophenyl acetic acid imidoester (15),

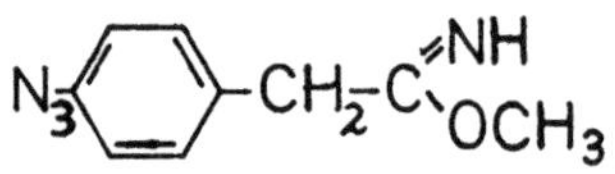

although other reactions have been found (15). Small changes in the structure have a strong influence upon the yield of the reaction. We have, for example, compared the cross-linking of alkaline phosphatase to rabbit IgG by the imido-ester shown above, and by p-azidobenzoic acid imidoester, which lacks the methylene group. To this purpose, the enzyme was incubated with a 20fold molar excess of the imidoesters for four hours at pH 7.5, and the excess of the reagents was dialyzed out thoroughly. The enzyme, which showed no loss of specific activity in both cases, was then irradiated in the presence of the equal amount of the IgG (w/w) at a protein concentration of 50 mg/ml with a 20 watt mercury lamp for 10 minutes. The products of the reaction were then separated by molecular sieve chromatography. In the case of the benzoic acid derivative, 35% of the IgG had been labeled by the enzyme, while the phenylacetic acid imidoester reached a yield of 61%.

Half-site Reaction and Cleavage

Two properties appear to be especially desirable for the reactions discussed here: The bifunctional yield should be high, i.e. the fraction of the reagent molecules attached to the macromolecule by only one covalent bond should be low. To facilitate the structural and functional analysis of the product, the use of reversible reactions or reagents cleavable under mild conditions without denaturation of the macromolecule is usually advantageous.

Many of the protein-reactive groups in these substances are activated carboxyl derivatives, as imido esters, acid azides, N-hydroxisuccinimide esters, or other highly reactive groups as isothiocyanates, isocyanates or bromoacetyl residues. Therefore, lysine side chains are probably among the first to be substituted on the cross-linked macromolecules, and with the highest yield. The reaction of the second group on the reagent, however, is evidently much influenced by the functional groups on the macromolecule in the usually small range it will reach. Thus, when bis-iodacetamido-azobenzene dicarbonic acid (5) was reacted with human hemoglobin A, the first iodoacetyl group formed a carboxymethyl derivative with the Cys 93 of the β chain, as expected. The second group carboxymethylated a sterically suitable lysine group, although at the chosen pH value of 6.8 the reaction of histidine side chains is approximately 10^3 fold faster with monofunctional iodoacetyl reagents than that of lysines.

Since most of the reactive groups in use are amenable to hydrolysis, this is the most important cause for half-site reactions. Its extent has been determined in only a very few cases. In fact, in most of the cross-linking experiments today, the amount of reagent actually attached to the macromolecules under investigation is not known. In the case of tartryl-bis-glycine azide (16), radioactively labeled in the glycine moiety, where the half-site attachment could be evaluated by periodate cleavage, it was found to amount to 28% of the covalently bound reagent. A similar value was obtained for an azobenzene-bis-isothiocyanate, where the half-site reaction could be monitored spectrophotometrically.

Half-sites may also occur when the second reactive group cannot reach a suitable partner, and its hydrolysis is either extremely slow - as in the case of bromoacetyl groups- or excluded by the structure of the macromolecule. This has been observed only rarely. However, during reaction of H_2DIDS (see above) with biological membranes, a fraction of the isothiocyanate groups was not hydrolyzed even during prolonged reaction times of several days, and their presence could be demonstrated with the aid of radioactive ethylene diamine.

Only a few structures for cleavable reagents have been proposed. Among the first were the azo dyes (5), as their cleavage by dithionite under physiological conditions does not cause any protein denaturation. Frequently, their hydrophobicity is a limiting factor in application. Aliphatic disulfides, as dimethyl-3,3'-dithio-bis-propionic acid imodoester (17) have been widely used in the investigation of the topography of ribosomal proteins. A certain drawback for other purposes is the possibility of accidental disulfide exchange reactions even at pH values around 6.0, as many systems will not permit reactions in the native state at more extreme conditions.

Vicinal diols, as the derivatives of tartaric acid (16), permit a mild cleavage by incubation of the cross-linked proteins with 10 mM periodate at pH 6. The reaction may be carried out after the first dimension of a two-dimensional gel electrophoresis, affording a "diagonal" analysis (16), as with the reductive cleavage of disulfide reagents. The slow hydrolysis of imidoester reagents is combined with this strongly hydrophilic structure in a reagent easily synthesized from dithiothreitol:

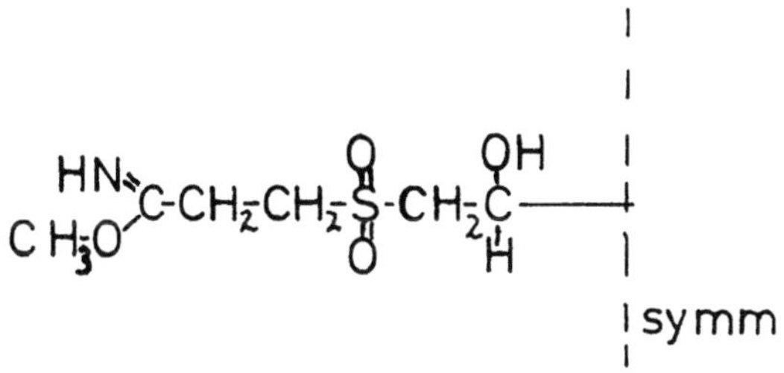

In many cases the cleavage reaction after cross-linking and isolation of the product not only enhances the detailed analysis of the points of attachment, but serves to clarify the effect of cross-linking upon the function of a protein itself. Thus, the cross-linking of F-actin leads to artificial dimers of the protein, that, in the isolated form, will not enter into the polymerization reaction from G- to F-actin. They are however able to serve as polymerization nuclei for native G-actin (13). When the dimers are cleaved by incubation in periodate, the resulting monomers will again polymerize with concomitant hydrolysis of bound ATP with normal specific activities. Therefore the cross-link did not cause a denaturation of the protein, but forced the dimer into a conformation prohibiting an elongation step in the polymerization.

Protein-Nucleic Acid Cross-Linking

All functional groups of the aromatic bases in nucleic acids suitable for substitution reactions are also present in amino acid side chains. Moreover, in all reactions that can be carried out under physiological conditions, protein side chains are more reactive than nucleic acids. The difficulty of finding selective reactions for linking proteins to RNA or DNA has usually been circumvented by using very fast mechanisms that can make use of a close contact between the two reaction partners. Most of these cross-links therefore have been obtained by irradiation and photolysis, either of nucleotides and proteins directly (18), or of aromatic azides or diazoketones already bound to the protein (15).

A few more selective procedures, however, are available for the purpose. Aliphatic α-bromoketones, at pH values below 6, react with cytosine, and especially adenine, to form the "ε"-heterocycle (19) with a speed at least comparable even to the substitution of sulfhydryls. In suitable reagents,

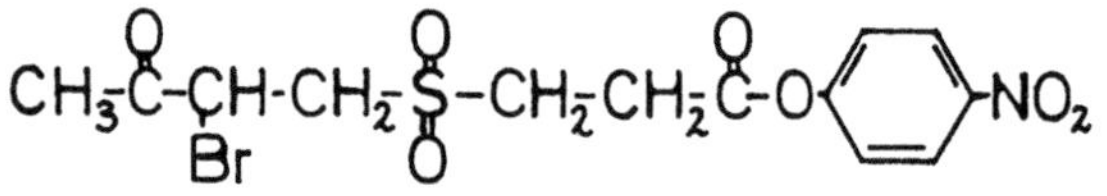

the derivative may then be linked to the protein in a second step above pH 7. In a model system, consisting of tRNA and polylysine, an overall yield of 30% cross-linking beteen the two components could be reached.

The compound was then used for the cross-linking of mRNA isolated from rabbit reticulocytes (20), with the aim of identifying the proteins attached to the polyA tail of this globin messenger. The cross-linked, purified messenger was digested with pancreatic ribonuclease A, and the purified polyA-nucleoprotein was again purified by density gradient centrifugation. Upon digestion with ribonuclease T_2+T_1, three polypeptides of molecular weight 40 000, 62 000, and 78 000 d were identified in gel electrophoresis.

A second selective reaction is offered by photolabile intercalating dyes, as psoralene. With the aid of ^{14}C-bromoacetyl-N-hydroxisuccinimidoester the derivative

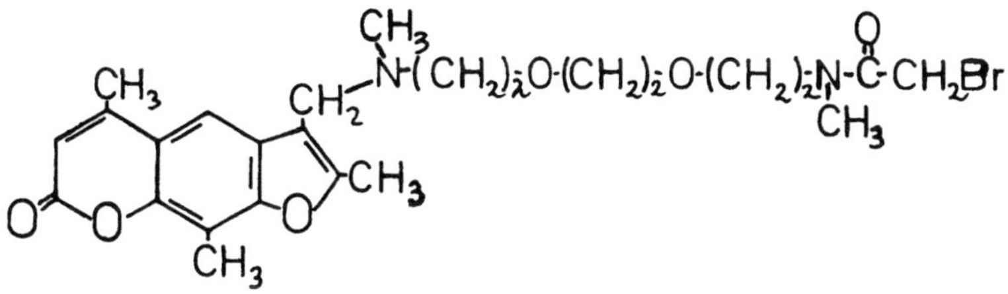

was synthesized from the corresponding amine (21). In mixtures of DNA and protein, the substance will readily intercalate into the nucleic acid, as can be observed by the strong change in fluorescence. It forms covalent links with the DNA upon irradiation at the absorption maximum at 250 nm with a low energy mercury lamp for 15 minutes in a fluid layer of 1 to 2 cm.

The reaction was applied to the "residual" DNA in nuclear ghost preparations (22), that is buried in the membrane and not accessible to DNase without its dissolution by detergent. DNA bound protein could be separated from the mixture by chromatography on Sepharose 4B columns in detergent solution, and preparative gel electrophoreses. After digestion with DNase, one of the lamina proteins (23), situated on the inside of the nuclear membrane, among others, was shown to be bound to the DNA.

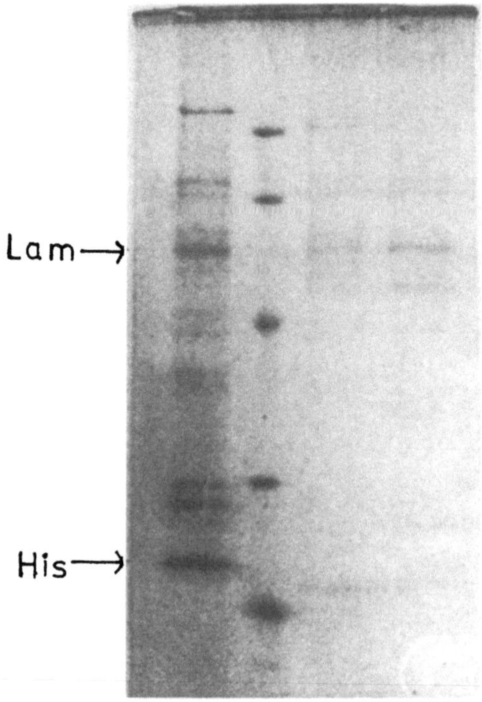

Fig. 1: DNase I digests of several fractions of proteins cross-linked to DNA in nuclear envelope ghosts, and isolated from discrete DNA-containing bands in agarose electrophoreses. Left lane: Total nuclear envelope proteins. Middle left lane: Molecular weight markers. Lam: lamina protein band. His: histone band.

References

1. Nutritional, Biochemical and Chemical Consequences of Protein Cross-linking: Advances in Exp. Medicine and Biology, Vol. 86A (1977).

2. Fasold, H., Bäumert, H., Fink, G.: Advances in Exp. Medicine and Biology, Vol. 86A, 207-213 (1977).

3. Zahn, H.: Angew.Chem. 67, 56-59 (1955).

4. Wold, F.: Methods Enzymol. 11, 617-640 (1967).

5. Fasold, H.: Biochem.Z. 339, 482-486 (1964).

6. Fasold, H., Klappenberger, J., Meyer, C., Remold, H.: Angew.Chem. 83, 875-882 (1971).

7. Marfey, P.S., Uziel, M., Little, J.: J.Biol.Chem. 240, 3270-3275 (1965).

8. Thomas, J.O., Kornberg, R.D.: Proc.Natl.Acad.Sci. U.S. 72, 2626-2630 (1975).

9. Fasold, H.: Interconvertible Enzymes, Springer Verlag N.Y. (1976), p. 60-64.

10. Hulla, F., Gratzer, H.: Bioch.Biophys.A. 291, 430-435 (1973).

11. Staros, J.V., Richards, F.M.: Biochem. 13, 2720-2726 (1974).

12. Lepke, S., Fasold, H., Pring, M., Passow, H.: J.Membr. Biol. 29, 147-154 (1976).

13. Fasold, H., Bäumert, H., Thrun, A., Kenmoku, A., Faulstich, P.: In press.

14. Ji, T.H., Ji, J.: Anal.Bioch. 121, 286-289 (1982).

15. Fink, G., Fasold, H., Rommel, W., Brimacombe, P.: Anal.Bioch. 108, 394-399 (1980).

16. Lutter, L.C., Ortanderl, F., Fasold, H.: FEBS Lett. 48, 288-295 (1974).

17. Wang, K., Richards, F.: J.Biol.Chem. 250, 6622-6626 (1975).

18. Barrios, M., Blobel, G., Fisher, P.A.: J.Biol.Chem. 258, 4548-4555 (1983).

19. Secrist, J.A., Barrio, J.R., Leonard, N.J., Weber, G.: Biochem. 11, 3499-3506 (1972).

20. Lebleu, B.: Methods Enzymol. XXX, 613-621 (1974).

21. Isaacs, S.T., Shen, C.J., Hearst, J.E., Rapoport, H.: Biochem. 16, 1058-1064 (1966).

22. Kondor-Koch, C., Riedel, N., Valentin, R., Fasold, H., Fischer, H.: Eur.J.Biochem. 127, 285-289 (1982).

23. Gerace, L., Blum, A., Blobel, G.: J.Cell Biol. 79, 546-561 (1978).

MICRO-ANALYSIS OF POLYPEPTIDE STRUCTURE USING COLOR REAGENTS

Jui-Yoa Chang

Pharmaceuticals Research Laboratories. CIBA-GEIGY Ltd. Basel
CH-4002 Switzerland

Introduction

Color reagents have been developed for micro-analysis of polypeptide structure. Those reagents include dimethylaminoazobenzene isothiocyanate (DABITC), dimethylaminoazobenzene sulfonyl chloride (DABS-Cl), dimethylaminoazobenzene iodoacetamide (DABIA), and dimethylaminoazobenzene maleimide (DABMA). They were specifically used for amino acid analysis (DABS-Cl), quantitative N-terminal analysis (DABITC), manual sequence analysis (DABITC), micro-peptide isolation (DABITC) and specific isolation of cysteine containing peptides (DABIA and DABMA). DAB chromophore derivatized amino acids or peptides are detected in the visible region with the detection limit of less than 1 pmol.

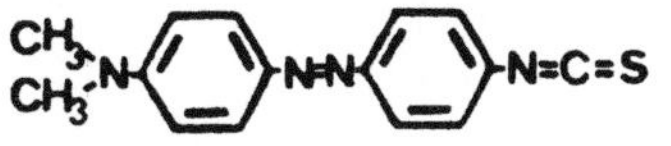

DABITC

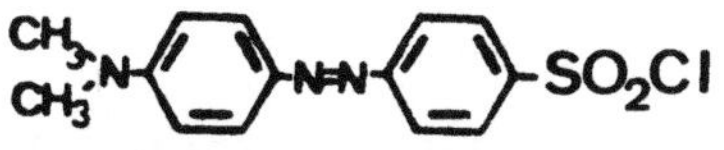

DABS-Cl

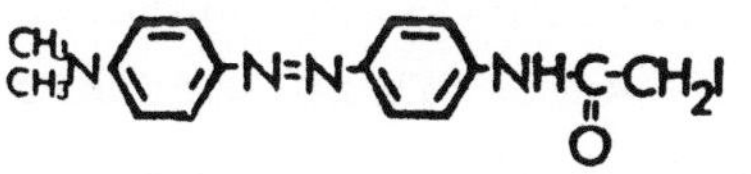

DABIA

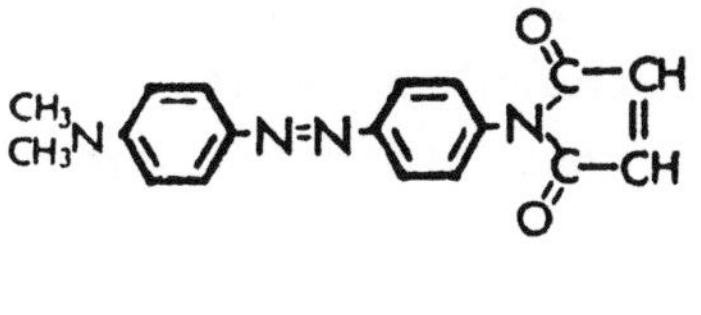

DABMA

Modern Methods in Protein Chemistry, Volume 2

1. Amino Acids Analysis (DABS-Cl) (1,2)

The combination of precolumn derivatization technique and HPLC is a potentially and practically more sensitive technique than the conventional analyser (3) for amino acid analysis. Several reagents are available for precolumn derivatization of amino acids (5-10). DABS-Cl (4) is the only color reagent currently available with which derivatized DABS-amino acids are detected in the visible region (436 nm) with limit of less than 1 pmol. DABS-Cl also reacts with both primary and secondary amino acid and DABS-amino acids are extremely stable.

1.1 DABS-Cl

DABS-Cl is commercially available from Fluka, Pierce, Adrich and highly purified DABS-Cl (double recrystallied) is now available from Pierce. A stock solution of DABS-Cl (4nmol/µl in acetone) was first prepared. One hundred microliters are pipetted into Ependorf tubes, dried under water pumps vacuum and then stored at -20°C. Each reagent tube is to be dissolved in 100 µl of acetonitrile shortly before dabsylation reaction. Standard DABS-amino acids can be prepared according to the details described in ref. 1.

1.2 Derivatization of Unknown Samples

Protein/peptide hydrolysates (0.2-1 µg) was dried in vacuum, added with 10 µl of sodium bicarbonate-carbonate buffer (pH 9) and 20 µl of DABS-Cl solution (4nmol/µl in acetonitrile) dabsylation was allowed to proceed at 50°C for 15-20 min. Without extraction or drying, the sample was directly diluted to 200-1000 µl with dilution buffer (ethanol/phosphate, pH 6.5, 25 mM , 1/1, v/v) for HPLC analysis. Conditions for dabsylation are very versatile. Acetone can be used to replace acetonitrile, although acetonitrile gave higher yields of glu, asp, thr and ser. With temperature 40-70°C are equally effective as long as standard and unknown samples were derivatized under the same condition.

1.3 HPLC Separation of DABS-amino acids

One HPLC condition which has been routinely used in our lab for DABS-amino acids separation is described in the legend of Fig. 2. C-18 columns from Merck, Vydac and Whatmann are all found to give complete separation of DABS-amino acids as that shown in Fig. 2. Amino acid composition analysis of unknown samples have shown that DABS-Cl method is as reliable as the conventional amino acid analyser.

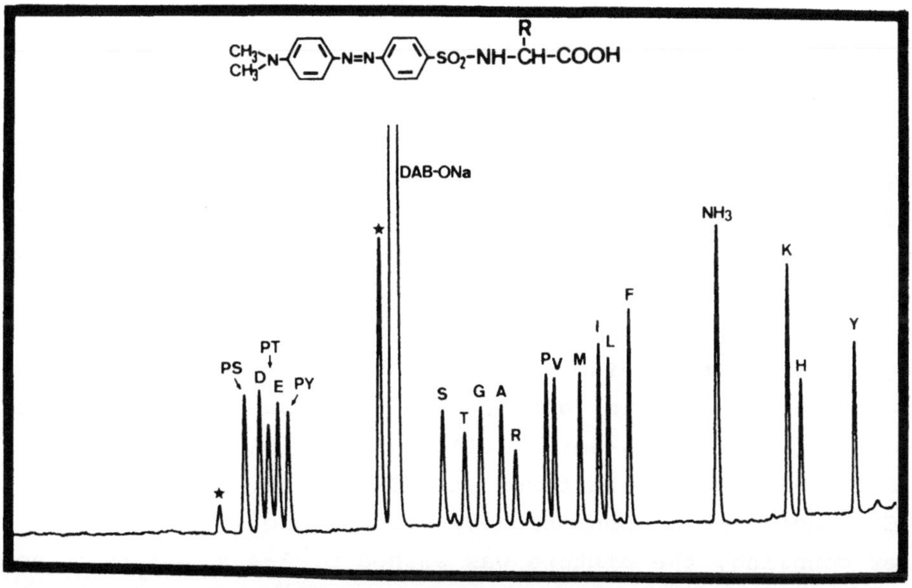

Fig. 2 Separation of standard DABS-amino acids on a Merck Lichrosorb RP-18 column. Approx. 7.5 pmol of each standard was injected. Solvent A was phosphate buffer (25 mM, pH,6.5) containing 4% DMF. Solvent B was acetonitrile. The gradient was 12 -30%B, in0-20 min, 30-60%B in 20-35 min, kept at 60%B from 35-45 min, then 60-12%B in 45-50 min. Column temperature 50°C. The flow rate was 1 ml/min. Detector, 436 nm, 0.01 AUFS. PS, PT and PY represent DABS derivatives of phosphoserine, phosphothreonine and phosphotyrosine respectively. Taken from Chang, J. Chromatogr. 1984, 295 193-200.

2. Quantitative N-Terminal Analysis (DABITC) (11-13)

DABITC presents advantages for quantitative N-terminal analysis in that (a) All N-terminal amino acids can be recovered quantitatively as DABTH-amino acids. (b) The color difference of DABTH (red), DABTC (blue) and DABITC (purple) is distinct and provide precise identification of amino acids from contaminant and (c) DABTH-amino acids can be detected by both TLC and HPLC in the visible region at the sensitivity limit of lower than 1 pmol. Quantitative N-terminal analysis serves to determine the quantity, identity and purity of a polypeptide preparation and is also very useful for monitering polypeptide processing.

2.1 Quantitative N-terminal Analysis of polypeptides

Polypeptide (0.05-1 nmol) was dissolved in in 40 µl of water and added with 80 µl of DABITC (10 nmol/µl in pyridine). Coupling reaction was performed at 70°C for 1h. After DABITC coupling, one can add 5 µl of PITC and let coupling continue for another 15 min. However, this second coupling with PITC is not absolutely necessary. The addition of PITC is not to complete the coupling, it is to couple the hydrolysed excess DABITC reagent, 4-dimethylamino-4'-aminoazobenzene (DAAB) to form a blue thiourea products (abbreviated as U, see Fig. 3). In our HPLC system for DABTH-amino acid analysis, DAAB interferes with DABTH-Ala and U interferes with DABTH-Phe. So whether add PITC or not depend upon which by-product one wish to avoid. After coupling, the mixture was washed by 3-4 x 500 µl of heptane/ethyl acetate (2:1, v/v). The remaining aqueous phase and residues were dried in vacuo or nitrogen evaporator and added with 50 µl of Aqueous acid (4N HCl/acetic acid, 1:2, v/v). Cleavage reaction was allowed to proceed at 54°C for 45 min. The sample was dried and redissolved in 10-20 µl of 80% ethanol for TLC analysis of DABTH-amino acids or in 100-500 µl of 50% acetonitrile for HPLC analysis.

2.2 Analysis of DABTH-Amino Acids

DABTH-amino acids can be analysed by both TLC (Fig. 3) and HPLC (Fig. 4). For TLC analysis, distinct color difference between the DABITC, DABTC and DABTH products greatly faciliates the correct assignment of DABTH-amino acid. However, TLC technique is only qualitative and the detection limit is about 5 pmol.

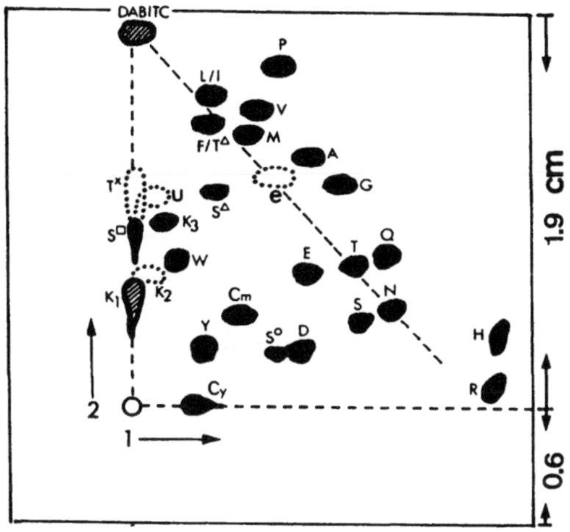

Fig. 3 Polyamide TLC separation of DABTH-amino acids. Solvent 1(first dimension), water/acetic acid (2:1, v/v). Solvent 2 (second dimension), toluene/n-hexane/acetic acid (2:1:1, by volume). The color appears as red (solid area), blue (dotted area) and purple (hatched area) after exposure of the thin layer plate to the HCl vapor. Taken from ref. 13.

3. <u>Micro-Peptide Isolation (DABITC)</u> (14,15)

DABITC derivatized peptide (DABTC-peptide) can be: (a) separated by RP-HPLC and detected in the visible region with the limit of 1 pmol and (b) recovered from HPLC and directly used for sequence analysis. Indeed, the DABITC method is the only precolumn derivatization technique for micro-peptide isolation

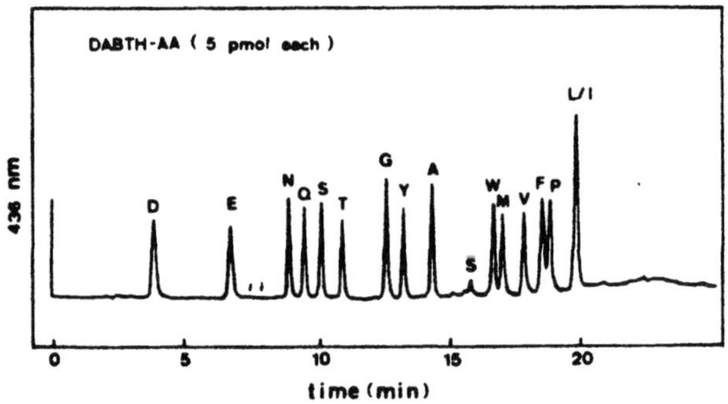

Fig. 4 HPLC separation of standard DABTH-amino acids. 5 pmol each of DABTH-amino acid was injected. Solvent A was 0.035M acetate, pH, 5.0. Solvent B was acetonitrile. The gradient was 45-70%B/0-10 min, kept at 70%B from 10-14 min, 70-80%B/14-16 min, kept at 80%B from 16-22 min, then 80-45%B/22-26 min. Column was Zorbax-ODS. Column temperature was 22°C. The flow rate was 1 ml/min. Detector, 436 nm, 0.005 AUFS.

which would allow the recovered peptide derivative to be used for subsequent sequence analysis. Furthermore, the purity, quantity and identity of the collected DABTC-peptide can be assessed through quantitative N-terminal analysis by incubation of DABTC-peptide with aqueous acid directly.

3.1 Derivatization of Peptides with DABITC

The derivatization procedure is identical to that of quantitative N-terminal analysis (section 2.1). After heptane/ethyl acetate (2:1, v/v) washing and drying, the derivatized peptide was dissolved in 20% of pyridine or 40% acetonitrile for HPLC analysis on the reversed phase column. Since the eluted peptides are detected in the visible region, a wide range of organic solvents and buffers can be used to achieve the optimum separation of DABTC-peptides. The pH of the aqueous buffer, however, should not be lower than 4.5. For analytical run, as little as 2-5 pmol of DABTC-peptide is sufficient (Fig. 5).

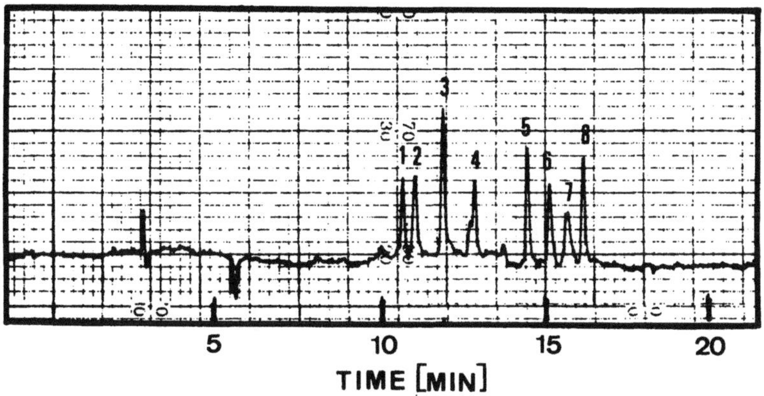

Fig. 5 Analytical separation of DABTC-peptides (4.5 pmol) derived from trypsin digestion of eglin c (see ref. 18). Seven peptides were expected. Peaks 4 ans 6 were derived from the same peptide through incomplete derivatization of DABITC at the Lys side chain. Solvent A was 10 mM phosphate, pH 7.0. Solvent B was acetonitrile. The gradient was 10%B-70%B in 20 min. Column was Bondapak C18 at room temperature.

3.2 N-Terminal Analysis of HPLC Eluted DABTC-Peptides

The eluted DABTC-peptides (50-100 pmol) were dried by lyophilization or nitrogen evaporation, dissolved in 50 µl of aqueous acid (4N HCl/acetic acid, 1/2, v/v), heated at 54°C for 45 min dried again and redissolved in 50% acetonitrile for HPLC analysis of DABTH-amino acids (Fig. 4).

3.3 Sequence Analysis of HPLC Eluted DABTC-Peptides

The eluted DABTC-peptides (0.5-2 nmol) were dried by nitrogen evaporation and then subjected to manual sequencing (13) or loaded on the automatic sequenator start with acid cleavage reaction.

2. Selective Isolation of Cysteine Containing Peptides (16,17)

Peptides containing amino acids which are specifically labelled with hydrophobic chromophore, can be selectively isolated by

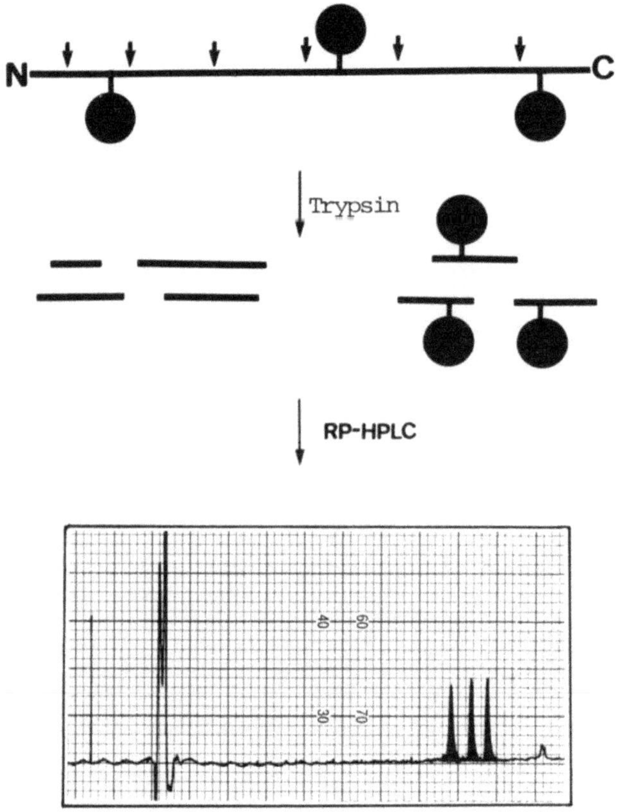

Fig. 6 Strategy for specific isolation of cysteine-containing peptides through hydrophobic labelling.

RP-HPLC. The strategy is outlined in Fig. 6. After specific labelling and enzyme digestion, two groups of peptides were generated. Those labelled peptide are colored and more hydrophobic. Those unlabelled are colorless and more hydrophvlic. Those two groups of peptides can be easily separated by RP-HPLC. This principle can be applied to isolate any peptide that contain amino acid residues which can be specifically modified with hydrophobic chromophore, such as lysine containing peptides and methionine containing peptides. The following section will only deal with the specific isolation of cysteine containing peptides.

4.1 Selective Labelling of -SH Groups with DABIA

Free -SH groups of native protein or reduced disulfide linkages can be specifically labelled with DABIA. Reduced protein should be desalted (sephadex G-25, superfine, eluted with 1M CH_3COOH) before reaction with DABIA. Native proteins or reduced proteins (1-10 nmol) were dissolved in 500 μl of 5M guanidine chloride/ 0.5M tris-HCl buffer, pH 8.4, and added with 500 μl of DABIA (0.5 mg/ml in DMF). Reaction was performed at room temperature fro 1 h with magnetic stirring. The sample was acidified with 25 μl of TFA and the excess reagent and salts were removed by gel filtration (sephadex G-25, superfine, eluted with 50% acetic acid). DABIA labelled protein was collected as a color band. The sample was freezed dried. If transfer of sample to smaller tubes is required, the dried sample is first dissolved in 200 μl of 50% formic acid, and then 400-600 μl of water.

4.2 Enzyme Digestion of DABIA Labelled Protein

The labelled protein (1-10 nmol) was dissolved in 100-300 μl of 0.5 M ammonium bicarbonate solution and added with 5-20 μg of trypsin or chymotrypsin for 20 h at 37°C. The labelled protein may not be soluble in the buffer, but usually become soluble after digestion proceeded. After digestion, the peptide mixture can be directly injected for HPLC separation, or dried and redissolved in 40% acetonitrile.

4.3 HPLC Isolation of Cysteine-Containing peptides

DABIA labelled cysteine peptides were separated from none-cysteine peptide by RP-HPLC and detected in the visible region (436 nm). As little as 1-5 pmol of colored peptide can be detected. An example is shown in Fig. 7.

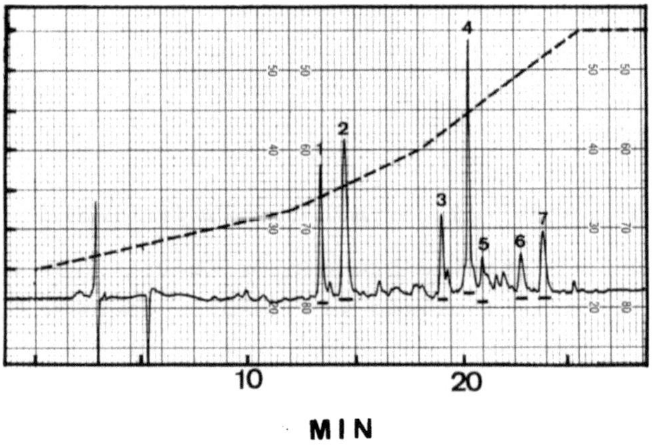

Fig. 7 Analytical separation of cysteine-containing peptides (30 pmol) derived from the chymotryptic digestion of DABIA-labelled immunoglobulin light chain 7S34.1 (see ref.16).

References

1. Chang, J.-Y., Knecht, R., Braun, D. G.: Methods Enzymol. 91, 41 (1983).

2. Chang, J.-Y., Knecht, R., Braun, D. G.: Biochem. J. 203, 803 (1982).

3. Spackman, D. H., Moore, S., Stein, W. H.: Anal. Chem. 30, 1190 (1958).

4. Roth, M.: Anal. Chem. 43, 880 (1971).

5. Simmons, S. S., Johnson, D. F.: Anal. Biochem. 82, 250 (1977).

6. Hill, D. W., Waters, F. H., Wilson, T. D., Stuart, J. D.: Anal. Chem. 51, 1338 (1979).

7. Lindroth, P., Mopper, K.: Anal. Chem. 51, 1667 (1979).

8. Jones, B. N., Paabo, S., Stein, S.: J. Liq. Chromatogr. 4, 565 (1981).

9. Gardiner, W. S., Miller, W. H.: Anal. Biochem. 101, 61 (1980).

10. Larsen, R. B., Grosso, D. S., Chang, S. Y.: J. Chromatogr. Sci. 18, 233 (1980).

11. Chang, J.-Y.: Anal. Biochem. 102, 384 (1980).

12. Chang, J.-Y.: Methods Enzymol. 91, 79 (1983).

13. Chang, J.-Y.: Methods Enzymol. 91, 455 (1983).

14. Chang, J.-Y.: Biochem. J. 199, 537 (1981).

15. Chang, J.-Y., Knecht, R., Ball, R., Alkan, S. S., Braun, D. G.: Eur. J. Biochem. 127, 625 (1982).

16. Chang, J.-Y., Knecht, R., Braun, D. G.: Biochem. J. 211, 163 (1983).

17. Chang, J.-Y.: FEBS Lett. 164, 307 (1983)

18. Knecht, R., Seemüller, U., Liersch, M., Fritz, H., Braun, D. G., Chang, J.-Y.: Anal. Biochem. 130, 65 (1983).

A MODULAR SYSTEM FOR MICROSEQUENCING OF PROTEINS, OLIGONUCLEOTIDE SYNTHESIS AND PEPTIDE SYNTHESIS

Rainer Frank, Marcus Trosin

Europäisches Laboratorium für Molekularbiologie
Postfach 10.2209, 6900 Heidelberg, West Germany

Introduction

Modern molecular biology requires both analytical methods such as microsequencing of proteins and peptides, and synthetic tools like peptide and oligonucleotide synthesis. Since all these methods have to be performed on microscale quantities, the executing systems are related with respect to the necessity of precise handling of very small amounts of liquids and gases. Therefore it seems advantageous to develop modular systems for the instrumentation of all the above mentioned methodologies. Due to the modular construction these machines are not only designed to our particular requisites, but also allow a high degree of flexibility with regard to systems updating, and adaptation to new chemistries. Furthermore they will serve as the working basis for the instrumentation of other relevant biochemical methods. Our modular system is subdivided into a common electronic programming unit and a solvent and reagent delivery system. The reaction system has to be individually designed for each method depending on the chemistry to be executed.

In this article the construction principles used in the design of a gas-phase protein sequencer and an oligonucleotide synthesizer are described. A short overview on the performance of both instruments is given. The design of a new peptide synthesizer which employs the same modular system is also presented.

Modern Methods in Protein Chemistry, Volume 2

The programming module

The programming module represents the government of an instrument and contains a processor unit and an interface unit (Fig. 1). The processor unit uses a microprocessor (Z80, 8 bits, 4 MHz), a 64 K memory, two floppy disc drives, a graphic interface and a board for real time control which are assembled on the processor bus (ECB-bus). For visual display of function parameters and running the machine the processor is linked to a 5" video monitor and 5 x 4 keyboard. The processor unit is connected through an I/O bus (8 bits parallel) to the interface unit, which supports the processor interface, two optoisolated I/O boards with 16 outputs each, the interface for a motor driven syringe. During design of the software and programming of individual cycles a terminal with an ASCII keyboard and a printer are linked to the programming module.

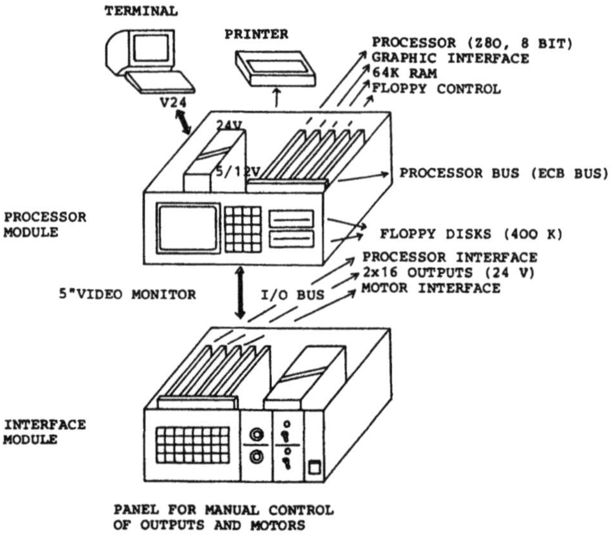

Fig. 1: Scheme of the programming module. The processor module and the interface module are linked by an I/O bus (8 bits parallel).

The microprocessor uses a CP/M 2.2 operating system. The main controlling program is written in BASIC, and ASSEMBLER is used for real time software.

For all three instruments we use the same program version permitting 99 different degradation or synthesis cycles to be executed during a run. The cycles are created by the operator designed for each chemistry. Each operation e. g. "wash column" is variable in duration and contains a maximum of 12 outputs. After writing the different cycles are stored on the floppy disc and are loaded in the processor memory immediately before execution of a degradation or synthesis run. Typing in of runs, cycles, and operations is further simplified by a set of menus.

The Reagent and Solvent Delivery Module

The major problem in the instrumentation of chemical processes is the construction of a reliable delivery system for corrosive gases, reagents and solvents. The valves in such a delivery system should exhibit the following characteristics: chemical inertness, no dead volume, vacuum tightness, pulse free delivery of microliter amounts of liquids, easy repair and maintenance and construction without any special equipment.

All of the above mentioned properties are found in the so called "zig-zag" valves (pneumatically activated diaphragm valves) developed by the Wittmann-Group in Berlin (1). The construction principle of the valves is shown in Fig. 2.

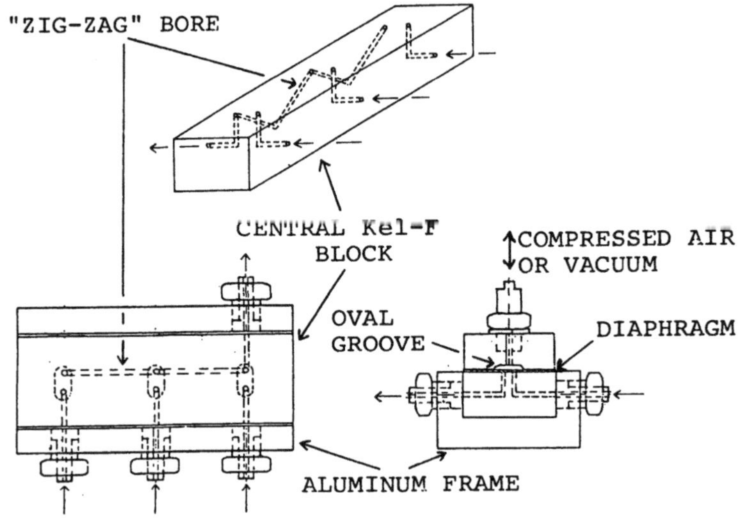

Fig. 2: Construction principle of a three unit delivery valve manifold. The "zig-zag" bore and the inlet bores have a diameter of 0.5 mm (gas-liquid phase sequencer and oligonucleotide synthesizer) and 0.8 mm (peptide synthesizer). The arrows on the Kel-F block indicate the inlets (right) and the outlet (left).

An aluminum frame surrounds the central Kel-F block which contains the individual valve units (inlets, diaphragms and heads) and the interconnecting "zig-zag" bore. On top of the Kel-F block the diaphragms are held in place by the valve heads. In comparison to the original design our valves are miniaturized to reduce the internal volume (0.5 mm inner bore) and the valve heads are modified to enable delivery of small amounts of liquids without any pulse. Depending on the number of liquids and gases that have to be transferred to the reaction systems, the valves are combined in manifolds from two to eight units. The valve manifolds are mounted as close as possible to the reaction chambers to minimize the internal volume of the system. The only materials in contact with reagents and solvents are Teflon, Kel-F and Pyrex glass. The same type of delivery valves is used to pressurize and vent the reagent and solvent bottles. To avoid any cross-contamination in the vapor phase all reagents and most of the solvents are separated from each other by inert valves.

The pneumatic system controlling compressed air and vacuum supply for activation of the delivery valves consists of three-way solenoid valves, assembled in a rack which can easily be removed from the instruments for repair and maintenance.

The Gas-Liquid Phase Protein Sequencer

Due to the sensitivity of the method and mechanical simplicity of the apparatus required we have chosen the gas-liquid principle (2) for the construction of our sequencer. This method employs gas phase reagents instead of liquids in the coupling and cleavage steps of the Edman degradation. Due to this modification the spinning cup assembly used in the liquid phase sequencer can be replaced by a simple reaction cartridge. The reaction cartridge forms a small reaction cell which contains a glass fibre disc. On this support the peptide or protein sample is fixed in a matrix of polybrene, a quarternary ammonium salt polymer (3).

The sequencer is constructed with the above described programming module and delivery system (Fig. 3) and works according to the same principles as published by Hunkapiller et. al. (4). Purified nitrogen delivered through low pressure regulators is used to pressurize the reagent and solvent reservoirs and to transfer the liquids and gases. The delivery of nitrogen and the partition of liquids and gases is controlled by the pneumatically activated diaphragm valves described in the previous chapter. After each delivery of a reagent or solvent the distribution manifold and the reaction cartridge are purged by a short nitrogen flush.

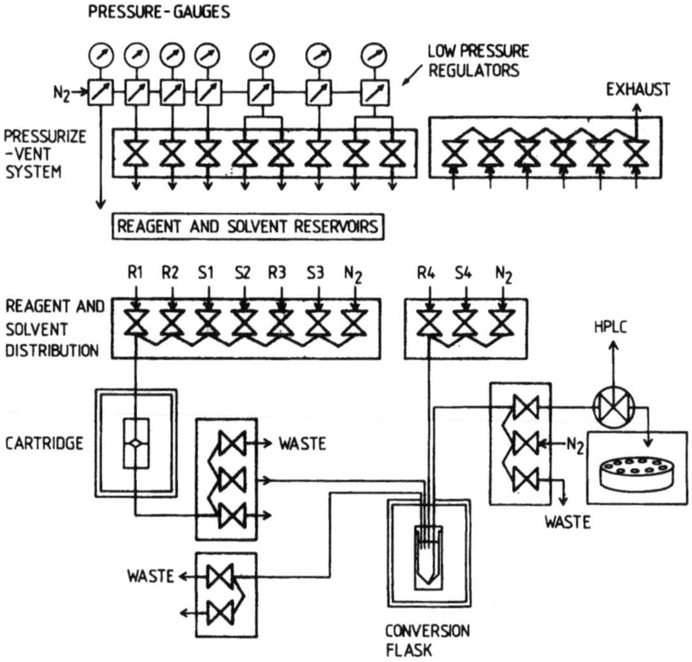

Fig. 3: Flow scheme of the reagent and solvent distribution system in the gas-liquid phase sequencer. For directing an aliquot of each >PhNCS-fraction to the HPLC system a pneumatically activated injection valve (Rheodyne) is used.

In our instrument we have replaced the vacuum stages, used in the original version to remove liquids from the reaction cartridge and conversion flask, with nitrogen flushes. Thus the elaborate and expensive vacuum system is completely eliminated. Additionally the relatively large conversion flask has been miniaturized to an internal volume of 0.8 ml allowing the extraction of >PhNCS-derivatives from the flask in 100 - 200 ul of solvent. This small volume is an important prerequisite for the on-line identification of the >PhNCS-amino acids by HPLC. The reaction cartridge is made from Pyrex glass and ultrasonically machined by Hellma GmbH, Müllheim, Baden.

The chemicals we use for microsequencing (Tab. 1) are of the highest purity commercially available (Applied Biosystems) or are purified according to the published procedures (4,5). However, for sample amounts of 3 nmoles and more we found the "sequence analysis" grade of various suppliers (Fluka, Pierce, Merck) to be reliable.

Table 1: Reagents and solvents used in the gas-liquid phase sequencer.

	Reagent/solvent	Volume per cycle (ml)
R 1	5 % Phenylisothiocyanate in n-heptane	0.05
R 2	10 % Trimethylamine in water	( )
R 3	Trifluoroacetic acid	( )
R 4	1 N HCl in methanol	0.1
S 1	n-Heptane	1.0
S 2	Ethylacetate	1.8
S 3	1-Chlorobutane	1.0
S 4	Methanol	1.2

For identification and quantitation of the >PhNCS amino acid residues we use a reverse phase HPLC separation system (Fig. 4). All >PhNCS amino acids are resolved by a 30 min gradient permitting the identification of less than 5 pmol of each >PhNCS amino acid.

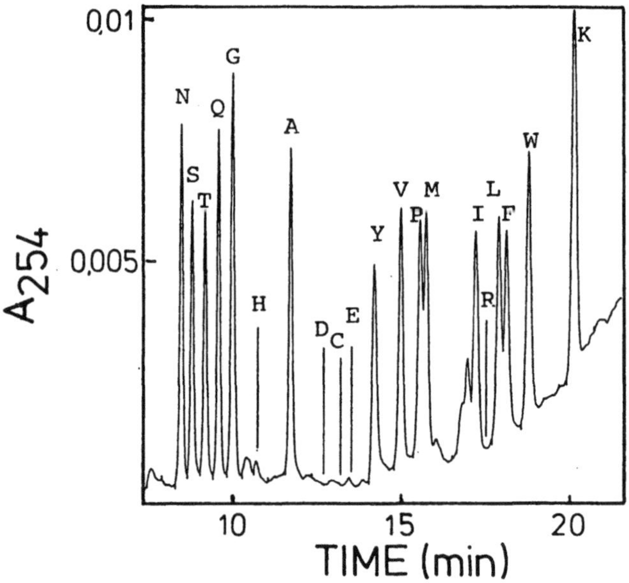

Fig. 4: Reverse phase HPLC separation of 100 pmol >PhNCS amino acid standard mixture. The chromatography was performed at 40 °C with a flow rate of 1 ml per min on a Hewlett-Packard 1084 B liquid chromatograph using a Du Pont Zorbax CN column (0.46 x 25 cm). The gradient program and the buffers are indicated below).

Time (min)	% B	
0	10	Buffer A: 20 mM NaAc pH 5.5
8	30	Buffer B: CH_3CN, CH_3OH (8:2)
16	40	
20	50	
22	80	
24	10	
30	10	/ new injection and start of the gradient

Some of the results of the first sequencing experiments are shown in Fig. 5 and 6. As far as we have evaluated the performance of the sequencer, we have found the lower limit for amino acid sequence analysis of proteins to be between 20 pmol and 50 pmol (10 to 20 residues) and for peptides between 50 pmol and 100 pmol. In test runs with sperm whale myoglobin we have found a repetitive yield of 95 % (500 pmol protein) and 91 % (30 pmol protein).

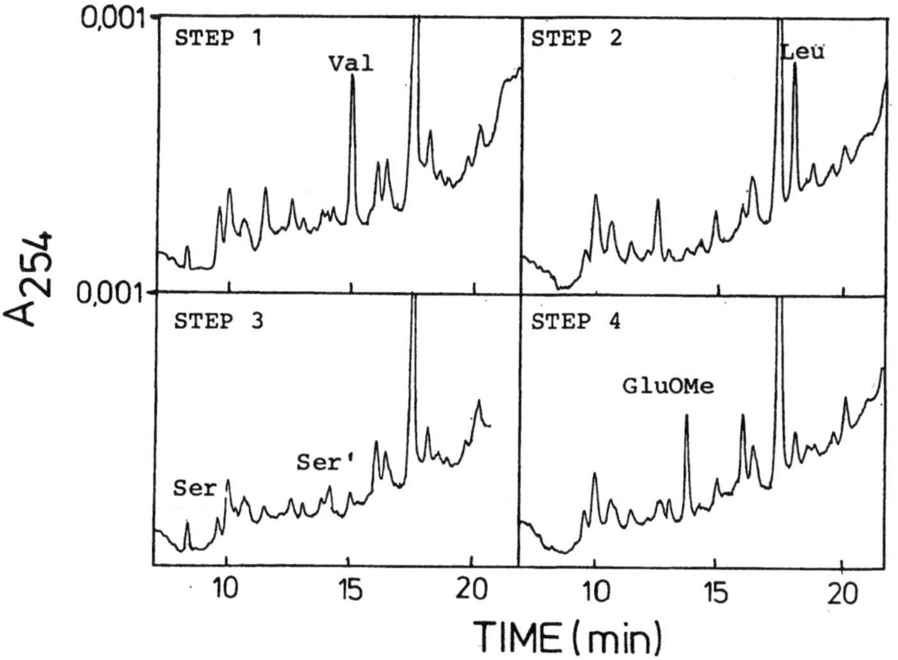

Fig. 5: NH_2-terminal amino acid sequence analysis of 30 pmol sperm whale myoglobin. 100 % of each sample were injected. The chromatographic conditions are the same as in Fig. 4.

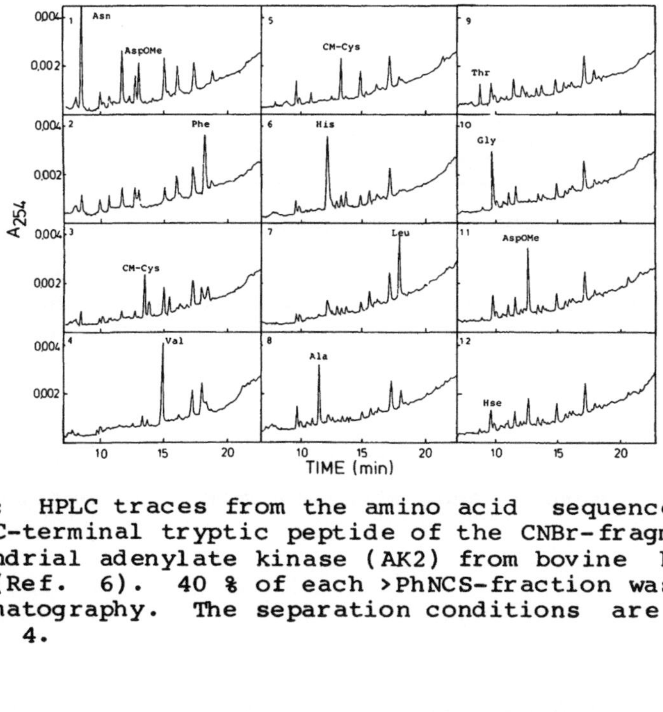

Fig. 6: HPLC traces from the amino acid sequence analysis of the C-terminal tryptic peptide of the CNBr-fragment C from mitochondrial adenylate kinase (AK2) from bovine heart (300 pmol) (Ref. 6). 40 % of each >PhNCS-fraction was subjected to chromatography. The separation conditions are described in Fig. 4.

The Oligonucleotide Synthesizer

For the construction of the oligonucleotide synthesizer we used the same programming module and reagent and solvent delivery system as for the gas-liquid phase sequencer. The chemistry executed by the instrument is the phosphoramidite chemistry developed by Caruthers (7). The oligonucleotides are synthesized on a controlled pore glass support (8) derivatized with a long chain alkylamino anchoring group to which the first nucleoside is coupled through its 3' hydroxyl group. The dimethoxytrityl nucleoside phosphoramidites are activated with tetrazole immediately before addition to the solid support.

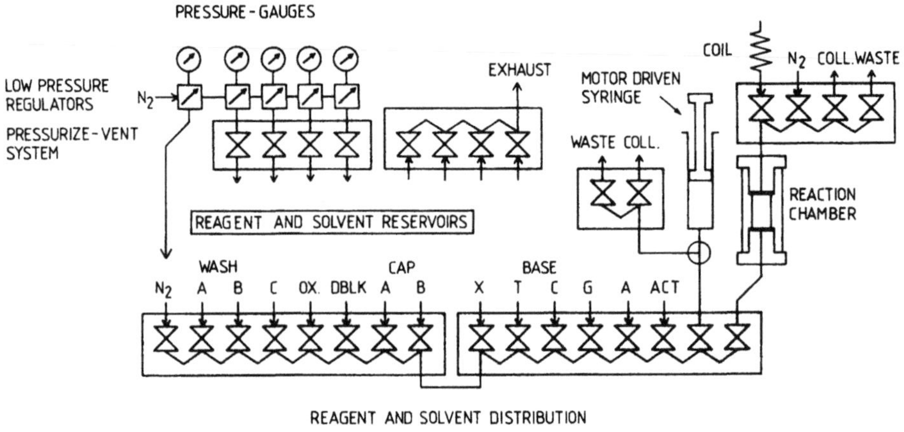

Fig. 7: Flow scheme of the reagent and solvent delivery system of the oligonucleotide synthesizer. The reaction chamber is made from Kel-F and closed on both ends with a Teflon membrane.

A special feature of our instrument is the motor driven syringe which meters and mixes monomer and activator and allows mixing of probes in all monomer combinations. After activation the syringe is used to recycle the coupling mixture (200 - 1000 ul) several times through the reaction column. This procedure forms a dynamic bed of the solid support and eliminates channeling and mixing problems especially when higher amounts of support (up to 500 mg) are used. All other reagents and solvents are transported by nitrogen pressure. Washing of the reaction column is performed by filling the column with solvents and driving out the solvent in the reverse direction with a flush of nitrogen. In each washing step this procedure is repeated several times.

Tab. 2: Synthesis cycle for solid phase synthesis of deoxyoligonucleotides using dimethoxytrityl nucleoside phosphoramidites.

Function	Reagent/Solvent	Vol. (ml)	Time (min)
Capping	A. 6.5 % 4-dimethylaminopyridine in tetrahydrofuran	0.25	2.0
	B. Acetic acid anhydride, 2,6-lutidine, tetrahydrofuran (1:1:8)	0.25	
Wash	Acetonitrile	3 x 1.0	1.5
Wash	1,2-Dichloroethane	2 x 1.0	1.5
Deblock	3 % dichloroacetic acid in 1,2-dichloroethane	1.5	2.0
Wash	1,2-Dichloroethane	2 x 1.0	1.5
Wash	Acetonitrile	3 x 1.0	1.5
Coupling	0.07 M solution of amidite in acetonitrile	0.2	3.0
	2 % tetrazole in acetonitrile	0.2	
Wash	Acetonitrile	3 x 1.0	1.5
Oxidation	1 % iodine tetrahydrofuran, 2,6-lutidine, water (8:1:1)	0.5	1.0
Wash	Acetonitrile	4 x 1.0	2.0

Depending on the quality of monomers and solvents used for synthesis we found a coupling yield between 96 % and 98 % allowing synthesis of DNA fragments up to 50 nucleotides long. (R. Frank, B. Connolly, unpublished results).

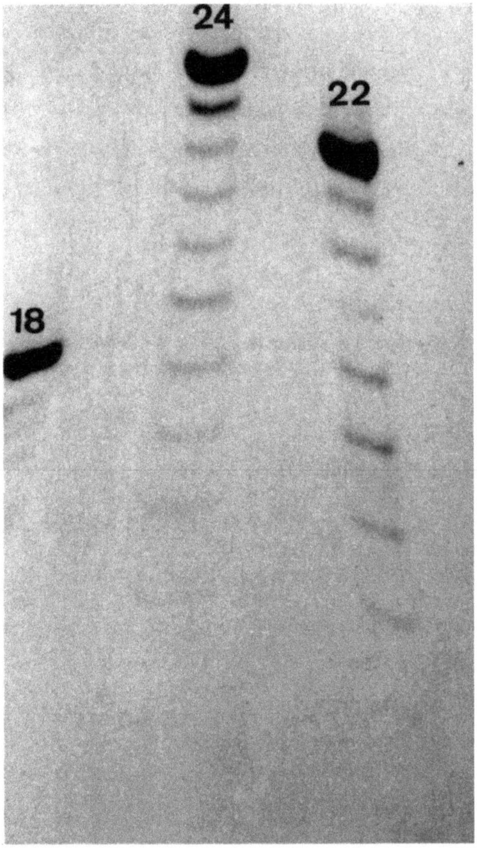

Fig. 8: Autoradiogram of ^{32}P labeled crude oligonucleotides 18, 24 and 22 bases long. The synthesis protocol is given in Tab. 2.

The Peptide Synthesizer

In contrast to oligonucleotide synthesis, in solid-phase peptide synthesis one has to handle a set of precursors with different reactivities, and depending on the sequence to be synthesized different protecting groups, deprotecting agents and forms of activitation.

To cover all these eventualities the synthesizer has to offer a high degree of mechanical flexibility. Therefore our synthesizer can handle five washing-, two activating- and two deblocking steps. The machine is equipped with the same pressurizing system for reagent and solvent bottles and the same liquid delivery system as is used in the oligonucleotide synthesizer. Whilst in the oligonucleotide synthesizer the solid support is treated in a batchwise manner the peptide synthesizer uses a continuous flow principle.

The continuous flow of liquids through the reaction column permits recirculation of the activated coupling mixture, quick washout of excess reagents and short deprotection times. To ensure high and reproducible flow rates and to perform the recycling of the coupling mixture, a valveless piston pump is used.

For transferring and metering of the protected amino acids and for preincubation with activator the same motor driven syringe is used as in the oligonucleotide synthesizer. In order to save valves and simplify programming the amino acids are stored in the order of the sequence to be synthesized in the turntable of a fraction collector which is controlled by the programmer. From this storage magazine which is equipped with a movable syringe needle the amino acids are transferred into the syringe and can be simultaneously mixed with an activator.

The activated coupling mixture is then delivered into the reaction chamber and recirculation is taken over by the piston pump.

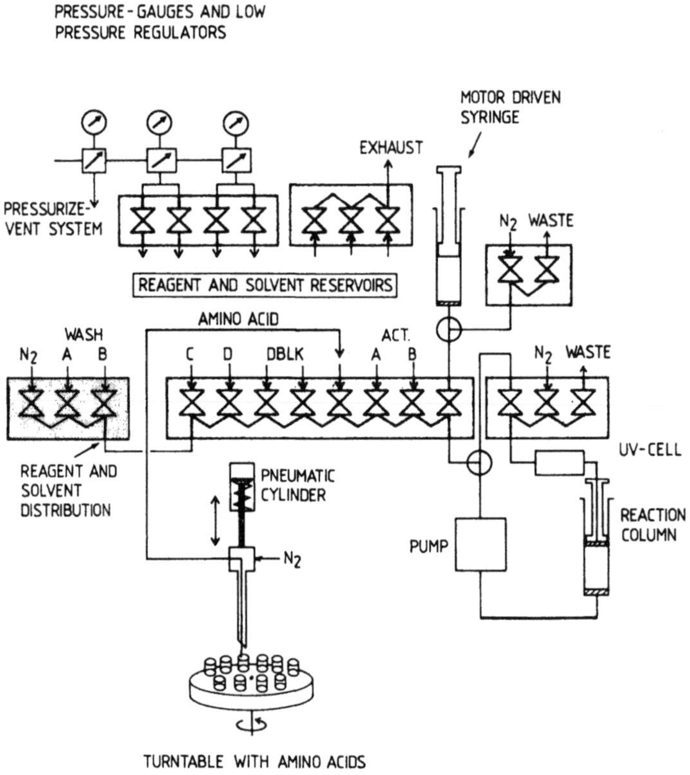

Fig. 9: Flow scheme of the reagent and solvent delivery system of the peptide synthesizer. The reaction column has a variable bed size and can be adjusted to the amount of support to be handled.

The mechanical design of the machine is based on the employment of a rigid support such as the polyamide-Kieselguhr composite resin (9) and the base labile N^{α}9-fluorenylmethoxycarbonyl (Fmoc-) protecting group (10).

In preceeding experiments with a semiautomated bench top prototype the polyamide resin has shown excellent flow characteristics and coupling efficiency (Leban, H. and Frank, R., unpublished results). The Fmoc-protecting group can be cleaved rapidly under mild conditions (50 % piperidine in dimethylformamide 20 min). The combination of the Fmoc-protecting group with an acid labile side chain protecting group allows the linkage of the peptide to the resin to be cleaved under mildly acidic conditions. Thus we can avoid the commonly used hydrogen fluoride treatment with its dangerous handling problems.

Due to the mechanical design and the programming flexibility the operator has the free choice of the acylation method e. g. employment of symmetrical amino acid anhydrides or esterification procedure with dicyclohexylcarbodiimide and 1-hydroxybenzotriazole (11). Control experiments with standard peptides have yet to show which method or combination of methods is best suited for our instrument.

Conclusion

The set of biochemical instruments we have constructed using a modular system represents an effective tool in molecular biology. The design of the programming module and of the reagent and solvent delivery system have proven to be reliable for both analytical and synthetic methods.

References

1. Wittmann-Liebold, B.: Modern Methods in Protein Chemistry, H. Tschesche, ed., Walter de Gruyter, Berlin. New York, pp. 229-266 (1983).
2. Hewick, R.M., Hunkapiller, M.W., Hood, L.E. and Dreyer, W.J.: J. Biol. Chem. 256, 7990-7997 (1981).
3. Tarr, G.E., Beecher, J.F., Bell, M. and Capra, J.D.: Anal. Biochem. 84, 622-627 (1978).
4. Hunkapiller, M.W., Hewick, R.M., Dreyer, W.J. and Hood, L. E.: Methods in Enzymology 91, Hiss, C.H.W. and Timasheff, S.N., eds., Academic Press, New York, pp. 399-413 (1983).
5. Esch, F.S.: Anal. Biochem. 136, 39-47 (1984).
6. Frank, R., Trosin, M., Tomasselli, A.G., Schulz, G.E., Schirmer, R.H.: Eur. J. Biochem. 141, 629-636 (1984).
7. Caruthers, M.H.: Chemical and Enzymatic Synthesis of Gene Fragments, Gassen, H.G., Lang, A., eds. Verlag Chemie, Weinheim, pp. 71-79 (1982).
8. Adams, S.P., Kavka, K.S., Wykes, E.J., Holder, S.B. and Gallupi, G.R.: J. Am. Chem. Soc. 105, 661-663 (1983).
9. Atherton, E., Brown, E., Sheppard, R.C. and Rosevear, A.,: J. Chem. Soc. Chem. Commun., 1151-1152 (1981).
10. Carpino, L.A. and Han, G.Y.: J. Org. Chem. 37, 3404-3408 (1972).
11. Gabriel, T.F., Jimenez, M.H., Felix, A.M., Midralewsky, J. and Meienhofer, J.: Int. J. Pept. Prot. Res. 9, 129-136 (1977).

ON-LINE DETECTION OF AMINO ACID DERIVATIVES RELEASED BY AUTOMATIC EDMAN DEGRADATION OF POLYPEPTIDES

Brigitte Wittmann-Liebold and Keith Ashman

Max-Planck-Institut für Molekulare Genetik
Abteilung Wittmann
Ihnestrasse 63-73
D-1000 Berlin 33 (Dahlem), Germany

Key-words

High performance liquid chromatography
Reverse phase isocratic separation
On-line detection of PTH-amino acid derivatives
Automated Edman degradation
Peptide and protein micro-sequence analysis

Summary

An on-line detection system for phenylthiohydantoin derivatives that are released by automatic Edman degradation of polypeptides has been installed in the Berlin liquid- and solid/gas-phase machines by connecting a simple high performance liquid chromatography system to the sequencers. Almost all PTH-amino acid derivatives can be separated within 45 minutes on a LiChrospher C_8-coated reversed phase column under isocratic conditions after automatic injection. This is controlled by the same microprocessor that governs the other parts of the sequencer program. The isocratic on-line identification enables quick and sensitive detection of the derivatives directly after their release, guarantees a stable base-line, low back-ground contaminations and reproducible retension times. Higher recoveries for labile PTH-amino acids were ob-

Modern Methods in Protein Chemistry, Volume 2

tained and the detection of modified residues facilitated. The isocratic solvent mixture is recycled for at least a week. Finally, the automatic detection system allows constant monitoring of the machine's performance.

Introduction

The design of a sequencer for the automatic stepwise degradation of polypeptides by Edman and Begg (1) greatly facilitated protein primary structural determinations and enabled an enormous increase in our knowledge of proteins. The automatic approach has turned out to be superior to any manually performed stepwise degradation due to its enhanced efficiency and increased repetitive degradation yield, if the sequencer performs optimally.

The automatic degradation as worked out by Edman and Begg only included the coupling of the polypeptide with phenylisothiocyanate (PITC), the cleavage reaction and the collection of the released thiazolinones; but it left the third part of the reaction set, the isomerization to the more stable phenylthiohydantoin derivatives to be carried out manually. Later, this part of the reaction was also automated for the modified liquid phase Beckman sequencer (2). Automatic conversion devices were also built into solid-phase machines (3-5) and the recently designed gas-phase sequencer (6).

Automatic conversion is a necessary step towards complete automation of the process as the released thiazolinones are labile in an oxygen atmosphere and undergo further destruction during drying. Often, amino acid "ghosts" rather than the authentic thiazolinones were subjected to manual conversion the day after their release in the sequencer. Further destruction takes place if the PTH-amino acid derivatives are not identified soon after their release on runs over the weekends or if the system employed for the final identification and quantification, e.g. a gradient HPLC or MS-system, breaks down. This effect is almost negligible where sufficient sample amounts can be applied for the polypeptide degradation; however, it becomes severe for sequencing of minute protein and peptide amounts where the amino acid derivatives are only released in low picomole quantities. Therefore, the installation of an automatic detection system into the sequencer which allows a quick and sensitive identification of the PTH-amino acids directly after their release greatly reduces this problem and is a further step towards complete automation in micro-sequencing.

In the present paper we describe such an automatic on-line detection system based on an isocratic HPLC-separation for PTH-amino acids as proposed 1980 (7). It has been installed into a liquid phase and a solid/gas-phase machine, both constructed in the institute's workshop in Berlin (8). The automatic detection system is simple and has been running for more than a year satisfactorily with both the sequencers. Its technical details and the advantages over the normal separate PTH-amino acid identification are summarized. The proposed on-line HPLC-injection is different from the one described recently for the installation into a solid-phase sequencer (9). Our version can be applied for all types of sequencers, it is simpler and inexpensive and needs no extra mechanical transport system.

Purification of reagents and solvents

<u>Phenylisothiocyanate</u> (PITC), purissimum grade (Fluka), was redistilled under reduced pressure (oil pump) three times and stored under nitogen in small ampoules at -20°C.
<u>Quadrol</u>, 1 M solution, was purchased from Beckman, Munich, diluted to 0.2 M with quartz-distilled water under nitrogen, and used without further purification.

Heptafluorobutyric acid (HFB) was technical grade (Fluorad, 3 M, Neuss) refluxed over BaO (10 g/200 ml) and redistilled. It was kept over Al_2O_3 (aluminia N, activity I, purchased from Woelm Pharma, Eschwege) under nitrogen for 20 h and redistilled twice from dithioerythritol (Pierce Comp.; 10 mg/150 ml), b.p. 120-121°C. The destillate was stored in ampoules under nitrogen at -20°C.
Trifluoroacetic acid (TFA), purum grade (Fluka) was redistilled from $CaSO_4 \cdot 0.5\ H_2O$ (dried at 500°C immediately before use) over a 30 cm column filled with glass rings, b.p. 72-73°C, and stored in 50 ml aliquots in sealed glass stoppered flasks under nitrogen at -20°C.
Benzene and heptane (both Uvasol grade, Merck) were kept over metallic sodium, refluxed over charcoal (10 g/lt) and distilled; they were redistilled under nitrogen over a 1 m column filled with glass rings, stored at 10°C under nitrogen and freshly prepared monthly.
Ethylacetate (pro analysis grade, Merck) was refluxed over charcoal and redistilled twice as given for benzene.
Methanol and 1,2-dichloroethane were Uvasol grade and acetonitrile LiChrosolv grade (all from Merck); these solvents were used without further purifications.

Technical devices

Liquid-phase sequencer parts and chemicals

The main parts of the sequencer, e.g. the microprocessor (version II), the vent-, press- and delivery system, and the reagent and solvent bottles were as described recently (8). The vacuum device was a smaller version of that given before (10). Cold traps and vacuum pumps were as in (10). The delivery system together with the connecting lines to the cup compartment and conversion device is illustrated in Fig. 1.

The reagents and solvents employed (see also Table 1) were:

R1, 5% PITC in heptane (at 0.05 bar);
R2, 0.2 M quadrol buffer, pH 9.0 (at 0.05 bar);
R3, anhydrous HFB (at 0.04 bar);
R4, 20% TFA in water (at 0.1 bar);
S1, ethylacetate/benzene (75:175, by vol, at 0.08 bar);
S2, not used;
S3, benzene (at 0.08 bar);
S4, 30% methanol in water (at 0.08 bar);
S5, methanol (at 0.08 bar).

The different nitrogen supplies were:

N1, for pressurizing the cup compartment (at 0.055 bar);
N2, for emptying the delivery line of the delivery block for dosage into the cup compartment (at 0.08 bar);
N3, for emptying the delivery line of the delivery block for dosage into the conversion flask (at 0.10 bar);
N4, nitrogen supply via capillary into the conversion flask (at 0.085 bar);
N5, nitrogen supply for the fraction collector (at 0.10 bar).

Solid/gas-phase sequencer parts

The details of this machine are the subject of a separate paper (11). In principle, the same parts, e.g. the vent-, press- and delivery system, the microprocessor and the solvent and reagent reservoirs as for the liquid-phase machine were used. The delivery of trimethylamine/water and TFA in the gas-phase was as described except that nitrogen instead of argon was used for the saturation (6).

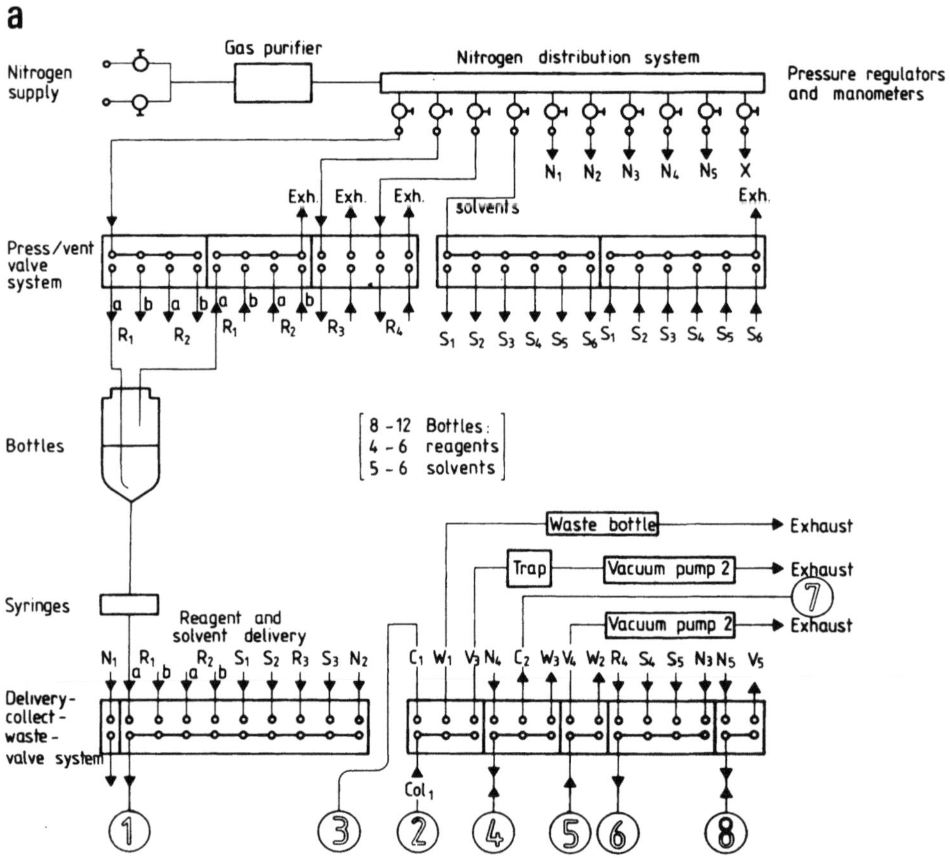

<u>Figure 1:</u> <u>Multiple Unit Micro-Sequencer</u>

Design of the combined liquid-, solid-, and gas-phase machine. <u>1a</u>: The nitrogen distribution system, the vent- and press-valve ports, the bottles (4-6 reagent and 5-6 solvent bottles) and the delivery-, collect- and waste valve system (all parts common to the combined version).
<u>1b</u>: The different reaction chambers, which provide alternatively sequencing in the spinning cup (left scheme), in the solid-phase column (middle part) and in the cartridge (right) are illustrated.

b

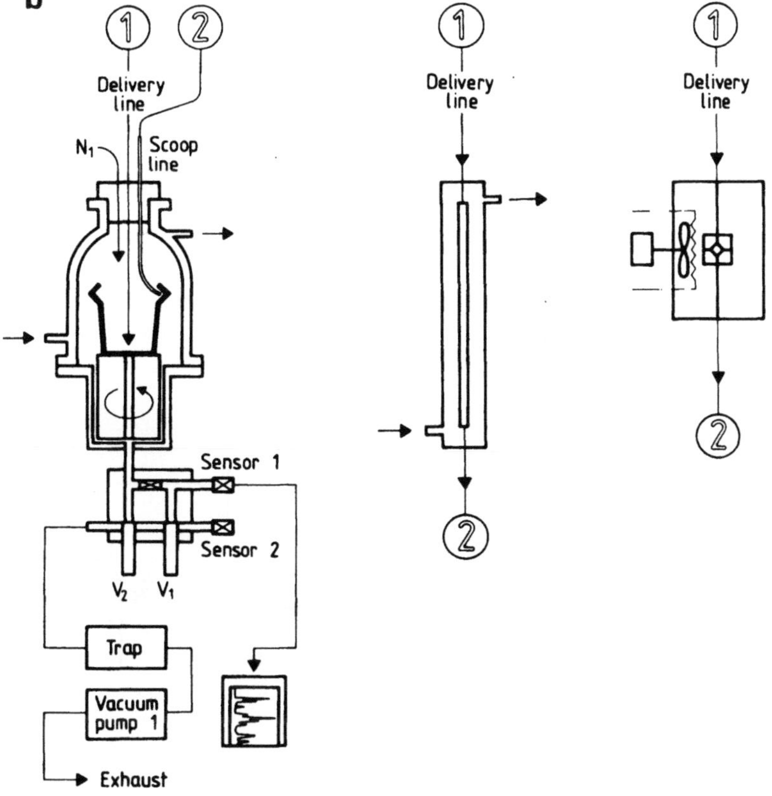

c

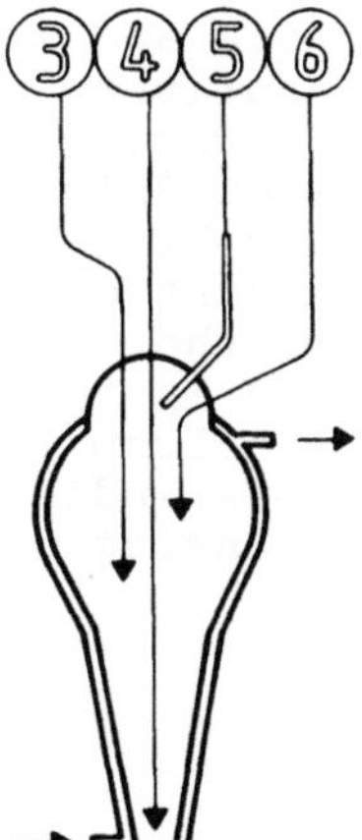

Fig. 1c: In- and outlet-lines of the conversion device (flask-type for liquid-phase sequencing; vessel-type for the other versions).

d

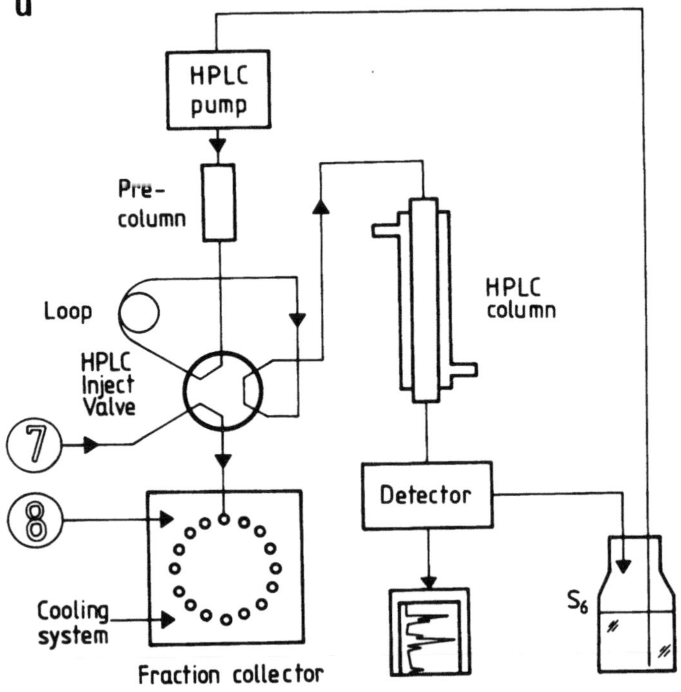

Fig. 1d: The connection to the on-line-HPLC system within the sequencer. The circled numbers show the connecting lines between: the delivery valve system and the reaction chambers (no. 1); from the reaction chamber towards waste 1/collect 1 (no. 2) and to the conversion flask (no. 3); from the conversion device to the collect 2 valve (no. 4), towards the HPLC-inject valve and the fraction collector (no. 7); the inlet from the delivery valve to the conversion flask (no. 6); and the vacuum/waste 2 outlet from the conversion flask (no. 5). The nitrogen inlet and vacuum outlet of the fraction collector is labelled by no. 8.
The configuration of the HPLC-injection valve as shown in Fig. 1d is that of the normal situation (no injection; HPLC-function 30 in "off"-position) and connects outlet no. 7 with that of the loop when function 30 is operated.
The syringes indicated in Fig. 1a employed usually in solid-phase sequencing were omitted.

HPLC-one-line equipment

The following equipment was used for the connection of the liquid-phase sequencer to the on-line HPLC-system: a Rheodyne automatic injection valve (governed by compressed air at 7 atm); a HPLC-pump, model 300/ B (Gynkothek, Munich); a fixed wave length detector at 254 and 313 nm (Waters, model 440) and a dual channel recorder (Kipp & Zonen). In case of the solid/gas-phase machine the HPLC-pump, model 64, and the fixed wave length detector at 254 nm were purchased from Knauer (Berlin).

Isocratic HPLC-separation

The separation of the PTH-amino acid derivatives is based on the isocratic solvent system for the identification of PTH-amino acids which was made separately from any sequencer degradation (7). The columns used for the on-line separations are: a) a column (250 x 4 mm, steel column purchased from Knauer), self-packed with Hypersil MOS, 5 µm (Shandon); and b) a LiChrospher 100-CH-8/II, 5 µm (250 x 4 mm), purchased from Merck). For the on-line separation the ternary solvent mixture proposed (7) was slightly changed to acetonitrile/water/dichloroethane/1 M sodium acetate, pH 5.3 (33:67:0.7:0.9, by vol). The water/acetate mixture (after the addition of 4 mg sodium azide in 10 µl of water) is degassed with stirring (water pump); then the acetonitrile/dichloroethane mixture is added and sonicated for 20 min. One liter of this solvent mixture is recycled for 7-10 days. After this period the column is rinsed with methanol/water (90:10, by vol) for 2 h and then reequilibrated with the above solvent mixture, freshly prepared. In order to protect the chromatography column a precolumn (filled with Hypersil MOS, 5 µm support, 50 x 4 mm) is installed between the outlet of the HPLC-pump and the injection valve. This precolumn traps solid material released from the sealings of the pump. It has to be replaced if the back pressure of the system becomes too high (after about a 6 - 8 weeks running time). The life time of the self-packed columns was about 9 months, and that of the LiChrospher column about three months under running conditions in the sequencers.

The flow rate in all experiments was 0.4 ml/min, the back pressure 36-40 bar at 57-62°C (maintained by thermostatted water which flows through an aluminium heating block of the column, built in the institute's workshop). The recorder speed was set at 2 mm/min. For degradations of 1.0 to 5.0 nmol protein the HPLC-detector of the liquid-phase machine was set at 0.02 AUFS if 1/4 of total PTH-amino acid amounts released from the sequencer were injected into the HPLC-column (in 100 µl solvent S4, 30% MeOH in water, from a total of 400 µl). With protein amounts of 5.0 to 10 nmol the range was set at 0.05 AUFS if the same aliquot was used for the direct identification.

Degradation cycles of 200-1000 pmole polypeptide obtained from the solid/gas-phase machine were identified by injecting 20 µl portions from a total of 100-140 µl S4, 20% methanol, the range set at 0.02-0.005 AUFS.

Liquid-phase sequencer program including on-line injection

In Table 1 the full sequencer program is listed which includes the main degradation program, that for the conversion device, for the on-line HPLC-injection, and that of the fraction collector.

Main Program

The program for the reaction in the cup and in the conversion flask, although written in a modern form provided by the

Table 1: Sequencer program for the degradation of proteins in the Berlin liquid phase sequencer with on-line detection

The program consists of STEPS, BLOCKS and the RANGE section (8). The STEPS contain the time of execution and the functions to be operated in this step. STEPS can be grouped to BLOCKS; in the RANGE section individual BLOCKS and/or STEPS are combined in the sequence of their execution during the degradation. Individual CYCLES can be created by changing the RANGE composition (examples see below). STEP times and functions can be changed upon execution of the program or before the run.

The program listed below was used for the degradation in the Beckman cup (i.d. 30 mm, straight version) and a smaller sized cup (i.d. 20 mm, see ref. 8); the speed was set at 3100 (large cup) and 2700 r.p.m. (small cup) throughout all steps. Applied amounts: 2.5 - 10 nmol of protein, without the addition of carrier.

Reagent and solvent functions (numbers in brackets)

Cup compartment:
R1: deliver (1), vent/press (31); 5% PITC in heptane
R2: deliver (2), vent/press (32); 0.2 M quadrol buffer
R3: deliver (3), vent/press (33); HFB
S1: vent/press/deliver (5); ethylacetate/benzene (75:175); 4.5 ml
S2: not used
S3: vent/press/deliver (6); benzene; 1 ml

Conversion device:
R4: vent/press/deliver (4); 20% TFA/water
S4: vent/press/deliver (7); 30% MeOH/water
S5: vent/press/deliver (8); MeOH

Nitrogen, waste, collect and vacuum supplies

N1 (9): nitrogen pressurizing cup compartment
N2 (10): nitrogen emptying delivery line, cup compartment
V1 (11): automatic change of nitogen and coarse vacuum (pulse of nitrogen for 1 sec/20 sec vacuum)
V2 (12): coarse vacuum (via restriction of the casing, see ref. 10)
V3 (13): fine vacuum
W1 (14): waste 1, exit cup compartment to waste
C1 (15): collect 1, exit cup compartment to conversion flask
N3 (21): nitrogen emptying delivery line of conversion flask
N4 (22): nitrogen through capillary into conversion flask
W2 (23): waste 2, exit conversion flask to waste
V4 (24): vacuum conversion flask
C2 (25): collect 2, exit conversion through HPLC-valve to fraction collector
W3 (26): waste 3, exit conversion flask via capillary to waste
V5 (27): vacuum fraction collector
N5 (28): nitrogen pressurizing fraction collector
St (29): step fraction collector
HP (30): operating HPLC-inject valve

Table 1.1

BLOCK	STEP	TIME (sec)	R1	R2	R3	Press	S1	S3	N1	N2	W1	C1	V1	V2	V3	R4	S4	S5	N3	N4	W2	V4	C2	W3	V5	N5	St	HP
			1	2	3	31-33	5	6	9	10	14	15	11	12	13	4	7	8	21	22	23	24	25	26	27	28	29	30
			CUP COMPARTMENT													CONVERSION FLASK									FRA.COL.			HPLC
R2	R21	4				32			9	10	14								21		23				27			
	R22	12				32			9		14								21		23				27			
	R23	4				32			9		14								21		23				27			
	R24	7		2		32			9		14								21		23				27			
	R25	4							9		14								21		23				27			
	R26	20							9	10	14								21			24			27			
	R27	10								10	14								21			24			27			
	R28	4							9		14								21			24			27			
R2V	R2V1	80								10				12					21			24			27			
	R2V2	120											11						21			24			27			
	R2V3	10								10										22		24			27			
	R2V4	80												12								24			27			
	R2V5	60							9										21							28		
R4	R41	4							9		14								21							28		
	R42	4							9		14										23					28		
	R43	19							9		14					4					23					28		
	R44	4							9		14										23					28		
	R45	32								10	14								21		23					28	29	
COU1	R1C1	4							9		14									22	23				27			
	R1C2	12				31			9		14									22	23				27			
	R1C3	20	1			31			9		14									22	23				27			
	R1C4	4							9		14									22	23				27			
	R1C5	20							9	10	14									22	23				27			
	R1C6	4								10	14									22	23				27			
	R2C1	4							9		14									22	23				27			
	R2C2	12				32			9		14									22	23				27			
	R2C3	8		2		32			9		14									22	23				27			
	R2C4	4							9		14									22	23				27			
	R2C5	20							9	10	14									22	23				27			
	R2C6	10								10	14									22	23				27			
	R2C7	4								10	14									22	23				27			
	CP1	700								10										22	23				27			
	C1V1	80								10				12						22	23				27			
	C1V2	120											11							22	23				27			
	C1V3	10								10										22	23				27			
	C1V4	80												12						22	23				27			
	C1V5	10							9											22	23				27			
	C1V6	150												12	13					22	23				27			
	C1V7	60							9											22	23				27			
	C1V8	4							9		14									22	23				27			
COU2	R1CW1	4							9		14									22	23				27			
	R1CW2	12				31			9		14									22	23				27			
	R1CW3	4				31			9		14									22	23				27			
	R1CW4	18	1			31			9		14									22	23				27			
	R1CW5	4							9		14									22	23				27			
	R1CW6	20							9	10	14									22	23				27			
	R1CW7	10								10	14									22	23				27			
	R2CW1	12				32				10	14									22	23				27			
	R2CW2	4				32			9		14									22	23				27			
	R2CW3	9		2		32			9		14									22	23				27			
	R2CW4	4							9		14									22	23				27			
	R2CW5	20							9	10	14									22	23				27			
	R2CW6	20								10	14									22	23				27			
	CP2	1000								10										22	23				27			
	C2V1	100								10				12					21			24			27			
	C2V2	120											11									24			27			
	C2V3	10								10												24			27			
	C2V4	120												12								24			27			
	C2V5	10							9													24				28		
	C2V6	200												12	13							24				28		
	C2V7	60							9											22								
	C2V8	20							9		14								21				25					
S4	S41	4							9		14								21		23					28		
	S42	18							9	10	14						7				23					28		
	S43	4							9	10	14										23							
	S44	25							9	10	14								21		23							
	S45	25							9	10	14									22	23							
	S46	30							9	10	14																	
	S47	16							9	10	14								21				25					30

Table 1.2

BLOCK	STEP	TIME (sec)	R1	R2	R3	Press	S1	S3	N1	N2	W1	C1	V1	V2	V3	R4	S4	S5	N3	N4	W2	V4	C2	W3	V5	N5	St	HP
			1	2	3	31-33	5	6	9	10	14	15	11	12	13	4	7	8	21	22	23	24	25	26	27	28	29	30
			CUP COMPARTMENT													CONVERSION FLASK									FRA.COL.			HPLC
S1	S11	4							9		14								21				25					
	S12	3					5		9		14								21				25					
	S13	4							9		14								21				25					
	S14	40							9	10	14								21				25					
	S15	4							9		14								21				25					
	S16	450					5		9		14								21				25					
	S17	40							9		14								21				25		27			
	S18	300							9	10	14											24			27			
S5	S51	50								10				12					21						27			
	S52	10							9										21		23					28		
	S53	15								10								8			23					28		
	S54	4								10											23					28		
	S55	30							9										21		23					28		
	S56	60							9		14								21				25					
	S57	20							9		14								21				25		27			
CYC1	R31	20				33			9		14								21						27			
	R32	4			3	33			9		14								21		23				27			
	R33	4							9		14								21		23				27			
	R34	25							9	10	14							8			23				27			
	R35	4								10	14										23				27			
	R36	50								10	14								21		23				27			
	CL1	700								10									21					26	27			
	CL2	20							9		14											24			27			
	CL3	20								10	14											24			27			
	CLV1	60								10				12								24			27			
	CLV2	160											11									24			27			
	CLV3	10								10												24			27			
	CLV4	80												12								24			27			
	CLV5	5								10												24			27			
	CLV6	200												12	13							24			27			
	CLV7	60							9											22					27			
	CLV8	10							9		14									22	23				27			
S3	S31	12							9			15								22	23				27			
	S32	65						6	9			15									23				27			
	S33	20							9			15									23				27			
	S34	100							9	10		15									23				27			
	S35	100							9	10		15								22	23				27			
	S3V1	60								10				12						22	23				27			
	S3V2	200											11							22	23				27			
	S3V3	60							9											22	23				27			
	S3V4	4							9		14									22	23				27			
CYC2	R3W1	12				33			9		14									22	23				27			
	R3W2	3			3	33			9		14									22	23				27			
	R3W3	4							9		14									22	23				27			
	R3W4	20							9	10	14									22	23				27			
	R3W5	50								10	14									22	23				27			
	CLW1	500								10										22	23				27			
	CLW2	20							9		14									22	23				27			
	CLW3	20								10	14									22	23				27			
	CLWV1	60								10				12						22	23				27			
	CLWV2	160											11							22	23				27			
	CLWV3	10								10									21			24			27			
	CLWV4	80												12					21			24			27			
	CLWV5	5								10									21			24			27			
	CLWV6	200												12	13							24			27			
	CLWV7	60							9										21						27			
	CLWV8	10							9		14								21		23				27			
S3W	S3W1	12							9			15								22	23				27			
	S3W2	60						6	9			15									23				27			
	S3W3	20							9			15									23				27			
	S3W4	100							9	10		15									23				27			
	S3W5	100							9	10		15								22	23				27			
	S3WV1	60								10				12						22	23				27			
	S3WV2	200											11							22	23				27			
	S3WV3	60							9											22	23				27			
	S3WV4	4							9		14									22	23				27			

Table 1.3

RANGE DECLARATION

PROGRAM A	PROGRAM B	PROGRAM C	PROGRAM D
COUPLING: 2x	COUPLING: 1x	COUPLING: 2x	COUPLING: 1x
CLEAVAGE: 2x	CLEAVAGE: 2x	CLEAVAGE: 1x	CLEAVAGE: 1x
R2	R2	R2	R2
R2V	R2V	R2V	R2V
R4	R4	R4	R4
COU1	COU2*	COU1	COU2*
COU2	S4	COU2	S4
S4	S1	S4	S1
S1	S5	S1	S5
S5	CYC1	S5	CYC1
CYC1	S3	CYC1	S3
S3	CYC2	S3	END CYCLE
CYC2	S3W	END CYCLE	
S3W	END CYCLE		
END CYCLE			

COU2* contains 1800 sec in STEP CP2

CYCLE DECLARATION

EXAMPLE: Ribonuclease A (bovine)

SEQUENCE:

```
1                 10                  20                  30
K E T A A A K F E R Q H M D S S T S A A S S S N Y C N Q M M
```

CYCLE 1	:	PROGRAM A	CYCLE 15-18	:	PROGRAM D
CYCLE 2	:	PROGRAM C	CYCLE 19-20	:	PROGRAM A
CYCLE 3	:	PROGRAM D	CYCLE 21-23	:	PROGRAM D
CYCLE 4- 8	:	PROGRAM A	CYCLE 24	:	PROGRAM B
CYCLE 9	:	PROGRAM C	CYCLE 25	:	PROGRAM A
CYCLE 10-11	:	PROGRAM B	CYCLE 26	:	PROGRAM D
CYCLE 12-13	:	PROGRAM A	CYCLE 27	:	PROGRAM B
CYCLE 14	:	PROGRAM C	CYCLE 28-30	:	PROGRAM A

microprocessor that gives much flexibility as detailed (8), is similar in its main events to that described earlier (12). It contains two coupling stages with PITC to guarantee completion of the reaction at basic residues as detailed (13), the appropriate drying steps, and a single wash with solvent S1 (ethylacetate/benzene, 75:175, by vol) thereafter. The program further consists of two cleavages with HFB to avoid carry-over at residues with bulky side chains as discussed (10,14), each of which is followed by the thiazolinone extraction with solvent S3 (absolute benzene), both extracts being collected in the conversion flask (to increase the yields of PTH-Asn and -Gln, see ref. 2,13).

The conversion is done with R4, 20% TFA in water (2,13), followed by a short drying stage. The PTH-amino acids are dissolved in 400 µl (liquid phase sequencer) or 100-140 µl (solid phase machine) of solvent S4 (used for the HPLC-injection, see below) and the flask is washed further twice with 800 (liquid phase) and 300 µl (solid-phase), respectively, of solvent S5 (methanol). The first extract of S5 is delivered into the fraction collector and the second discarded to waste (W3, see Table 1).

On-line injection

The connection of the conversion program with that of the HPLC-process is outlined in the scheme, Fig. 1. The collect line of the flask (outlet C2, function 25, see Table 1) is connected with the HPLC-inject valve and the fraction collector (connecting line no. 7 in scheme 1a and 1d). Therefore, the solvent S4 transports the PTH-amino acid from the conversion flask towards the fraction collector via the inject valve through line C2.

At all normal stages with no injection into the HPLC-column

the HPLC-solvent mixture pumped from the HPLC-pump is directed via the loop of the automatic injection valve towards the HPLC-column and towards the detector and flows back to the HPLC-solvent reservoir (injection function 30 in "off"-position, as drawn in the diagram, Fig. 1d). But after the completion of the conversion and the dissolution of the PTH-amino acid the automatic injection valve is operated (function 30 switched to "on"-position) opening the collect C2 line via the loop of the inject valve towards the fraction collector, hereby filling the loop (20 to 100 µl depending on the sequencer type) with the PTH-solution. In the moment when the loop is filled function 30 switches back to "off"-position, turning back the injection valve, and the aliquot placed in the loop is injected into the HPLC-column. The remainder of the PTH-amino acid solution is then purged into the fraction collector with nitrogen N3 (function 21, delivered from the conversion flask through line C2 via the injection valve, but not via the loop, towards the fraction collector).

In order to ensure that the loop is full with the PTH-amino acid solution at the moment of injection and to guarantee an exactly reproducible filling time all the connecting lines (e.g., line C2 and its valve) must be clear and completely free of liquid droplets before that event. This is done shortly before the dosage of the S4 solvent into the conversion flask takes place by blowing nitrogen N4 (via capillary, function 22) through the collect C2 line towards the fraction collector for about 100 sec. This guarantees fully reproducible filling times for the loop. Further, during the loop filling stage it is important that the PTH-amino acid solution is bubble free. This was achieved by using a 30 sec delay (all nitrogen and waste functions of the conversion flask switched to "off") after S4 is delivered into the conversion flask. By this means constant injection conditions were arrived at as has been tested with reference PTH-amino acid mixtures. This is further proven by the constant and low background obtained

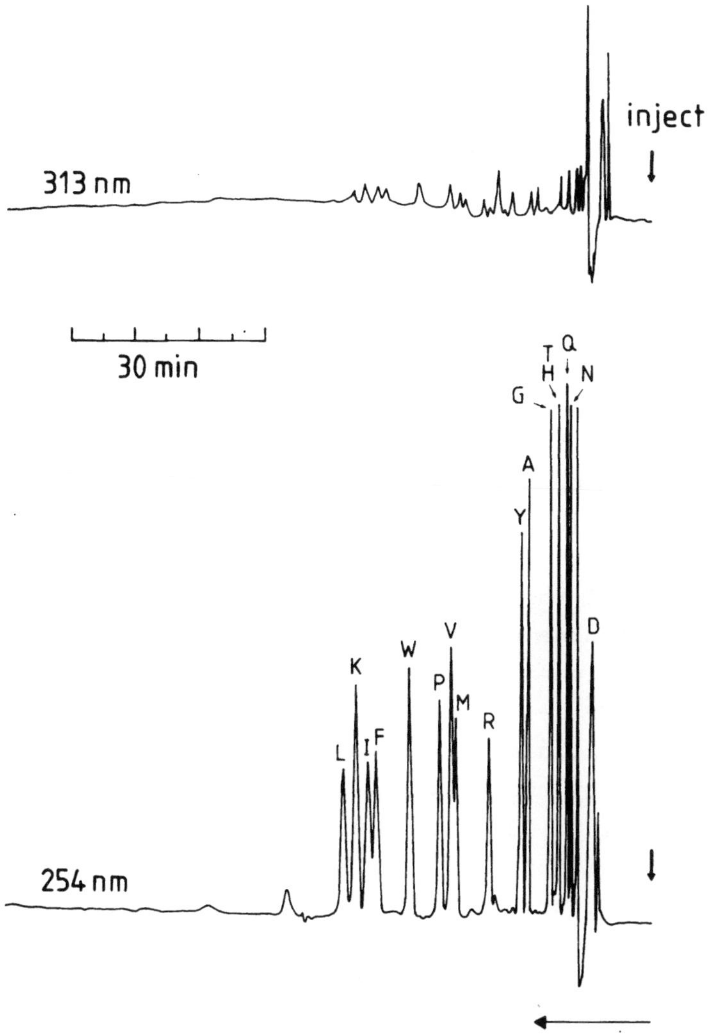

Figure 2: PTH-amino Acid Identification

Isocratic separation of standard PTH-amino acid derivatives by automatic on-line injection of the sequencer into the reversed phase HPLC-column, LiChrospher 100-CH-8/II, 5 µm, 250 x 4 mm; flow rate 0.4 ml/min; temperature 61°C; isocratic solvent system: acetonitrile, water, 1,2-dichloroethane, 1 M Na acetate, pH 5.3 (33:67:0.7:0.9, by vol); for more details, see Methods.

with all on-line detecting sequencer runs performed with proteins of known or unknown sequence.

Results with on-line detection

PTH-amino acid separation

In Figure 2 the on-line separation of standard PTH-amino acid derivatives under isocratic conditions is shown. The highest resolution was achieved with the LiChrospher-column (see Methods) in spite of the relatively high injection volume applied: all PTH-amino acids were separated except the pairs PTH-Glu/Asp, PTH-Gln/Ser and PTH-His/Thr. The latter two pairs have not to be resolved as the sequencer releases PTH-Ser and PTH-Thr in form of their dehydrated compounds, detectable at 313 nm. The threonine derivative further forms an additional peak at 254 nm which migrates beyond the last hydrophobic derivative on the HPLC-trace. Thus, the only poorly resolved pair remains the acidic derivatives, PTH-Asp and -Glu. These can be separated by changing the injection medium, solvent S4, from 30% methanol to 10-15%, or by decreasing the concentration of acetonitrile in the HPLC solvent mixture. However, then the peaks of the hydrophobic derivatives, mainly that of PTH-Phe/Ile/Lys/Leu/Norleu are broadened too much. Therefore, the conditions as chosen in Fig. 2 give the best result in the on-line mode. Further possibilities to resolve the acidic pairs are to employ the conversion in methanolic acid.

The main reasons for the good separation of the PTH-amino acid derivatives are the slow flow rate (0.4 ml/min) plus the water/methanol injection solvent mixture which concentrates the dilute sample on top of the column; e.g., sample volumes of up to 100 µl can be resolved. This would not be possible if only methanol or acetonitrile were used for the injection.

We further tested the possibility of using acetonitrile/water mixtures as the injection solvent. This would reduce the injection peak, which occurs directly after PTH-Asp/Glu on the HPLC-trace. However, it was noticed that the fractions stored in the fraction collector in this mixture plus one wash of solvent S5 dried very slowly under the vacuum applied to the fraction collector. As a result the remaining samples were partially destroyed during storage. For picomole quantities of most of the derivatives we observed drastic decreases in the yields when stored for 24 h in this mixture. On the other hand, storage in 30% methanol/water (first wash) plus methanol (second wash) does not seriously affect the yields of most of the amino acid derivatives besides the very labile ones. We also tested many degradation cycles by reinjecting the stored derivatives after several days and it was possible to identify them.

Compared with the support Hypersil MOS with which empty steel-columns were filled the LiChrospher support gives a much higher resolution. The self-packed column poorly resolved PTH-Met/Val, PTH-Phe/Ile and PTH-Lys/Leu. Further, PTH-Arg which can be recovered from the LiChrospher material in form of a rather sharp peak not eluted from the Hypersil material under the same chromatographic conditions.

Degradation results

A typical sequencer result from a protein of unknown sequence degraded with PITC using the on-line identification system is presented in Fig. 3. The protein, a DNA-binding protein with a molecular mass of approximately 16 000, was provided by Dr. C. Gualerzi. It was degraded in the Berlin liquid-phase sequencer, and 100 µl samples, i.e. one fourth of the total amounts released, were injected on-line in 30% methanol into the HPLC-column applying the program presented in Table 1.3/A.

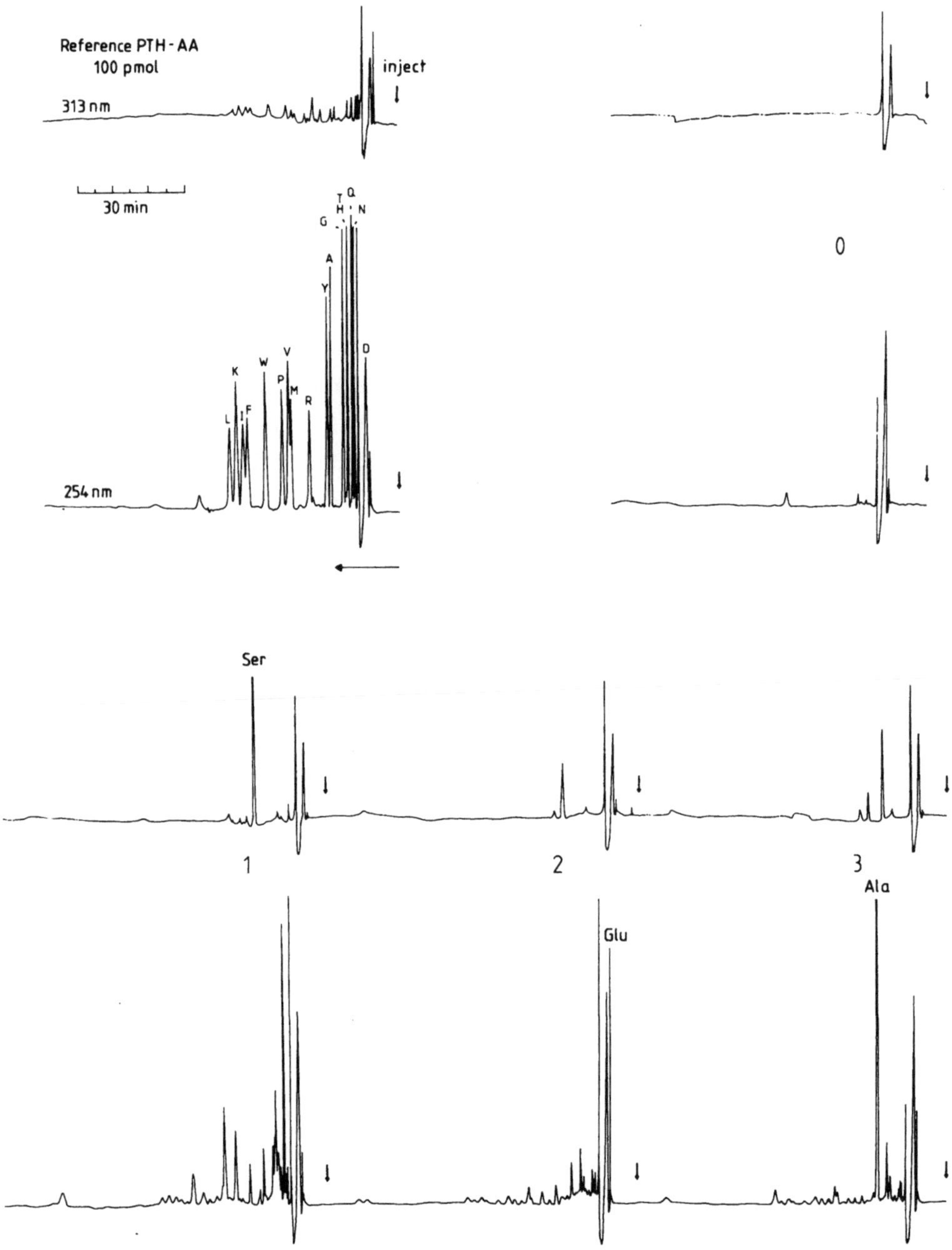
Reference PTH-AA
100 pmol
inject
313 nm
30 min
L
K
I
F
W
P
V
M
R
Y
A
G
T
H
Q
N
D
254 nm
0
Ser
1
2
3
Glu
Ala

Fig. 3.1

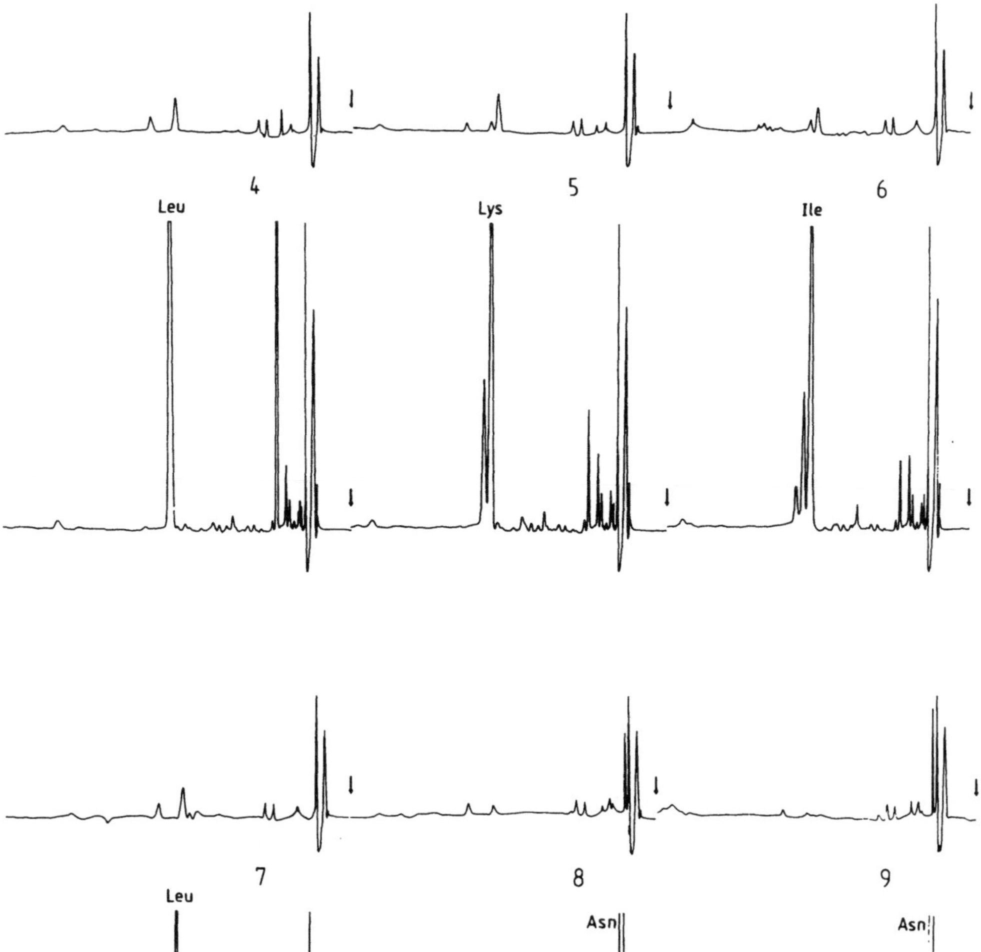

Fig. 3.2

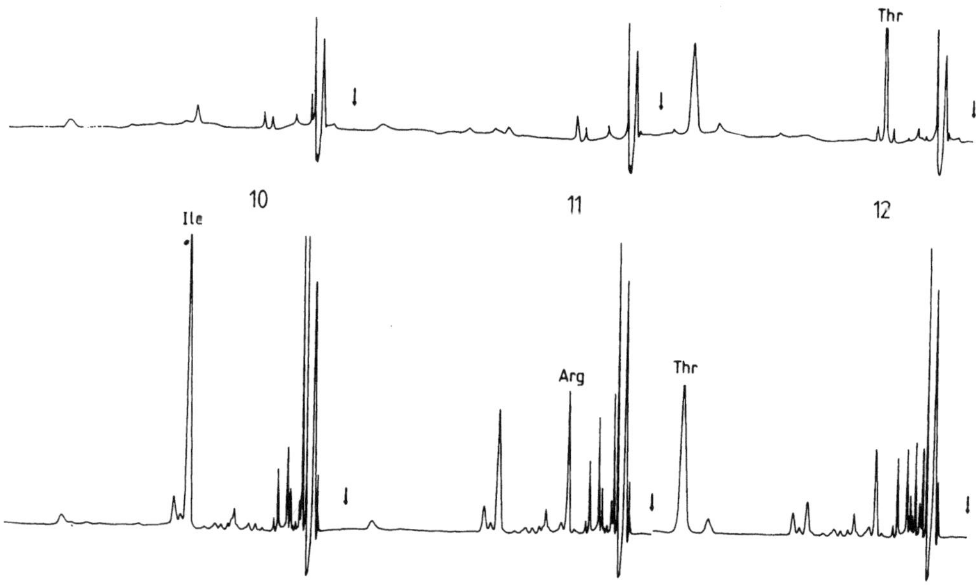

Figure 3: Sequencer Run of a DNA-binding Protein

The N-terminal sequence up to position 50 was derived by liquid-phase sequencing with on-line detection. 5 nmol protein (M_r 16 000) were degraded and one fourth injected in 100 µl 30% MeOH; sensitivity range 0.05 AUFS (at 254 nm) and 0.02 AUFS (313 nm). The figures show from left to right: the HPLC-trace of the standard PTH-amino acids (100 picomole each); the first (wash) cycle; and degradation cycle 1 to 12, Ser-Glu-Ala-Leu-Lys-Ile-Leu-Asn-Asn-Ile-Arg-Thr- (the main peaks are out of scale).

In case of this protein differentiation of all Glu- and Asp-steps was made by reinjecting aliquots of the remaining samples which were dried in the fraction collector tubes into a separate isocratic HPLC system, with a modified solvent system (acetonitrile/water/dichloroethane/1 M sodium acetate, pH 5.3 (31:69:0.7:0.9, by vol). All other steps were identified on-line; the easy identification of all hydrophobic residues and of arginines, e.g. Ile-6, Leu-7, Arg-11, and of all Ser- and Thr-steps, is remarkable.

Conclusions

In summarizing the described on-line identification system for the installation into sequencers it can be stated:

1. The on-line HPLC-system is inexpensive, easy to use and stable.
2. The sensitivity is increased compared to any detection systems used separately from the sequencer. This holds true for both, the relatively inexpensive isocratic or any of the sophisticated gradient systems employed, as the amino acid derivatives are directly injected into the HPLC-column shortly after their release from the sequencer. Therefore, losses are decreased and labile amino acids undergo less destructions.
3. Depending on the support employed almost all PTH-amino acid derivatives can be separated in one run of 45 min including the groups of PTH-Val/Met/Pro and α-PTH,-ε-PTC-Lys/PTH-Phe/Ile/Leu/Norleu. PTH-His migrates before PTH-Gly, PTH-Arg well separated between PTH-Tyr and PTH-Met, and PTH-Trp between PTH-Pro and PTH-Phe. 45 min suffices for the separation of all derivatives except for the second peak of threonine which elutes shortly before the new injection.

4. Due to the low flow rate used and the water containing injection solvent the sensitivity is very high. 10 pmol and less can be detected. The quantitation of all residues is easy and can be combined with an integrating system if desired.

5. The detection of modified amino acid residues is made easier, since within one run modified and unmodified residues of the same amino acid can be easily compared.

6. The low and constant background contaminations are identical from one degradation cycle to the next. This constancy is improved over any manual identification made after the degradation.

7. Due to the isocratic separation chosen the base line keeps constant for long periods of time and has not to be corrected daily. For highest sensitivity ranges of the detectors (less than 0.02 AUFS) the base line shift has to be readjusted daily. Here a detector with the possibility for an automatic reset function is useful, e.g. the detector purchased from Knauer, the reset of which can be governed by the microprocessor and operates once within one degradation cycle.

References

1. Edman, P., Begg, G.: Eur. J. Biochem. 1, 80-91 (1967).
2. Wittmann-Liebold, B., Graffunder, H., Kohls, H.: Analyt. Biochem. 75, 621-633 (1976)
3. Birr, C.: In: Solid Phase Methods in Protein Sequence Analysis, Proc. First Internat. Conf., Pierce, Rockford, Ill. (1975), pp. 115-129.
4. Bridgen, J.: In: Methods Enzymol. 47, C.H.W. Hirs, Timasheff, S.N., eds. (1977), pp. 385-391.
5. Machleidt, W., Hofer, H.: In: Methods in Peptide and Protein Sequence Analysis, C. Birr, ed., Elsevier, North Holland, Amsterdam (1980), pp. 35-47.

6. Hewick, R.M., Hunkapiller, M.W., Hood, L.E., Dreyer, W.J.: J. Biol. Chem. 15, 7990-8005 (1981).

7. Lottspeich, F.: Hoppe-Seyler's Z. Physiol. Chem. 361, 1829-1834 (1980).

8. Wittmann-Liebold, B.: In: Modern Methods in Protein Chemistry, H. Tschesche, ed., de Gruyter, Berlin, New York, pp. 229-266 (1983).

9. Machleidt, W.: In: Modern Methods in Protein Chemistry, H. Tschesche, ed., de Gruyter, Berlin, New York, pp. 267-302 (1983).

10. Wittmann-Liebold, B.: In: Polypeptide Hormones, R.F. Beers, Jr., E.G. Basselt, eds., Twelfth Miles Internat. Symp., Raven Press, pp. 87-120 (1980).

11. Wittmann-Liebold, B., paper submitted.

12. Wittmann-Liebold, B.: In: Chemical Synthesis and Sequencing of Peptides and Proteins, T.Y. Liu, A.N. Schechter, R.L. Heinrikson and P.G. Condliffe, eds., Elsevier North Holland, Inc., Amsterdam, New York, pp. 76-110 (1981)

13. Wittmann-Liebold, B., Geissler, A.-W., Marzinzig, E.: J. Supramolec. Struct. 3, 426-447 (1975).

14. Brandt, W.F., Henschen, A., von Holt, C.: Hoppe-Seyler's Z. Physiol. Chem. 361, 943-952 (1980).

PROGRAMMABLE ELECTRONIC CONTROL OF SEQUENTIAL PROCEDURES OF ANALYTICAL DEVICES

Use of Commercially Available Low Price Home Computer for a Liquid Phase Sequencer

Joachim Friedrich, Heiko Thieme

Max-Planck-Institut für experimentelle Medizin
D-3400 Göttingen

1. Introduction

In order to update a Beckman Liquid Phase Sequencer 890 B the question of electronic control was considered. This should replace the punch card system totally and facilitate handling the sequencer. Instead of installing a commercially available controller, it was decided to design a new system.
Such a system should become as versatile as possible i.e. it should also be suitable for connection to devices other than the sequencer and it should take over as many functions as possible for automatization. Of course, these functions should all be fully reliable.
Nevertheless the time learning to handle the system should be as small as possible and the price should be very low.
With the appearance of low cost microcomputers, it seemed to be possible to fullfill all these criteria above.
The current result is a controller which contains the home computer Commodore C= 64, one monitor, one tape and two self built interfaces as the hardware components (Figure 1). The operating system, the software part, on which user written programs will then be based, is itself written in machine language.
The reliability has been tested. The versatility has been expanding since installation. The use in routine is very easy. Programming is also

Modern Methods in Protein Chemistry, Volume 2

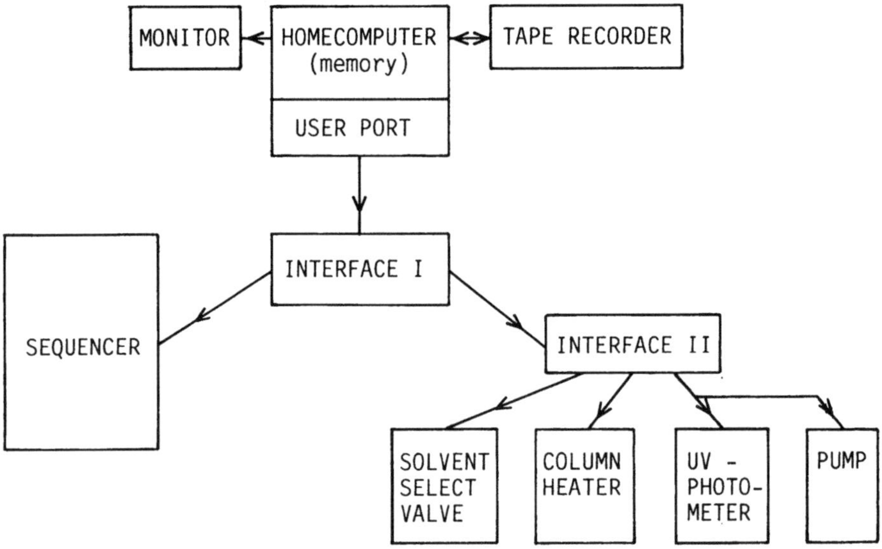

Fig. 1: The current version of the this laboratory control system. It consists of two main parts: the sequencer itself and devices of the HPLC-PTH-AA-detection system connected by interface II.

rather easy, but still requires the manual. The price is low regarding home computer and interfaces. However, the creation of the operating system, being written in machine language, was time consuming and so would be costly for single use.

The controller (and its earlier version) has been working routinely and quite satisfactorily since February 1983.

This communication describes the principles of designing, realization and use of that controller. This includes programming and aspects of further developments. The use of the controller in the sequencer laboratory is illustrated.

2. Designing of the Controller

The meaning of sequential procedure - as in the case of the Liquid Phase

Sequencer - is a step by step procedure. It does not consider gradient procedures and data processing so far. Therefore, the control method is designed in the following way:

A group of steps are combined in one program or, optionally, to subroutines. This group, or program, is repeated cycle-wise. Such a procedure is called a run.
All cycles can be assigned to the same program but it is possible for every cycle to have its own program (for double coupling, proline-specific cleavage etc.).
In every step relays are activated: one or two combinations of relays are closed e.g. to open a specific set of valves. This may be applied to the Liquid Phase Sequencer:

R1 WASTE, R1 BOTTLE PRESSURIZE, NITROGEN TO CELL and WASTE OPEN .

At the same time one or more texts are printed on the screen. Combinations and texts are valid for a step-specified time until the next step or cycle is begun up to the end of the last cycle.
This must have

a set of cycles, each of which is assigned to one program.
a set of one or more programs of steps (optionally including subroutines).
a set of combinations of relays to be closed.
a set of texts to be printed on the screen and to be used during programming.

Since realization is based on a (6502-/6510-) machine language operating system and since machine language allows high speed and a great variability, management

of daily events and of events at any date and time

is also implemented (1).

3. Realization

It is necessary to have one home computer connected to an interface e.g.

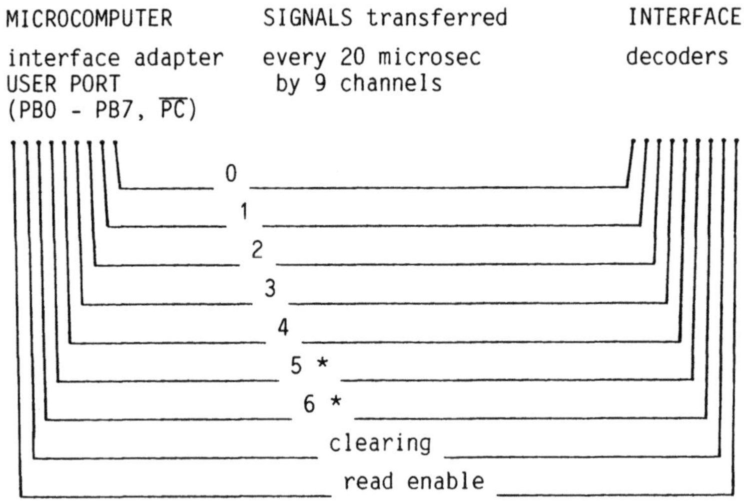

Fig. 2: Signal transfer. Seven of nine channels are used in the current version. Channels 0 - 4 represent 2 to the 5 = 32 possible relays. *: 5 and 6 are not used. Read enable means a delay of 1 microsec to ensure, that any code (of channels 0 - 6) is decoded (to one relay) only after all channels have their clearly defined signal states (either ground or 5 V(+)). Clearing is a function applying to all relays.

by the user port for serial transfer of signals (Figure 2), and the interface for decoding and setting the appropriate relays, which are then connected to the device (Liquid Phase Sequencer). Communication between controller and user is effected via the home computer's keyboard and a (colour) monitor. A tape recorder (or disk) is used for loading and installing the operating system into the C= 64 and also for saving and loading user written programs.

The operating system mediates the wishes of the user to the controller

to activate the devices, especially the sequencer,
to make and store a set of combinations and
to make and store a set of programs.

It enables the controller, to receive the orders, to show the results of

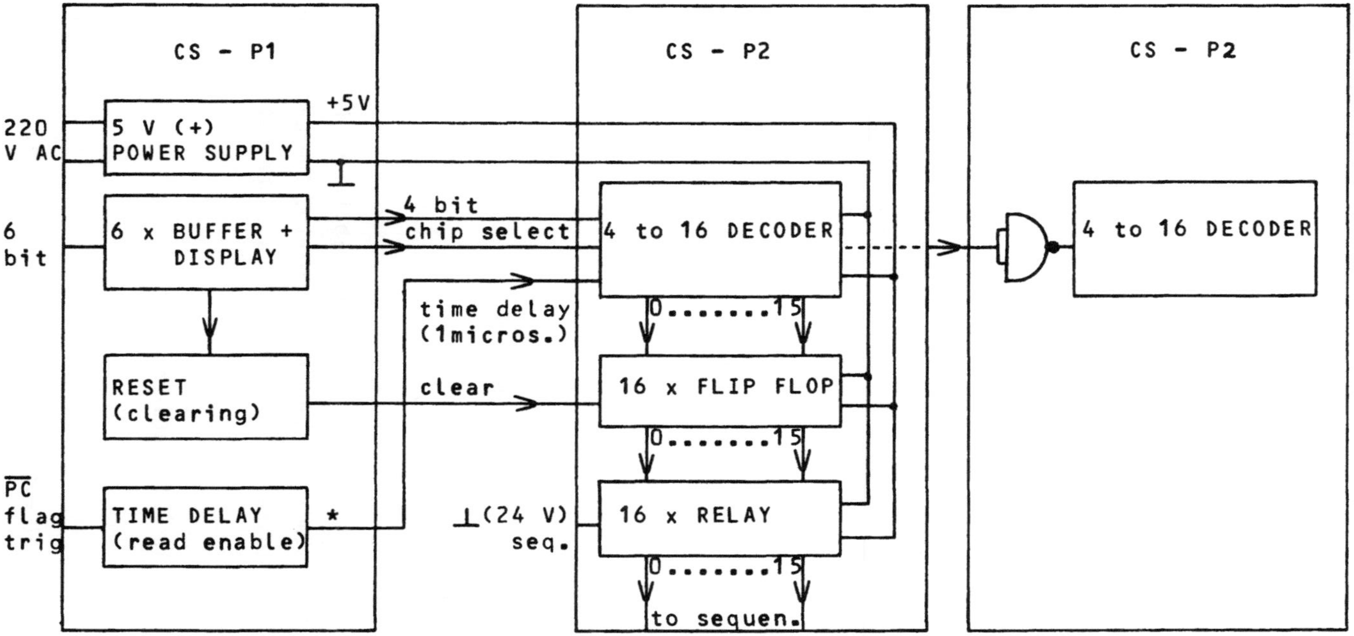

Fig. 3. Scheme of Interface I. For further details see Figure 6. *: The "time delay" block was necessary in a first version of the controller; it is not needed with the Commodore C= 64.

programming and assigning and to show the actual state of a run, the date and the time of day.
Unless there are misfunctions, the user is not aware that it is itself program. It comprises a storage of about 15 kbytes, a quarter of the memory, which could be used on the C= 64. The memory space, which is reserved to contain all the data, comprises 25 kbytes.
The operating system allows decoding for 128 different relays but in the case of the system described here only 32 are used. The interface therefore contains two cards each with one decoder, 16 flip-flops and 16 relays and one card for power supply and the reset/clearing function (deactivation of all relays) (Figure 3).
A combination of relays will be set serially and the setting time per relay is about 20 microseconds. Including preceding clearing, the mean setting time of one or two combinations takes about 500 microseconds (Figure 2).
All created programs, combinations and texts are stored in the micro-computer itself for direct access. During handling no tape or disk is needed.

4. Use of the Controller

The operating system (together with the special programs for sequential procedures, with relays combinations and texts) is loaded from tape or disk in the beginning - theoretically it need never be loaded again - and is run.
A few orders are sufficient to manage all possibilities (Table 1). In routine procedure only two orders have to operate: the cycles must get to "know", what to do (Figure 4), and one cycle must be called to start. The orders are, e.g.:

"c:1 10 >Standard-Programm"

"c:1"

The first order means that every cycle no. 1 through no. 10 is defined as a "Standard-Programm"-cycle. The second order means that the run is

Tab. 1: List of Orders

u, d	further entries	set time (german: "Uhrzeit"), date
c:	"	assign and start cycles
x:	"	program combinations and texts (explaining texts and markers)
p:	"	program programs and subroutines
c, p		read assigned cycles, actual step
i-, i+	(")	delete, insert one step
u:	"	program time ("Uhrzeit-") program
d:	"	program date program
lp, lx		load programs, combinations
sp, sx		save programs, combinations

```
c:1  10 ❯Standard-Programm
---------------------------------------------
00 00: 00 00 ❯Trocknungs-Programm
01 10: 01 00 ❯Standard-Programm




 6. September 1984                 15.30.20 Uhr
```

Fig. 4: Assigning cycles. On the screen the cycles assigned appear grouped. In this case cycle no. 0 (from no. 0 to no. 0) was assigned to "Trocknungs-Programm" before cycles no. 1 through no. 10 were assigned to "Standard-Programm". Normally the last cycle of the last assignment terminates a run. Numbers after colons mean addresses of program storage.

started, beginning at cycle no. 1.
The run, now started, will be finished at the end of cycle no. 10 (the last of all defined). Thereafter, the system waits for the next run. During the sequential procedure it is possible to write, to change or to read programs, combinations or texts, and the procedure can be supervised directly by some special keys:

- to hold, to interrupt and to continue
- to go faster or slower (changing step time)
- to go to the next or to the preceding step

By means of the screen the user can be informed about cycle no., program name, subroutine name, step no. and remaining time of the step. A first and/or a second combination name - as explaining texts - are printed on the screen. By simply ordering

"p"

one gets all information (the numbers of all setted relays) about the actual step. Date and time are also shown on the screen (Figure 5).

The procedure for protein sequencing in our laboratory is as follows:

Polybrene(R), manually precycled five times in the presence of butylamine for de-activation, is dried in the cup of the Beckman Liquid Phase Sequencer and precycled once again. Consequently, one cycle of the "Trocknungs-Programm" and one cycle of the "Standard-Programm" are needed. The following orders are entered:

"c:0 ›Trocknungs-Programm"

"c:1 ›Standard-Programm"

The sample is then loaded into the cup for drying. If it is more than 0.5 ml, we use a program beginning with low speed. Again cycle no. 0 is assigned to a program for drying, cycle no. 1 through the desired maximum number are assigned to a standard program for coupling and cleavage.
This is our routine procedure unless double coupling, proline-specific

p	order
01 005 19 45 2 sec	s.below
PITC ins Cup (HIGH DRIVE) 000 001 016 018 + + 009	name and numbers of relays of first combination
Konversion (N2 durch Flask), 50'C 020 021 022	name and numbers of relays of second combination
Standard-Programm	
K O P P L U N G	
Cyc: 03 Prog: 01 Schr: 005 Zeit 002	
001 e+0 sec	actual remaining step time
6. September 1984 19.24.21 Uhr	date and time

Fig. 5: Picture of the screen during sequential procedures.In the second line the step's address of program storage, the first, the second combination no. and total step time appear.

cleavage etc. are required. Any cycle may be assigned to its own program, one after the other with automatic endmarking, or, if the maximum cycle number has been defined, cycle-wise. The order in the case of a proline-specific cycle could be:

"c:11 >PROLIN*", followed e.g. by:

"c:12 20 >Standard*" and

"c:21 >Lauf beendet" ("run finished").

The "*"'s serve here as an abbreviation sign.
The last cycle for coupling and cleavage is followed by one "cycle" of waiting with vacuum to the fraction collector. (Until now we have not used an autoconverter, for which the system is prepared.) Waiting is

stopped by a function key to go to the next and last step (nitrogen to the fraction collector), the samples of the derivatized aminoacids (extracted by chlorobutane containing hexylamine and ethanethiol as scavengers) are subjected to manual conversion and identification with autosampler and integrator by an isocratic system according to Lottspeich, adapted to self packed narrow bore columns for higher sensitivity (2,3). The use of the controller applied to the HPLC-system is described in more detail below.

5. Programming

Programming occurs in four main steps. Firstly, texts are created. They serve as markers for different programs or as markers for different subroutines.

Secondly, combinations are created. 128 relays, three lag functions (lag times of 1 sec, 1/10 sec and 1/100 sec) and one endmarker can be chosen. Additionally the combinations can be given an explaining text, which will be printed on the screen during the procedures, or used during further programming. (This feature resembles FORTH-programming: one is allowed to create one's own "orders" - i.e. programs, subroutines and combinations - with their own "names".)

Thirdly, if wished, subroutines are developed. Any location of the program/steps storage is available. 4096 steps altogether can be programmed. The first subroutine step should call a subroutine marker. Thereafter, one or two combinations, the time unit of the step (default value 1 sec, choices from 1 millisecond up to 1000 seconds are possible) and the sum of the units (0 through 999) are programmed. The minimum step time is zero/one millisecond, the maximum step time 999000 seconds (about 12 days). A subroutine is ended by a stop sign.

Fourthly, after all desired subroutines are prepared, the main program/s is/are created. Any location of the program/steps storage is also available. The first step of the main program should call a program marker. The following steps may solely call one subroutine after another, or they may call combinations, as subroutine steps do, having their step

times. A main program is also ended by a stop sign.
Among the steps of a main program, "null steps" - steps without any function - can be interspersed. Thus, a programmer can make blocks in a program. Furthermore, as "null steps" are just the stop sign of subroutines, these parts can also be called as subroutines if properly marked.
At the end of programming combinations, texts and all steps can quickly and easily be read and corrected.
Finally, the routines are begun by assigning any cycle to any main program and by starting with any cycle. One may start with the interruption possibility or with the interface turned off in order to monitor the run procedure before the real run is started. Even then, any programmed data can be changed.
As long as the power supply is on all data remain stored.

Programming is one of the great advantages of this controller in comparison to the punch card system. All programs (programs for drying the samples, the standard program, the program for double coupling and the program for proline-specific cleavage) are now written and stored. If subtle changes have to be made, they are done rapidly and without any problem, especially without adhesive tape or punching a new card. This is useful when a new program or an alternative procedure are to be tested.
Meanwhile all programs, except those for drying, are composed of subroutines. By this set of subroutines a new program may be written within one minute.

6. Control of Further Devices

Just as two combinations of relays can be used in parallel in sequential procedures e.g. one for coupling and cleavage and one for autoconversion (Figure 5), one time program for daily events and one date program can also operate simultaneously. For these programs any of the relays can be reserved.
This is used for HPLC-control with respect to buffer change, column oven, HPLC-pump and UV-photometer. The daily routine in the sequencer labor-

atory begins with automatic change of buffer and heating the column, before the assistant reaches the laboratory. Soon thereafter pump and photometer are automatically switched on. The further control is then done by the autosampler. The integrator itself reacts to the autosampler in a programmed manner. With the aid of a BASIC-program it records the chromatograms of samples and test standard, each in two wavelengths twice with different sensitivities. The chromatogram of one wavelength (313 nm for dehydro-serine and -threonine) is switched on for only a few minutes during an analysis.
At the end of the analyses the buffers are automatically changed back for purging the column. Thereafter column heater, HPLC-pump and photometer are switched off. This all may happen after the assistant has left the laboratory.

7. State and Price of Current Version

The system is connected to the Liquid Phase Sequencer by 19 lines via relays with the circuits completely separated; as is the case of the four connections to the buffer select valves, to the column oven and to HPLC-pump and UV-photometer. The autosampler and the integrator operate independently (Figure 1). The 19 lines to the Liquid Phase Sequencer are fused to the connections for the four cards of the sequencer, which were used (and can be used, furthermore, without limitation) for the punch card system (sequencer lines (J1, J2, J3) - (4, 6, 9, 10, 13, 14) and J4-9). The four lines to the HPLC, which control voltage values of 12 V DC and 220 V AC, activate relays of the second interface (Figures 6 and 7).

The prices are:	home computer Commodore C= 64	ca. 200 $ / 600 DM
	tape recorder	ca. 50 $ / 150 DM
	monitor	ca. 150 $ / 350 DM
	interface to the sequencer	ca. 500 $ / 1300 DM
	operating system: may be less than	700 $ / 2000 DM
	depending on number of installations	
	interface to HPLC system	ca. 80 $ / 200 DM

8. Future Developments

As exemplified in the sections above, a small laboratory control system has been created. Future developments could be based on the inbuilt possibilities of the Commodore C= 64:

1) User Port. Not only 128 but 256 different values could be presented on the user port. Two decoders are sufficient - one for group/card selection, one for function/relay selection. Some values could have special functions, e.g. for general clearing and clearing blockwise.

2) Paddle Connections. With these it is possible to use two inbuilt A/D converters for four different measurements with values in 256 steps. Temperature control, vacuum control, photometer input for simple data storage etc. in conjunction with running procedures could be realized.

3) Serial Port. The serial port contains the IEC-bus to floppy disk and printer. Recording, printing out of programs, combinations and texts, data storage on disk are possible in terms of multitasking procedures.

4) Graphics. Quite another picture or the results of chromatographs could be presented on the (colour) screen. Graphics could be used for self explaining guides by means of menus to overcome the need of a manual.

5) Data/Address Bus. Here a series of new possibilities could be opened for outputs (setting) as well as for inputs (receiving data). This could imply:

- control of over 1000 relays
- control of gradient systems or speed control of motors to be driven (reagent dosages)
- highly accurate photometer data input for integration and storage

and anything else one would like to have.

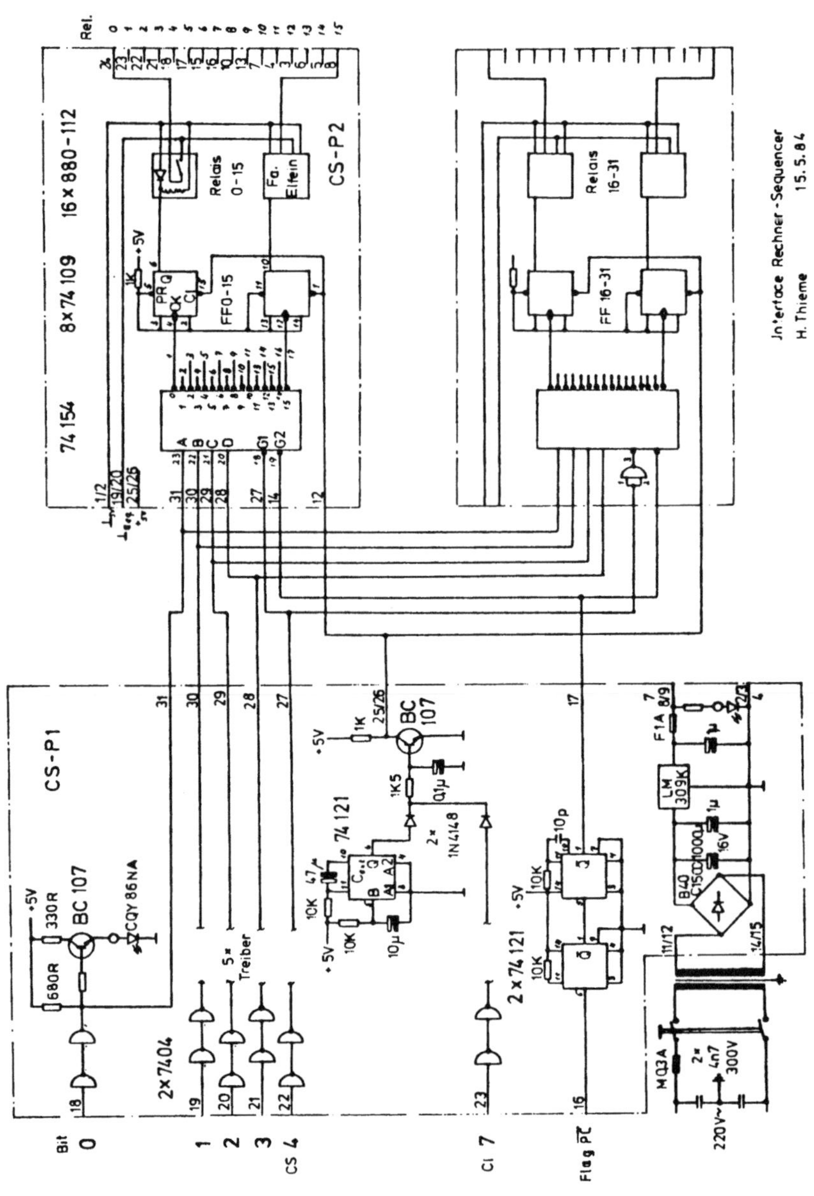

Fig. 6. Interface I.

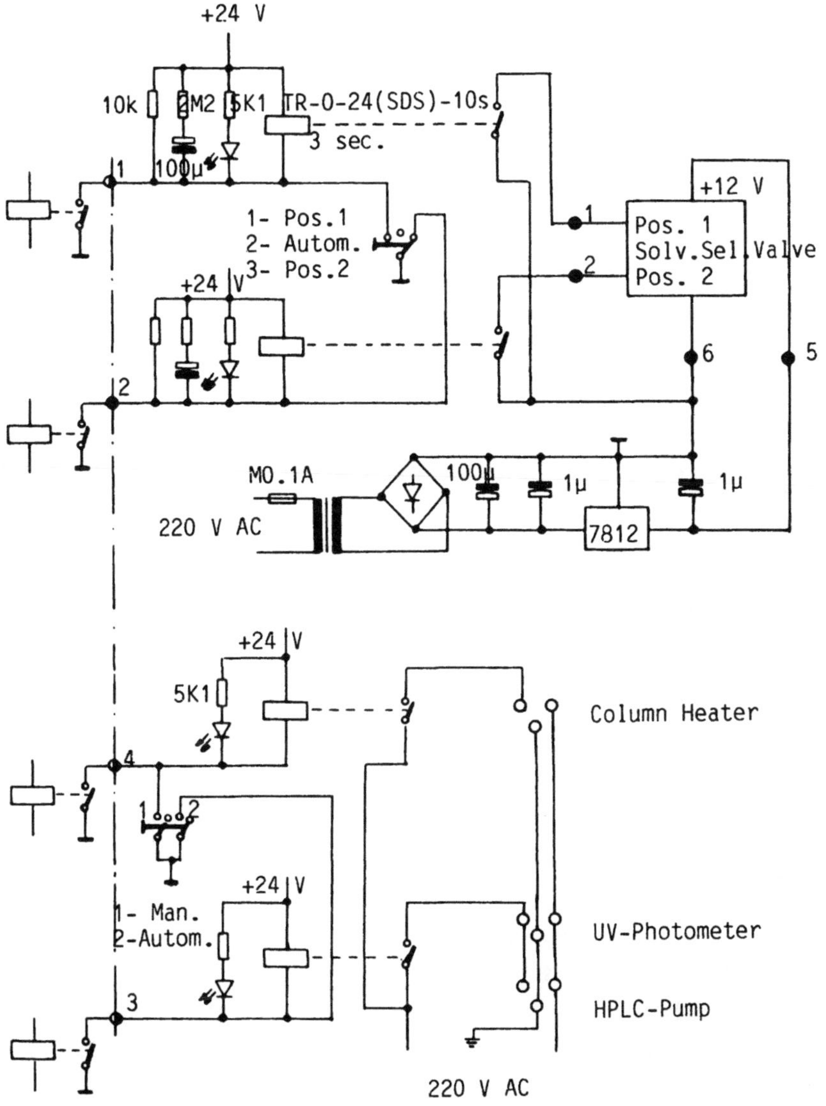

Fig. 7: Interface II

For direct use 256 addresses of 8 bits/256 values, which could be combined by A/D- and D/A-converters, are available.

6) Multitasking. Finally - as this has already been realized in the beginning - a set of several devices (for sequential, for gradient procedures as for data processing and equilibrium systems) could be controlled simultaneously and independently. But firstly, it is assumed to be better to have one separate controller for every device and secondly, the memory capacity would begin to set limitations.

7) Other Microprocessors/Microcomputers. With a well structured operating system there should not be a great problem to change, for example, to 32 bit microprocessors of similar machine language as to Motorola's 680X0. This seems to be one processor of choice for realization of the points mentioned above. (For comparison: The 6510-microprocessor of the C= 64 is an eight bit processor.)

9. Summary and Perspectives

A sequencer laboratory control system has been set up and has been in operation since the beginning of 1983. Activating a Beckman Liquid Phase Sequencer 890 B and controlling the detection system for PTH aminoacid derivatives, it has the following specifications:
One run of up to 128 cycles can be started. For every cycle, one of all created programs including subroutines are available. Every step allows setting of two relays combinations simultaneously and/or printing of texts. Every combination contains up to 32 relays out of 128 and includes explanatory text. Step time durations are from zero/one millisecond up to about 1 million seconds. The reserved storage allows programming of over 4000 steps for programs and subroutines and of 128 combinations and markers.
The system's ability is likely to expand. If it is desired, all can be changed and new possibilities can be realized. With exact knowledge of the techniques, the reliability of functions can be controlled directly The more installations there are, the lower the costs are (regarding the

operating system) and the more opportunities are provided to develop higher reliabitity and higher versatility and to facilitate its use.
This form of self-designed sequencer control is not the only one. There are at least five others known to the author (e.g. 4,5,6). There seems to be special advantages associated with each of these. These advantages may be similar to those aspects pointed out above. This might have coincided with low cost microprocessors/-computers, while universally adaptable laboratory control systems - at least for this price - were not commercially available. Here some changes will surely occur in the future.
The author is not primarily interested in commercial aspects. A fundamental system has been created which can and will be adapted in coming years, always in connection with special laboratory situations.

10. Acknowledgement

B. Wehle and H. Giersiefen (Cologne) are gratefully acknowledged for their help and ideas during installations of the controller, as well as K. Hirsch and J. Taylor for reading and preparation of text.

11. References

1. Commodore 64 Programmer's Reference Guide, Commodore Business Machines, Inc. (1982)

2. Lottspeich, F.: Hoppe/Seyler's 361, 1829-1834 (1980)

3. Lottspeich, F.: this volume

4. Machleidt, W. and Hofner, H.: in Methods in Peptide and Protein Sequence Analysis (Birr, Chr., ed.) pp 35 - 47, Elsevier/North Holland, Amsterdam (1980)

5. Wittmann-Liebold, B.: this volume

6. Frank, R.: this volume

IMPROVED ISOCRATIC SEPARATION OF PHENYLTHIOHYDANTOIN AMINO ACIDS DOWN TO THE FEMTOMOLE LEVEL

Friedrich Lottspeich

Gentechnologische Arbeitsgruppe Mikrosequenzierung am
Max-Planck-Institut für Biochemie
D-8033 Martinsried bei München

In 1980 we reported a simple, quick and reliable high performance liquid chromatography method for the isocratic separation of the common phenylthiohydantoin (PTH) amino acid derivatives (1). The isocratic separation mode has been chosen because it is superior to the gradient elution mode with respect to base line stability, reproducibility, ease of quantification and equipment costs. Furthermore, no time is needed for equilibration. The only PTH amino acids not well resolved with this system was the pair PTH-Gln/PTH-Ser. To distinguish between PTH-Gln and PTH-Ser was not difficult because of the far smaller yield of PTH-Ser, the fact that PTH-Gln is always accompanied by PTH-Glu and the small retention time difference existing. Problems arose only when Ser and Gln were found together in the same step of the Edman degradation.

However, the development of new stationary phase materials and better packing techniques makes it now possible by using an isocratic system to separate all the common PTH amino acid derivatives in a single run within 15 min.

An early improvement over the originally obtained separation was the use of LiChrospher C-18 (5µ) rather than LiChrosorb (see Fig. 1). The elution conditions were essentially the same as in the original paper (1) and are given in Table 1. The peaks were much sharper and therefore the automated quantification was facilitated. A variety of well packed C-18 materials (5µ) were tested since this time and most of them gave a similar separation pattern. Only to position PTH-His and PTH-Arg the molarity

Modern Methods in Protein Chemistry, Volume 2

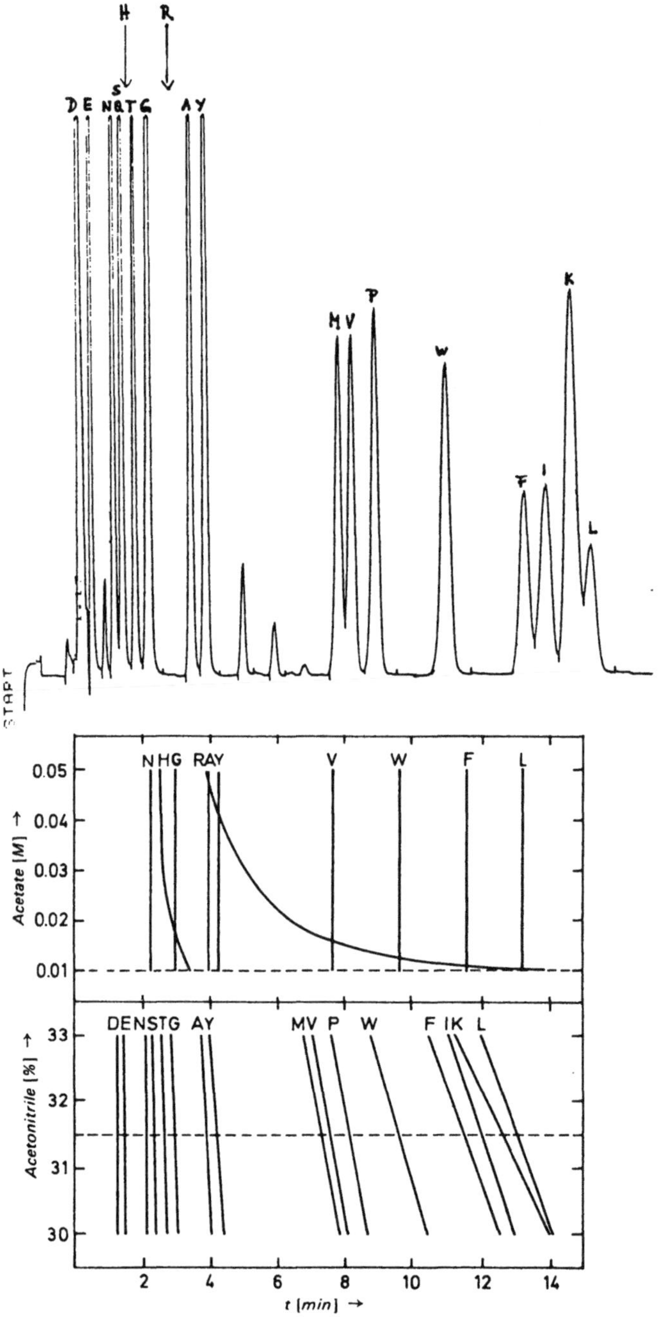

Fig.1. Upper part: HPLC separation of PTH amino acids under the conditions given in Table 1. Column: LiChrospher CH-18 (5μ), 250 x 4 mm.
Lower part: Effect of the molarity of the sodium acetate buffer and the acetonitrile concentration on the retention times of PTH amino acids.

of the buffer has to be adpated, essentially as described in (1) (see Fig. 1, lower part).

The availability of columns with a newly developed highly selective and very efficiently packed stationary phase material, LiChrospher CH-8 SUPER (Merck, Darmstadt), offers for the first time the possibility to separate all the common PTH amino acid derivatives (including PTH-Gln and PTH-Ser) by an isocratic elution mode in less than 15 min. The separation is shown in Fig. 2. The separation conditions are given in Table 1. The only

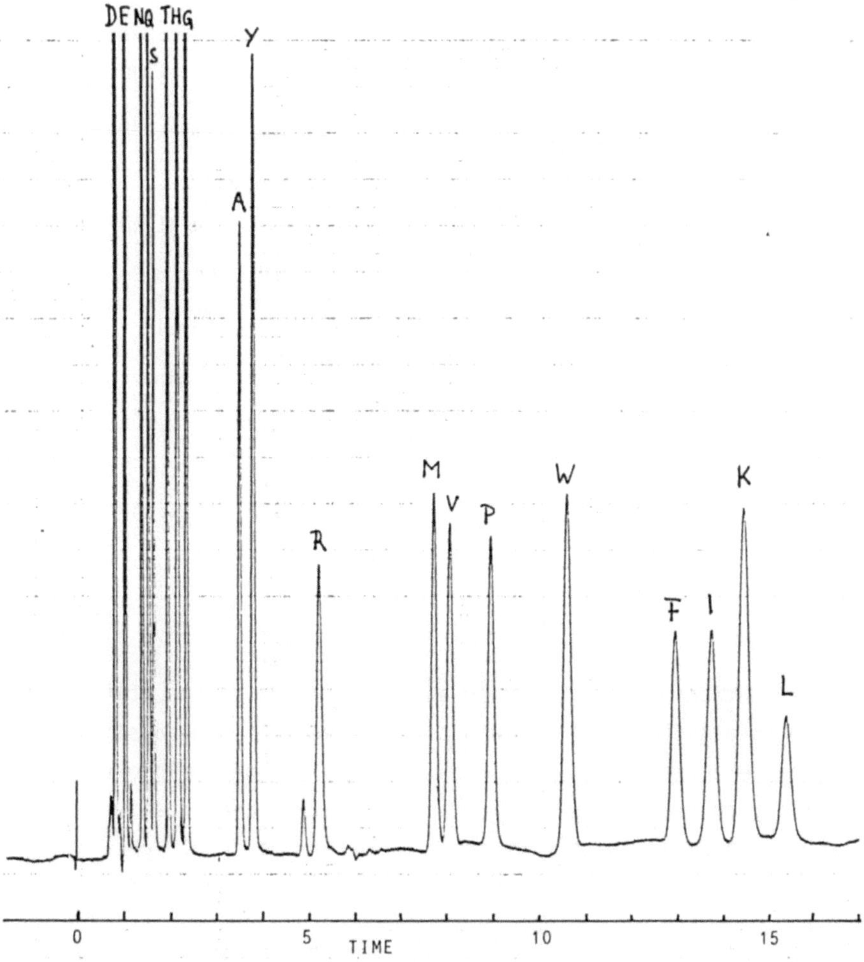

Fig. 2. HPLC separation of PTH amino acids. Column: LiChrospher CH-8 SUPER, 250 x 4 mm. Conditions as given in Table 1.

differences to the previous reported method are the higher flow rate (2.0 mL/min) to obtain a reasonable short analysis time, and a reduced temperature of 60°C instead of 62°C with the LiChrospher CH-18 column. The highly reliable detection of PTH amino acids down to the 3-5 picomole level (Fig. 3, upper line) with a very simple HPLC equipment makes this system very useful for routine PTH amino acid analysis.

Table 1. Conditions for the HPLC-separation of PTH amino acids

Column:	LiChrospher CH-18 (5μ)	LiChrospher CH-8 SUPER	LiChrospher CH-8 SUPER
Dimension:	250 x 4 mm	250 x 4 mm	250 x 2 mm
Buffer:	68.5 % 0.01 M sodium acetate, pH 5.0 31.5 % acetonitrile (LiChrosolv) 0.5 % 1.2 dichloroethane (Uvasol)		
Flow:	1.5 mL/min	2.0 mL/min	0.5 mL/min
Temperature:	62°C	60°C	60°C
Detection limit:	3 pmol	2 pmol	1 pmol
Analysis time:	14 min	15 min	15 min
Equipment:	LKB 2150 HPLC pump pulsation damping unit LKB 2155 Oven LKB 2151 Variable wave length detector, detection wave length: 265 nm		

The introduction of the gas phase sequencer makes it possible to obtain sequence information on the low picomole level. This necessitates to lower the detection limit of PTH amino acids to the femtomole level. One possibility to reach this goal is to use small diameter size columns (Microbore columns). With 2 mm inner diameter one should get an improvement in peak height by a factor of 4 compared to a normal 4 mm inner diameter column. With a 1 mm column an increase by a factor of 16 should be possible.

We have tried chromatographies with 2 mm Supersphere CH-8 columns

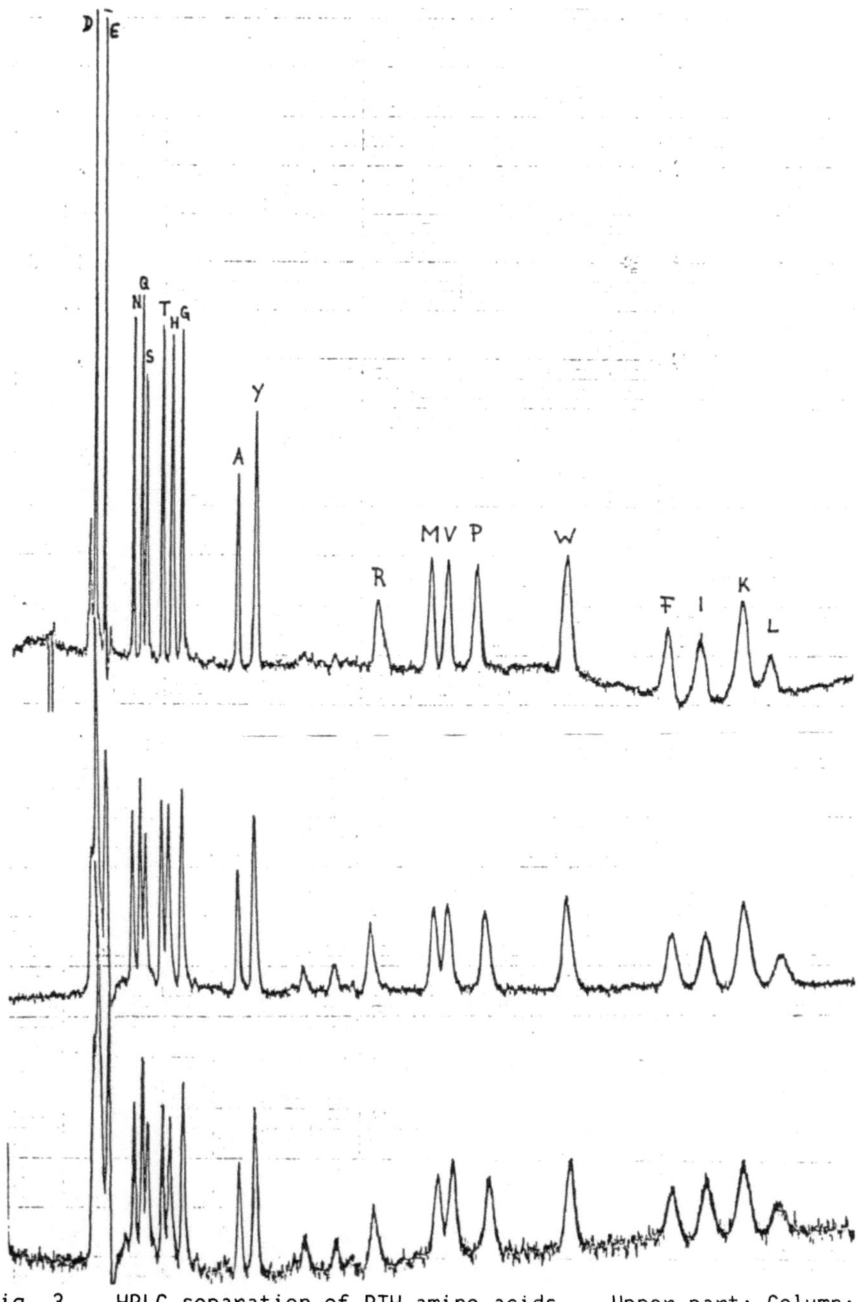

Fig. 3. HPLC separation of PTH amino acids. Upper part: Column: LiChrospher Super 250 x 4 mm, 4 pmol of each PTH amino acid, A.U.F.S.: 0.0025. Middle part: Column: LiChrospher Super 250 x 2 mm, 1 pmol of each PTH amino acid, A.U.F.S.: 0.0025. Lower part: Column: LiChrospher Super 250 x 2 mm, 500 fmol of each PTH amino acid, A.U.F.S.: 0.00125.

(Merck, Darmstadt) and to our surprise we could get the theoretical increase in detectability without any loss of selectivity (Fig. 4). Even with injection volumes as large as 10 µl the loss of resolution was minimal (Fig. 5). The elution conditions are given in Table 1.

In addition to this considerable improvement in the detection limit there was the very desirable reduction by a factor of 4 in the amount of solvents needed. These two facts will make microbore high performance liquid chromatography in near future the most useful detection method for PTH amino acids. However, the microbore columns used in this study were presently not commercially available, and the long term stability of these columns was not evaluated.

Acknowledgement

We gratefully accepted the LiChrospher SUPER columns as a gift of Dr. E. Eisenbeiss, Merck, Darmstadt.

Reference

1. Lottspeich, F.: Hoppe-Seyler's Z. Physiol. Chem. 361, 1829-1834 (1980)

APPLICATION OF COMPUTER PROGRAMS FOR THE STRUCTURAL ANALYSIS OF PROTEINS

Georg E. Schulz
Institut für Organische Chemie und Biochemie der Universität
D-7800 Freiburg i. Br., Albertstr. 21, F.R.Germany

Our structural knowledge of proteins is based on X-ray crystallography, which became only possible with the advent of electronic computers. Therefore, protein structures and computers are intimately connected with each other. Today, one may distinguish the following computer program applications in this field :

1). X-ray diffraction analysis by the multi-isomorphous replacement method
2). Structural refinement on the basis of X-ray data or energy potentials,
3). Visualizing and manipulating structures using a three-dimensional display system
4). Analysis of amino acid sequences

   a) sequence alignments:
      I. protein - protein
      II. peptide - protein
      III. peptide - peptide
      IV. special cases
   b) prediction of secondary structures
   c) hydrophobicity- and helix amphipathy profiles
   d) antigenicity profiles.

Here, I shall not elaborate on the first three subjects, since they are well-covered elsewhere (1,2,3). The fourth subject demonstrates that not only the geometric X-ray diffract-

Modern Methods in Protein Chemistry, Volume 2

ion data need computer evaluation: Computer programs are also required for an efficient application of amino acid sequence information.

Protein - Protein Sequence Alignment

Soon after establishing the first sequences it has been recognized that they change with time. During evolution these changes accumulate, rendering sequences which have emerged from a common ancestor more and more dissimilar. Thus, it became an important endeavour to find homologous proteins on the basis of sequence similarities.

For this purpose Needleman & Wunsch suggested a scheme scoring amino acids which were identical in two compared sequences (4). Since such a scheme can only be successful with rather closely related proteins, it was quickly superseded by a more elaborate scoring scheme based on amino acid residue similarities (5). A further refinement, allowing to detect even more distantly related proteins, was introduced with the application of natural exchange frequency scores (6).

An example of such a sequence comparison is given in Fig.1, where the sequences of a cytosolic (AK1) and a mitochondrial (AK3) adenylate kinase have been aligned (7). For this purpose a segment of 7 residues of the AK3 sequence (Fig.1, horizontal) is overlaid at all possible positions of the sequence of AK1 (Fig.1, vertical). A similarity value is assigned to each pair of amino acids in the overlaid segments according to the matrix of natural substitution frequencies given in Fig.3c (see below). The sum of the 7 similarity values of the segment is the score at the respective overlay position. Overlaying all possible segments of AK3 and AK1 yielded about 10,000 scores, the distribution of which is

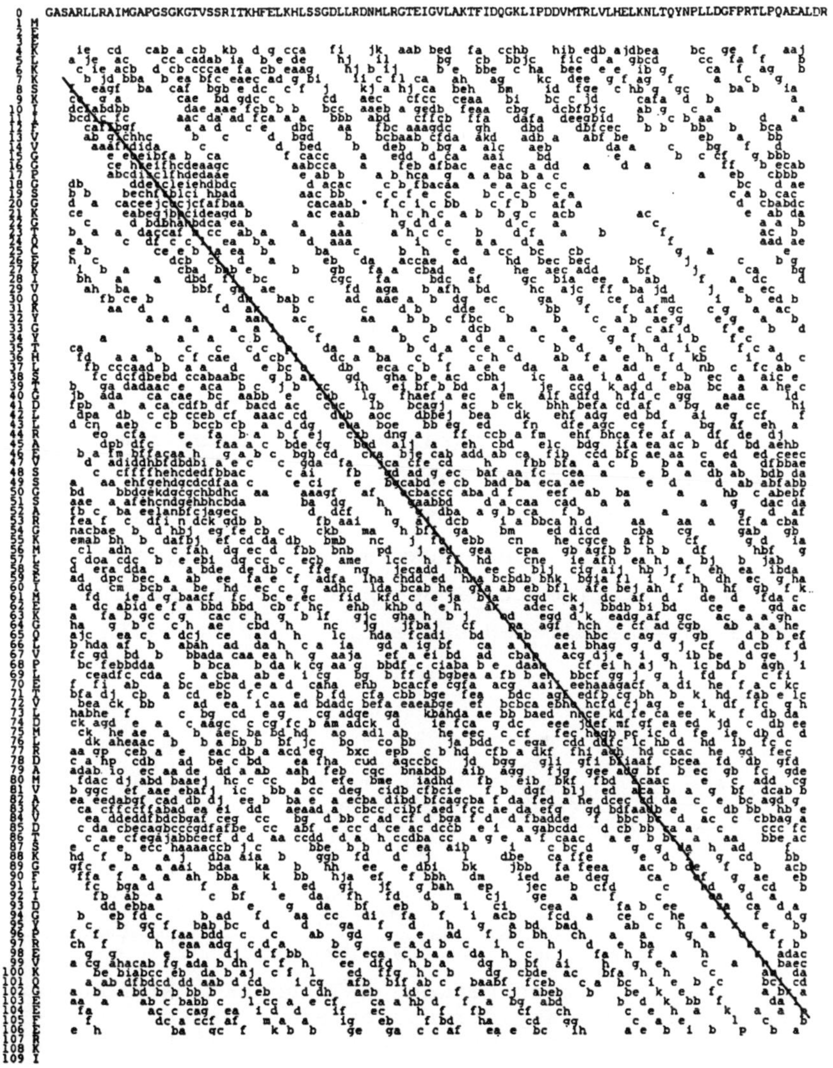

Fig. 1 Sequence comparison of the first 102 residues of mitochondrial adenylate kinase (AK3, horizontal) with the first 109 residues of the cytosolic enzyme (AK1, vertical) using the natural exchange frequency matrix MDM-78 (6). The segment length is 7. The scores are printed in an analog manner (alphabetically). The line indicates the best alignment as shown in Fig. 2.

approximately Gaussian. As a consequence, it is meaningful to calculate the standard deviation of the scores from their average and to express each score by its distance from the average as measured in standard deviation units. These values have been displayed in a manner following Staden (8), so that aligned stretches of sequence can be readily recognized as ridges of high scores, i.e. late letters of the alphabet.

The best alignment is given by the pathway from the N- to the C-terminus with the highest overall score and the lowest number of insertions & deletions. No additional insertion or deletion was allowed if it failed to increase the overall score of 10 adjacent segment overlay positions by more than 20%. Besides a segment length of 7, the program was run with lengths of 3, 5 and 13. These auxilary runs confirmed the alignment, demonstrating its independence of segment length. The scores along the aligned sequences are plotted in Fig.2.

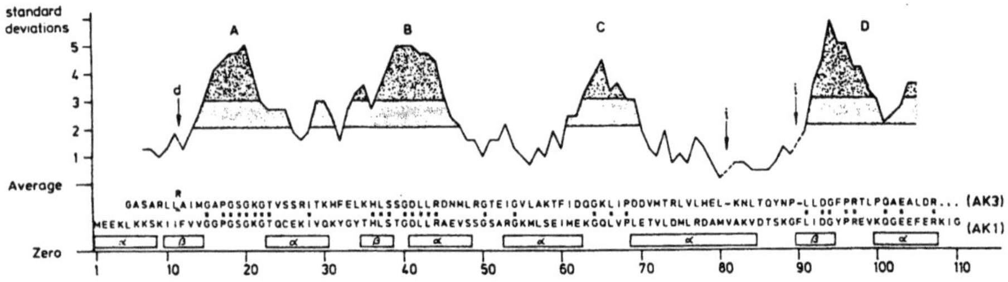

Fig.2 Local similarities between AK3 and AK1. Residue numbers and secondary structures are those of AK1 (9). The ordinate starts at a score of zero. The distribution of all scores is approximately Gaussian. The average of the distribution and the standard deviations from the average are given.

The regions A, B, C, D are unquestionably homologous since the local scores exceed 3 standard deviations. For establishing the alignments between these regions as few as possible insertions & deletions were introduced. Val-81 and Phe-90 of

AK1 had to be declared insertions in order to connect regions C and D. No insertion & deletion was assigned in the other two less well aligned stretches on both sides of region B, because none of them would have increased the local score of 10 residues enough (more than 20%) to make it reasonable. At the N-terminus a deletion was applied because it improved the score appreciably. Still, this deletion remains to be corroborated by a three-dimensional structure comparison. As the insertions are framed by regions of high similarity, their existence is established; only their positions may undergo small shifts.

Since the sequence comparison showed that AK3 and AK1 are homologous, it is safe to assume that both enzymes have very similar chain folds (10). Accordingly, the known chain fold of AK1 (9) can be used as a geometric reference for discussing amino acid differences.

In the light of the three-dimensional structure the most and the least conserved parts of the sequence assemble at specific locations of the globular molecule : the best conserved part is at the active center of the enzyme, whereas the least conserved part is at the opposite side of the molecule. The latter part of the molecule differentiates between cytosolic and mitochondrial enzyme and is therefore most likely recognized by the system which transports AK3 from the cytosol into the mitochondrion. This example demonstrates that sequence data may lead to detailled structural interpretations if they can be aligned to a known structure.

Peptide - Protein Sequence Alignment

During sequence analysis data turn up piecemeal, one peptide after the other. For placing them into correct order and deriving the whole sequence peptides have to be overlapped. Not all of these overlaps come by easily. Therefore, a homologous sequence can be very helpful, if it is used as a template for peptide ordering. Using homologous sequences with more than 50% identical amino acids such an alignment may reach the degree of certainty, rendering overlaps superfluous.

In some instances an X-ray structure analysis preceeded the sequence analysis (11). Here, one can apply the same rules as for homologous proteins if a so-called map sequence has been read from the electron density map. Depending on map quality, such sequences are up to about 50% correct. Experience shows that they are useful if more than 25% of the residues are right, which can be achieved with an electron density map at 3Å resolution. Such X-ray-based alignments have the advantage that no additional residues (insertions) may occur between adjacent peptides, since these would have been recognized in the map.

When aligning peptides to an electron density map (11) I began with a rather gross similarity matrix (Fig. 3a) designed according to expected map reading mistakes. A more refined matrix designed on the same basis (Fig. 3b) was added later. Soon, I realized that the reading errors are correlated to the physical properties of side chains. Therefore, I introduced the natural exchange frequency matrix (Fig. 3c), and thereafter, a fourth matrix designed with the experience of committed reading errors combined with the natural matrix (Fig. 3d).

Surprisingly, peptide fitting experience with these matrices showed that no matrix was better or worse than the others. Clear alignments usually emerged with all 4 matrices. Only rarely an alignment was significant with one matrix and insignificant with the others. Therefore, I adopted the empirical routine to use always all 4 matrices for fitting peptia)

	A	C	D	E	F	G	H	I	K	L	M	N	P	Q	R	S	T	V	W	Y
A	10	0	0	5	0	10	0	0	5	0	0	0	5	0	5	5	5	5	0	0
C	0	20	5	0	5	0	5	5	0	0	5	5	0	0	0	5	5	0	5	5
D	0	5	10	5	0	0	5	0	5	0	0	5	0	5	5	5	5	0	0	0
E	5	0	5	10	0	0	0	0	5	0	0	5	0	5	5	5	0	0	0	0
F	0	5	0	0	20	0	15	5	0	10	10	0	5	5	0	0	0	5	15	15
G	10	0	0	0	0	10	0	0	5	0	0	5	5	0	5	5	0	0	0	0
H	0	5	5	0	15	0	20	5	0	5	10	5	0	5	5	0	0	0	10	15
I	0	5	0	0	5	0	5	15	0	10	5	0	0	5	0	0	5	10	5	5
K	5	0	5	5	0	5	0	0	10	0	0	5	0	5	10	5	0	0	0	0
L	0	0	0	0	10	0	5	10	0	10	5	5	0	5	0	0	0	10	0	0
M	0	5	0	0	10	0	10	5	0	5	20	0	0	5	0	0	5	0	5	10
N	0	5	5	5	0	5	5	0	5	5	0	10	5	5	5	5	5	10	0	0
P	5	0	0	0	5	5	0	0	0	0	0	5	10	0	0	5	5	5	0	0
Q	0	0	5	5	5	0	5	5	5	5	5	5	0	10	5	0	5	0	0	5
R	5	0	5	5	0	5	5	0	10	0	0	5	0	5	10	5	5	0	0	5
S	5	5	5	5	0	5	0	0	5	0	0	5	5	0	5	10	10	5	0	0
T	5	5	5	0	0	0	0	5	0	0	5	5	5	5	5	10	10	10	0	0
V	5	0	0	0	5	0	0	10	0	10	0	10	5	0	0	5	10	15	0	0
W	0	5	0	0	15	0	10	5	0	0	5	0	0	0	0	0	0	0	'30	15
Y	0	5	0	0	15	0	15	5	0	0	10	0	0	5	5	0	0	0	15	20

b)

	A	C	D	E	F	G	H	I	K	L	M	N	P	Q	R	S	T	V	W	Y
A	17	3	12	9	2	10	3	6	4	7	4	11	4	7	3	14	12	13	0	2
C	3	26	5	9	12	1	11	10	12	9	13	7	10	9	13	4	5	5	9	9
D	12	5	20	12	2	4	4	8	6	8	4	15	9	13	5	14	11	11	1	3
E	9	9	12	20	4	3	5	13	8	13	11	14	10	15	6	5	12	12	2	4
F	2	12	2	4	25	0	14	11	8	10	13	7	11	4	10	3	4	5	13	15
G	10	1	4	3	0	28	0	3	1	3	0	3	2	3	1	4	3	6	0	0
H	3	11	4	5	14	0	24	8	9	7	11	4	13	11	9	5	4	4	11	14
I	6	10	8	13	11	3	8	20	9	15	12	10	11	14	9	6	9	10	4	10
K	4	12	6	8	8	1	9	9	21	10	11	6	6	8	14	6	4	5	7	12
L	7	9	8	13	10	3	7	15	10	19	12	13	10	13	9	7	10	13	3	9
M	4	13	4	11	13	0	11	12	11	12	26	6	9	11	12	4	7	8	8	11
N	11	7	15	14	7	3	4	10	6	13	6	17	8	13	5	12	11	12	3	4
P	4	10	9	10	11	2	13	11	6	10	9	8	26	12	5	6	8	5	7	12
Q	7	9	13	15	4	3	11	14	8	13	11	13	12	18	6	5	9	10	3	5
R	3	13	5	6	10	1	9	9	14	9	12	5	5	6	22	5	3	4	8	13
S	14	4	14	5	3	4	5	6	6	7	4	12	6	5	5	16	13	13	1	3
T	12	5	11	12	4	3	4	9	4	10	7	11	8	9	3	13	18	14	2	4
V	13	5	11	12	5	6	4	10	5	13	8	12	5	10	4	13	14	18	1	8
W	0	9	1	2	13	0	11	4	7	3	8	3	7	3	8	1	2	1	30	12
Y	2	9	3	4	15	0	14	10	12	9	11	4	12	5	13	3	4	8	12	23

c)

	A	C	D	E	F	G	H	I	K	L	M	N	P	Q	R	S	T	V	W	Y
A	48	10	33	33	-5	43	16	25	18	11	19	32	41	26	15	41	42	32	-28	-5
C	10	80	-21	-23	-13	-4	-4	7	-24	-30	-22	-6	2	-24	-6	30	8	11	-48	33
D	33	-21	69	64	-26	36	37	6	31	-10	4	51	20	46	17	33	29	9	-38	-13
E	33	-23	64	68	-24	32	37	10	29	-4	9	44	24	55	19	30	26	12	-40	-13
F	-5	-13	-26	-24	121	-18	12	40	-23	48	32	-5	-16	-17	-15	-2	-1	18	34	100
G	43	-4	36	32	-18	78	9	4	13	-11	2	33	25	18	4	41	30	16	-40	-22
H	16	-4	37	37	12	9	95	6	30	9	9	46	28	59	46	22	17	8	2	29
I	25	7	6	10	40	4	6	75	11	54	52	12	10	10	10	16	31	67	-21	21
K	18	-24	31	29	-23	13	30	11	77	1	34	40	19	37	64	28	30	6	-5	-14
L	11	-30	-10	-4	48	-11	9	54	1	89	67	1	5	12	0	2	13	49	12	21
M	19	-22	4	9	32	2	9	52	34	67	94	13	9	20	26	14	24	48	-12	6
N	32	-6	51	44	-5	33	46	12	40	1	13	50	25	38	30	37	34	13	-12	9
P	41	2	20	24	-16	25	28	10	19	5	9	25	89	32	28	39	33	18	-26	-19
Q	26	-24	46	55	-17	18	59	10	37	12	20	38	32	70	43	25	22	11	-18	-10
R	15	-6	17	19	-15	4	46	10	64	0	26	30	28	43	91	27	21	5	52	-12
S	41	30	33	30	-2	41	22	16	28	2	14	37	39	25	27	46	43	20	5	2
T	42	8	29	26	-1	30	17	31	30	13	24	34	33	22	21	43	56	33	-22	3
V	32	11	9	12	18	16	8	67	6	49	48	13	18	11	5	20	33	73	-32	5
W	-28	-48	-38	-40	34	-40	2	-21	-5	12	-12	-12	-26	-18	52	5	-22	-32	203	28
Y	-5	33	-13	-13	100	-22	29	21	-14	21	6	9	-19	-10	-12	2	3	5	28	131

d)

	A	C	D	E	F	G	H	I	K	L	M	N	P	Q	R	S	T	V	W	Y
A	14	7	8	8	4	13	8	8	9	6	8	12	14	11	5	13	12	10	1	2
C	7	20	5	5	2	6	3	9	4	5	12	6	4	5	2	11	9	11	1	1
D	8	5	15	15	2	11	9	6	10	5	6	14	10	14	7	12	10	7	1	1
E	8	5	15	20	3	11	8	7	10	6	6	12	11	14	7	11	10	8	1	1
F	4	2	2	3	20	2	7	11	3	12	11	3	2	3	4	4	5	7	10	15
G	13	6	11	11	2	29	6	6	7	4	5	11	12	10	4	12	10	7	1	1
H	8	3	9	8	7	6	20	7	11	6	6	13	6	12	10	9	9	6	10	7
I	8	9	6	7	11	6	7	25	7	10	10	7	6	8	5	8	10	15	4	6
K	9	4	10	10	3	7	11	7	24	6	8	12	8	11	15	10	9	8	2	3
L	6	5	5	6	12	4	6	10	6	25	15	5	4	6	4	6	7	10	3	6
M	8	12	6	6	11	5	6	10	8	15	20	7	6	8	10	8	9	10	4	6
N	12	6	14	12	3	11	13	7	12	5	7	17	9	12	10	13	12	8	2	2
P	14	4	10	11	2	12	6	6	8	4	6	9	14	11	4	11	10	8	1	1
Q	11	5	14	14	3	10	12	8	11	6	8	12	11	21	13	11	11	9	2	2
R	5	2	7	7	4	4	10	5	15	4	10	10	4	13	20	7	6	5	3	3
S	13	11	12	11	4	12	9	8	10	6	8	13	11	11	7	13	13	9	2	3
T	12	9	10	10	5	10	9	10	9	7	9	12	10	11	6	13	18	11	2	2
V	10	11	7	8	7	7	6	15	8	10	10	8	8	9	5	9	11	22	2	3
W	1	1	1	1	10	1	10	4	2	3	4	2	1	2	3	2	2	2	30	20
Y	2	1	1	1	15	1	7	6	3	6	6	2	1	2	3	3	2	3	20	25

Fig.3 Similarity matrices used for aligning peptides to an electron density map - and thereafter - for aligning peptides to homologous proteins.

des to X-ray data, and later on (using the same program), also to homologous protein sequences. In this procedure, a signal gets quite a high confidence rating if it appears with all matrices. The rating is appreciably lower if one matrix misses the fit.

After a number of sequence fits it turned out that peptides with less than 10 residues have to be approached with caution. Even if a signal emerges in all matrices with a significance of e.g. 4 standard deviations above the average signal height, the placement is not certain. This parallels the experience with protein - protein alignments (Fig.1), where insertions & deletions should only be assigned between stretches of 10 or more aligned residues in order to avoid mistakes.

The peptide fitting program has also been used for aligning longer sequences. In this case the sequences were split into segments of 20-30 residues being placed independently. With this procedure the confidence rating increases if consecutive segments are placed consecutively.

Peptide - Peptide - Search

As mentioned above the alignment of peptides with lengths shorter than 10 residues yields spuriously high significances. Therefore, relating two proteins merely on the basis of a single peptide fit is hazardous. Although recognizing identical hexapeptides, or, a nonapeptide with 7 identical residues elicits great surprise, a simple estimate calms the waves: Among the easily available protein sequences one has to expect many identical tetra- and pentapeptides as well as appreciable numbers of hexa- and heptapeptides as shown in Table 1. These frequencies have been supported by screenings of known sequences (12). In conclusion, the observation of identical short peptides has to be approached with great caution.

Table 1 Estimated numbers of identical peptides in 300 proteins with 200 residues each.

peptide length	expected number
4	20 000
5	1 000
6	50
7	2
8	0
...	...

Special Cases of Sequence Homology

There exist some special cases, however, where short peptides are most likely to indicate structural homology. The clearest example is a highly conserved ßαßαß- unit of dinucleotide binding proteins shown in Fig.4. The corresponding sequence alignments are given in Fig.5. No high significance heralds these fits using a general alignment program. Most of them have been found by visual inspection.

The most obvious feature in Fig. 5 is the pattern of conserved glycine residues with smallish amino acids between them. They contact the pyrophosphate moiety of the dinucleotide (Fig.4). In all cases there are residues like Val and Ile with a high propensity of participating in a ß-pleated sheet at the 4 positions prior to the first Gly of the pattern. This part of the chain is a ß-strand, which is in general the first strand of a 4- or more stranded parallel sheet. At the C-terminal side of the Gly pattern the chain forms an α-helix. Here one

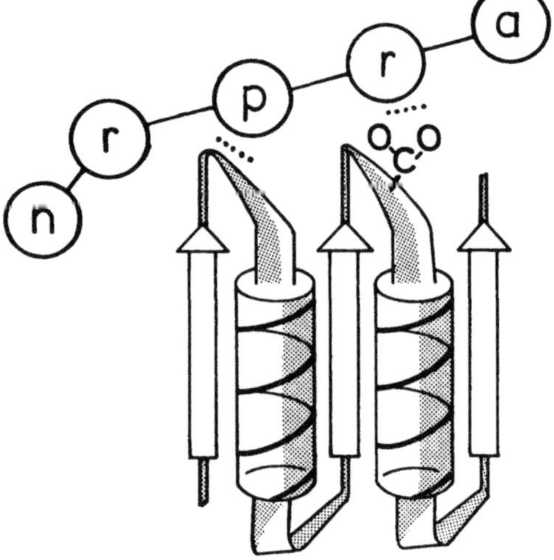

Fig.4 Sketch of a ßαßαß- chain fold which is well conserved during evolution (13,14) together with the relative location of a bound dinucleotide (NAD, NADP or FAD). The residues contacting the pyrophosphate (p) and the adenine(a) ribose (r) hydroxyls are indicated.

observes small residues in about every fourth position, i.e. where the side chain contact the ß-sheet. These small side chains allow an intimate contact between helix and sheet.

After this helix the chain continues in a second, parallel ß-strand containing residues with generally lower ß-propensities than the first strand. This ß-strand usually ends in an Asp or Glu residue at a position which is 23 to 25 residue positions after the first Gly of the pattern. The carboxyl functions of these Asp or Glu invariably bind to the 2′and/or 3′hydroxyls of the adenine riboses (Fig.4). When NADP is the bound dinucleotide, the 2′hydroxyl is phosphorylated and therefore negatively charged. Here, the one observes an Arg at the equivalent position instead.

```
                                bbbbb aaaaaaaaaaaaa  bbbbbb
** GR      FAD    22 ..DYLVIGGGSGGLASARRAAELGARAAVLESH..
** PHBH    FAD     4 ..QVAIIGAGPSGLLLGQLLHKAGIDNVILERQ..
 * LPDHp   FAD     8 ..DVTVIGSGPGGYVAAIKAAQLGFKTVCIEKN..
 * LPDHc   FAD     7 ..QVVVLGAGPAGYSAAFRCADLGLETVIVERY..
 * MR      FAD   100 ..QVAVIGSGGAAMAAALKAVEQGAQVTLIERG..
   GENE1    ?    238 ..KVAVIGAGISGLVVAQELLHAGVDDVTIYEA..
   DAAO   (FAD)    2 ..RVVVIGAGVIGLSTALCIHERYHSVLQPLDV..
   NADDH  (FAD)    7 ..KIVIVGGGAGGLEMATQLGHKLGRKKKKAKI..
** GR     NADP   189 ..RSVIVGAGYIAVEMAGILSALGSKTSLMIRH..
 * MR     NADP   272 ..RLAVIGSSVVALELAQAFARLGSKVTVLARN..
 * LPDHc   NAD   176 ..RLLVMGGGIIGLEMGTVYHALGSQIDVVEMF..
   NADDH  (NAD)  172 ..NIAIVGGGATGVELSAELHNAVKQLHSYGYK..
** LDH     NAD    23 ..KITVVGVGAVGMACAISILMKDLADEVALVD..
** ADH     NAD   194 ..TCAVFGLGGVGLSVIMGCKAAGAARIIGVDI..
** GAPDH   NAD     2 ..KIGINGFGRIGRLVLRAALSCGAQVVAVNDP..
   GLUDH  (NAD)  246 ..TFAVQGFGNVGLHSMRYLHRFGAKCVAVGES..
```

Fig.5 Alignment of dinucleotide-binding enzymes at the conserved βαβ-unit. The structures of the sequences marked by an asterisk are known from unambiguous sequence comparisons. Two asterisks denote known X-ray structure. The observed secondary structure (b = beta, a = alpha) as well as the sequence number of the first given residue are indicated. Well conserved residues are marked by bars. The abbreviations are: GR = glutathione reductase, PHBH = p-hydroxy benzoate hydroxylase, LPDH = lipoamide dehydrogenase (p = pig, c = E.coli), MR = mercuric reductase, GENE1 = gene 1 protein of crown gall tumor virus, DAAO = D-amino acid oxidase, NADDH = respiratory NAD dehydrogenase, LDH = lactate dehydrogenase, ADH = alcohol dehydrogenase, GAPDH = glyceraldehyde-3-phosphate dehydrogenase, GLUDH = glutamate dehydrogenase.

The alignment of Fig.5 is based on the observed similarities between the known three-dimensional structures : chain folds and dinucleotide positions. In a number of cases, however, only the sequences are known and applied for the alignment. As long as the whole chain can be correlated (sequences with asterisks in Fig.5) the homology is well established. For D-amino acid oxidase and others, however, the sequence homology extends only over the segment depicted in Fig.5. In these instances homology is derived from the fact that the proteins bind a dinucleotide together with the observation that most dinucleotide-binding proteins have the discussed structural features in common. In the case of gene-1 of crown gall tumorvirus no dinucleotide cofactor has been established yet. Here, the homology is merely based on a rather short segment of the chain and has to be approached with great caution. The observation sufficed, however, to initiate the search for a dinucleotide bound to the gene-1 protein hoping to reveal the function of this tumor protein.

A special case of much more tentative nature is given in Fig.6 : a glycine pattern slightly different from the one of Fig.5 has been observed with mononucleotide-binding proteins. The alignment of Fig.6 is founded on an even shorter chain segment, and, there is only one three-dimensional structure namely the one of adenylate kinase known. In adenylate kinase

```
                bbbbbb    aaaa

** AK1      ...IIFVVGGPGSGKGT...
 * AK2      ...RAVLLGPPGAGKGT...
 * AK3      ...LRAIMGAPGSGKGT...
   ATB      ...FHSFGGGTGSGFTS...
   BTB      ...THSLGGGTGSGMGT...
   MYN      ...SMLITGESGAGKTE...
   MYR      ...SILITGESGAGKQV...
   AF1      ...RELIIGDRGVGKTV...
   BF1      ...KVGLFGGAGVGKTV...
   P21      ...KLVVVGAGGVGKSA...
```

Fig.6 Tentative alignment of mononucleotide-binding proteins at the flexible loop of cytosolic adenylate kinase (residues 10 through 23), which is the only established three-dimensional structure (9). See Fig.5 for notations. Abbreviations are: AK1 AK2, AK3 = cytosolic, outer-, inner mitochondrial adenylate kinases; ATB, BTB = α-, ß-subunits of tubulin; MYN, MYR = myosin of nematode, rabbit; AF1, BF1 = α-, ß-subunit of F1 ATPase of E.coli; P21 = p21 human bladder protein containing the mutation site of the oncogenic variant.

the glycines are located in a highly conserved flexible loop of the chain (7) which engulfes a phosphoryl group of the mononucleotide, i.e. the chain segment is located at a very special position (15). As in Fig.5 the residues at the N-terminal side of the glycine pattern have high ß-propensities and form the first strand of a more extended parallel ß-sheet, and the residues at the C-terminal side form an α-helix. Most interesting is the similarity between adenylate kinase, where structural movements of this chain segment have been established (16), and myosin, which certainly undergoes a large structural change when transforming chemical (splitting of ATP) to mechanical energy.

Prediction of Secondary Structures

Since sequences change faster than chain geometries during evolution (10), they also vary faster than secondary structures. Accordingly, the combination of sequence alignments with secondary structure predictions seems advisable, in particular for marginal cases with sequence similarities of low significance. This aspect has already been approached above when drawing attention to high ß-propensities in Figs.5 & 6.

Moreover, this combination has been applied at a larger scale (17). The improvement, however, was rather limited. This may be explained by the fact that α-, ß- and turn-propensities, which dominate all predictions, are implicitly contained in the natural exchange matrices (Fig. 3c) and therefore already dealt with. Furthermore, prediction success is limited to only two thirds of all residues at best (10). Thus, an unreflected incorporation of this method may cause additional errors.

Hydrophobicity- and Helix Amphipathy Profiles

Hydrophobicity- (18) or hydropathy- (19) profiles of sequences are likely to reveal and distinguish exterior and interior chain parts of proteins. Therefore, these profiles can be used as additional indicators for structural homology. However, since the hydrophobicity values are implicitly included in the natural exchange matrix (Fig.3), their importance in sequence comparisons is rather limited. These profiles are well suited for locating hydrophobic sequence parts which are likely to span membranes as observed im membrane-bound proteins and in signal sequences.

Helix amphipathy profiles are not contained in the natural exchange matrix because they are of a non-local nature. They have been suggested as indicator for helices at protein surfaces and thus for helix prediction (20,21). In my hands, however, these profiles were not versatile because they gave spurious helix signals too frequently. They were valuable, however, as a means for characterizing clearly predicted α-helices, differentiating between helix locations at the surface and in the interior (22).

Antigenicity Profiles

Apart from establishing evolutionary and structural connections sequences can be used for synthesizing antigenic probes. Three major applications have emerged:

1) Finding and isolating the protein corresponding to an established DNA sequence,
2) Locating a rare protein in the cell,
3) Production of a cheap vaccine in form of a peptide mimicking the antigenic protein.

In all these applications oligopeptides are synthesized and

coupled to a carrier molecule as for instance bovine serum albumin, which is subsequently used for antibody induction. In the first two cases antibodies are produced in rabbits, isolated, and then used for binding to the proteins containing the corresponding oligopeptide. In the last case the produced antibodies provide an immune defense against the organism carrying the antigenic protein.

For this purpose one has to locate those chain segments which confer antigenicity to the protein. These chain parts can be located using an antigenicity plot (23). It turned out, however, that antigenicity correlates closely with hydrophilicity so that it can be derived from the hydrophobicity plot mentioned above. As a further requirement the isolated oligopeptide should assume the same structure as in the intact protein. This is only likely if local forces govern the structure of the chain, a situation which correlates with successful secondary structure prediction (10).

Since conformational fidelity and antigenicity are particularly important for vaccine production, one has to expect that oligopeptides work as vaccines only in rare cases. In foot and mouth disease virus an oligopeptide of the coat protein has been selected on the basis of the antigenicity profile and secondary structure predictions (22). This protein contains an α-helix identified by all five applied prediction methods. Furthermore, the antigenicity plot has a strong peak at one helix end, and the helix shows a very clear helix amphipathy signal indicating its location at the surface. This oligopeptide has been synthesized and tested with appreciable success so that its application as a vaccine can be anticipated after further technical development.

Using immunogenic oligopeptides for identifying the corresponding proteins has been met with much more success (24):

In this method the same conditions apply, but not as stringent as with vaccines. Considering the booming development of genetic engineering this method seems to have a very bright future.

References

1. Blundell, T.L., Johnson, L.N.: Protein Crystallography, Academic Press, London 1976
2. Konnert, J.H., Hendrickson, W.A.: Acta Crystallogr. A36, 344-350 (1980)
3. Rice, D.W., Schulz, G.E., Guest, J.R.: J.Mol.Biol. 174, 483-496 (1984)
4. Needleman, S.B., Wunsch, C.D.: J.Mol.Biol. 48, 443-453 (1970)
5. McLachlan, A.D.: J.Mol.Biol. 61, 409-424 (1971)
6. Dayhoff, M.O.: Atlas of Protein Sequence and Structure, vol.5, suppl.3, p.354, Natl.Biomed.Res. Foundation, Silver Springs MD 1978
7. Wieland, B., Tomasselli, A.G., Noda, L.H., Frank, R., Schulz, G.E.: Eur.J.Biochem., in press (1984)
8. Staden, R.: Nucl. Acids Res. 10, 2951-2961 (1982)
9. Schulz, G.E., Elzinga, M., Marx, F., Schirmer, R.H.: Nature (London) 250, 120-123 (1974)
10. Schulz, G.E., Schirmer, R.H.: Principles of Protein Structure, Springer Verlag, New York 1979
11. Krauth-Siegel, R.L., Blatterspiel, R., Saleh, M., Schiltz, E., Schirmer, R.H., Untucht-Grau, R.: Eur.J.Biochem. 121, 259-267 (1982)
12. Kabsch, W., Sander, C.: Proc. Natl. Acad.Sci. USA 81, 1075-1078 (1984)
13. Rao, S.T., Rossmann, M.G.: J.Mol.Biol. 76. 241-256 (1973)
14. Wierenga, R.K., Drenth, J., Schulz, G.E.: J.Mol.Biol. 167, 725-739 (1983)
15. Pai, E.F., Sachsenheimer, W., Schirmer, R.H., Schulz, G.E.: J.Mol.Biol. 114, 37-45 (1977)
16. Sachsenheimer, W., Schulz, G.E.: J.Mol.Biol. 114, 23-36 (1977)
17. Argos, P., Hanei, M., Wilson, J.M., Kelly, W.N.: J.Biol.Chem. 258, 6450-6457 (1983)

18. Rose, G.D.: Nature, 272, 586-590 (1977)

19. Kyte, J., Doolittle, R.F.: J.Mol.Biol. 157, 105-132 (1982)

20. Schiffer, M., Edmundson, A.B.: Biophys.J. 8, 29-39 (1968)

21. Eisenberg, D., Weiss, R.M., Terwilliger, T.C.: Nature, 299, 371-374 (1982)

22. Pfaff, E., Mussgay, M., Böhm, H.O., Schulz, G.E., Schaller, H.: EMBO J. 1, 869-874 (1982)

23. Hopp, T.P., Woods, K.R.: Proc.Natl.Acad.Sci. USA 78, 3824-3828 (1981)

24. Palfreyman, J.W., Aitcheson, T.C., Tailor, P.: J. Immunol. Meth., in press

DNA AND RNA SEQUENCE ANALYSIS

Ulrich Krawinkel
Institut für Genetik der Universität Köln, Weyertal 121,
D-5000 Köln 41

1) Summary

The chain termination sequencing method in combination with the M13 cloning system is highly suitable to analyse the sequences of small and large DNA fragments. Some examples of the large variety of random and non-random sequencing protocols are presented. A technique for sequencing mRNA is discussed with its applications and limitations.

2) Introduction

Determination of long stretches of nucleotide sequence is required to study the structure of genes encoding macromolecules. All techniques for sequence determination of DNA and RNA give rise to a nested set of radioactive oligonucleotides sharing one end and differing in length by single base increments at the opposite end. As it is possible to obtain the fragments ending in each of the four bases a single size fractionation by polyacrylamide gel-electrophoresis with subsequent autoradiography permits direct reading of the sequence (for review: Davis, 1982). The most widely used techniques to determine DNA and RNA nucleotide sequences either utilize chemical degradation or enzyme-directed copying of single-stranded templates. In the Maxam

Modern Methods in Protein Chemistry, Volume 2

and Gilbert degradative procedure (Maxam and Gilbert, 1977) chemical reactions, each specific for one of the four desoxynucleotides of DNA, cause strand cleavage at positions where an A, a C, a G or a T is present. This technique may indirectly be applied to determine RNA sequence by utilizing reverse transcription to generate cDNA which then is subjected to chemical degradation.

The chain termination sequencing method devised by Sanger et al. (1977) depends on the primed synthesis of a radioactive complementary copy of a single-stranded DNA-template by DNA-polymerase. Base specific chain termination is achieved by adding to the chain-extension reaction one of the (2'-3') dideoxy derivates (ddNTPs) of the natural deoxynucleotides. The dideoxy analogues do not have the 3'OH group which is necessary for 3'-5' phosphodiester bounding, chain growth thus stops whenever a dideoxy nucleotide is incorporated. By choosing an appropriate dideoxy/deoxy ratio incorporation of dideoxy nucleotides will occur at random along the chain at each possible site. By addition of one of the dideoxy analogues to each of four separate chain-extension reactions copies are accummulated which are identical at the 5' end but differ from each other in length by the increments corresponding to the distance between successive nucleotides of one particular kind (A, C, G or T).

RNA sequence determination by the chain termination method essentially follows the same principle with some modifications (for review: Rocca-Sera, 1984). RNA-dependent DNA polymerase from avian myeblastosis virus (AMV-reverse transcriptase) is used instead of the Klenow fragment of DNA-polymerase I. The dideoxy nucleotide fraction in the reverse transcription reaction has to be augmented in order to meet the high preference of AMV-reverse transcriptase to use deoxynucleotides as a substrate instead of dideoxynucleotides.

As techniques to determine DNA and RNA sequences already have been exhaustively reviewed this paper will discuss strategies devised to rapidly obtain sequences of genes encoding macromolecules. Emphasis is laid on employment of the chain termination sequence technique. The strategies discussed here may be adopted with slight modifications for chemical degradation sequencing.

3) The M13 cloning and sequencing system

The chain termination sequencing technique requires pure single-stranded DNA from the region of interest to act as a template for DNA-polymerase. The most convenient way to obtain pure single-strands is to clone the fragment of interest into the bacteriophage vectors M13 (Messing et al., 1977; Gronenborn and Messing, 1978; Schreier and Cortese, 1979; Messing et al., 1981; Messing and Viera, 1982; Norrander et al., 1983, Kieny et al. 1983) fd (Hermann et al., 1980) or pEMBL (Dente et al., 1983). Mature phages contain single-stranded DNA ((+) strand) which easily can be purified. The pEMBL family of vectors has the property of being encapsidated as single stranded DNA upon superinfection with phage F1.

The M13 cloning system is the one which commonly is used in combination with chain-termination sequencing. The M13 cloning vehicles have been engineered to contain the part of the E.coli ß-galactosidase gene, which encodes the α-peptide and to complement ß-galactosidase activity when grown on E.coli hosts carrying a lacZ gene with the M15 deletion

(Gronenborn and Messing, 1978). The function of the lacZ gene product serves as a marker system for discriminating vector DNA (the intact α-peptide complements ß-galactosidase function, thus leading to the formation of blue plaques on indicator plates) from recombinant DNA (lacZ gene segment encoding α-peptide is inactivated by insertion of foreign DNA, the plaques formed are colourless).

Multiple cloning sites have been introduced into the N-terminus of the lacZ gene segment of M13 thus allowing a great variety of restriction fragments to be cloned and sequenced. The more recent developments of the M13 vectors contain cloning sites which occur only once within the vector. Therefore, it is possible to direct the orientation of restriction fragments exhibiting two different restriction termini ("forced cloning"). Such fragments can be sequenced from both sides when pairs of cloning vehicles are chosen which contain a particular array of cloning sites in oppostive orientation, such as M13mp18 and M13mp19 (Norrander et al., 1983). An application of "forced cloning" is presented in chapter 4 of this article.

All DNA fragments inserted into a cloning site have the same lacZ gene sequences flanking their 5' and 3' ends. The clones thus can be primed for the sequence reaction by a primer molecule which hybridizes on the (+) strand of the vector to the lacZ gene sequence directly adjacent to the end of the inserted DNA fragment. A sequence in the reverse orientation can be primed by a primer hybridizing to the lacZ gene sequence at the opposite end of the insert on a newly synthesized (-) strand (Hong, 1981). An outline of M13 sequencing techniques using oligonucleotide primers is shown in Fig. 1.

```
         ***----
         ***----------
         ***--------------
         ***---------------------
         ***--------------------------
         ***--------------------------------
M13(-)strand---G-----G---G------G----G-----G-###
---------===---C-----C---C------C----C-----C-===M13(+)strand
               ------------------------------###
                     ------------------------###
                         --------------------###
                                -------------###
                                     --------###
                                           --###
```

Fig. 1 Principle of DNA-sequencing by the chain termination method. Synthesis of the M13(-)strand in the presence of ddNTPs, ddGTP in the case presented here. (###) sequence primer. (***) reverse sequence primer. (===) lacZ gene sequence encoding the alpha peptide.

M13 clones also can be used to generate single-stranded hybridisation probes particularly suitable for the detection of RNA (Hu and Messing, 1982; O'Hare et al., 1983).

4) Random and non-random sequence approaches

One limitation of sequencing techniques is that only 400 nucleotides from the priming site can be reliably read from gels. Therefore, analysis of a large DNA fragment involves the use of several restriciton enzymes to generate small overlapping fragments that are cloned into M13 vectors for subsequent sequence analysis. The entire DNA sequence is assembled by matching of overlapping sequences (Messing et al., 1981; Sanger et al., 1980). Small overlapping DNA-fragments for sequence analysis can also be obtained by sonification and subsequent end-repair to generate clonable fragments with blunt ends (Fuhrman et al., 1981). A completely random approach to analyse large DNA fragments is presented by Deininger (1983). Sonification was chosen to sub-fragment the 17.2 kb EcoRI C-fragment of Epstein-Barr-Virus. The resulting sub-fragments are end-repaired with T4-DNA polymerase to generate blunt-ended DNA-fragments which are cloned into the Hind II site of common M13 vectors. In order to prepare subclones of relatively homogeneous and specific size a vigorous size fractionation step is included in the cloning procedure. Anderson (1981) determined the entire sequence of a 4251 bp DNA fragment by establishing a M13 library of small sub-fragments generated by DNase I digestion in the presence of Mn^{2+}.

A disadvantage of these random sequencing approaches is unnecessary redundancy in analysis as some regions may be analysed several times before the entire DNA-sequence can be assembled. On the other hand, the duplication of data may ordinarily be quite useful as it is desireable to determine a sequence several times on both strands of the DNA to completely eliminate sequence errors. The slow accumulation of new data towards the end of a random sequencing project can be overcome by employing non-random cloning and sequencing procedures.

For instance, the redundancy of analysis can be avoided when subclones are generated by "walking" along a DNA-fragment using subclones which already have been sequenced as a probe to identify clones carrying in opposite orientation the overlapping adjacent fragment (Hu and Messing, 1982). Another approach to sequence large DNA-fragments in a stepwise fashion is priming the sequence reaction (n + 1) with an oligonucleotide primer complementary to a site located at the end of the previously determined sequence (n). Other non-random techniques have been developed to analyse the sequence of large DNA fragments independently of suitable restriction sites. These techniques are based on generation of a set of progressive deletion mutants of a clone carrying the entire DNA fragment to be analysed. Poncz et al. use exonuclease BAL31 to generate deletion mutants (1982), Hong (1982) presents a method which utilizes DNase I in the presence of Mn^{2+}, Henikoff (1984) generates deletion mutants by unidirectional digestion of the insert sequence with exonuclease III: the polylinker region of the vector between the insert and the priming site is cut with two restriction enzymes. The site adjacent to the priming region leaves a four base 3' protusion which protects the remainder of the vector from exonuclease III attack. The site adjacent to the insert leaves a 5' protruding end giving access to unidirectional digestion of the insert region.

5) Direct determination of genomic sequences

M13 clones for DNA-sequence analysis generally are generated from DNA fragments which are clonally amplified in bacterial or animal viruses. Blankenstein et al. (1984) devised a method to clone immunoglobulin V_H-region genes located on selected restriction fragments from genomic DNA directly into M13 vectors for subsequent nucleotide sequence analysis. Genomic DNA was digested with restriction enzymes A and B. Fragments exhibiting site A at one end and site B at the other end were rigorously size-fractionated and inserted by "forced cloning" into M13 phages digested with enzymes A and B. Site A is adjacent to the priming site of the M13 vector thus directing the orientation of the inserted fragment for a sequence analysis starting at site A. Transformation to generate a M13 library of genomic DNA fragments was performed in microtiter plates. Microtiter-wells containing phages carrying an immunoglobulin V_H-region gene were identified by means of a specific probe in a dot-blot assay. Combination of this technique with random cloning approaches may soon allow to generate complete genomic libraries in M13 phages.

6) RNA-sequencing

Determination of mRNA sequence utilizing reverse transcriptase and chain termination inhibitors has been employed in various experimental systems (Bina-Stein et al. , 1979; Hamlyn et al., 1978; Kaartinen et al., 1983; Kitamura and Wimmer, 1980; McGeoch and Turnbull, 1978; Rocca-Sera et al., 1983a; Rocca-Sera et al., 1983b; Sasavage et al., 1980; Tonnelle et al., 1983; Zimmern and Kaesberg, 1978). The method is outlined in Fig 2.

```
5'UT  /   coding region            /    3'UT      polyA

------T-----T---------T--------T-----T-------T----AAAAA mRNA
                      -------------------------###
                               ----------------###
                                     ----------###
                                             --###

            -------***
      -------------***
-------------------***
```

Fig. 2. Principle of mRNA sequencing by the chain termination method. cDNA synthesis in the presence of ddNTPs, ddATP in the case presented here. (###, ***) sequence primers.

Essentially, two requirements have to be fulfilled for successfully sequencing mRNA.

(i) The mRNA to be analysed should be relatively abundant, i.e. it should represent at least 5% of the total cellular mRNA. It is in addition desirable to maximize the amount of intact mRNA by affinity chromatography on oligo dT-cellulose. About 10-20 µg of an immunoglobulin V-region mRNA are needed to determine the complete sequence of the V-domain.
(ii) A specific primer is required to initiate the cDNA synthesis in a mRNA template by reverse transcriptase. The most widely used way to obtain a specific primer is to chemically synthetize the required oligonucleotide. In fact it is quite easy now to achieve the chemical synthesis of 10-20 base oligonucleotides. The primer usually iniates DNA-synthesis in a region which is conserved in a particular family of mRNAs, for instance the joining-region of immunoglobulin mRNAs (Sablitzky and Rajewsky, 1984).

As the use of one primer usually allows sequence determination of 250-300 bases, the synthesis of a second "downstream" primer is required to derive a complete sequence of a protein molecule (see Fig. 2).

The major advantage of mRNA sequencing over DNA-sequencing of cDNA cloned in plasmids and over protein sequencing is its rapidity. The constraints of the method concern the limitation to abundant mRNAs, and the dependence on primers which restrict the method essentially to the analysis of mRNAs sharing common sequences. It also should be mentioned that the fidelity of mRNA sequence reading does not reach the standard of DNA-sequencing yet.

A combination of primer-intitated cDNA synthesis on a specific mRNA species and conventional DNA sequencing is developed and employed in several laboratories now (D. Baltimore, T. Imanishi-Kari, G. Brownlee, M. Weigert, pers. comm.). Specific cDNA is isolated from a gel and subjected to chemical degradation sequencing according to the Maxam-Gilbert protocol. This variation of mRNA sequencing is said to improve the readability of sequence gels.

7) Processing sequence data

DNA and RNA sequencing methods yield rather complex data at such a rate that computer facilities must be used to interpret sequence results, in particular when the M13 system is employed on large DNA fragments. Computer programs are available for translation of sequences in all possible phases, searching particular or all possible restriction sites, writing out the complementary sequence, search for possible secondary structures of DNA, comparing sequences, establishing sequence data banks and many other features (Staden, 1977, 1978, 1979, 1980; Queen and Korn, 1980; Devereux and Haeberli, 1983).

8) References

1. Anderson, S.: Nucl. Acids Res. 9, 3015-3027 (1981).

2. Bina-Stein, M., Thoren, M., Salzman, N. Thompson, J.A.: Proc. Natl. Acad. Sci. USA 76, 731-735 (1979).

3. Blankenstein, T. Zoebelein, G., Krawinkel, U.: Nucl. Acids Res. in press.

4. Davies, R.W. in Gel Electrophoresis of Nucleic Acids (Rickwood, D. and Hames, B.D.. eds.), IRL Press, Ltd., Oxford and Washington D.C., 1982 pp. 117-172.

5. Deininger, P.L.: Anal. Biochem. 129 216-223 (1983).

6. Dente, L. Cesareni, G., Cortese, R.: Nucl., Acids Res. 11 1645-1655 (1983).

7. Devereux, J., Haeberli, P.: Program Library of the University of Wisconsin Genetics Computer Group 1983.

8. Fuhrmann, S.A., Deininger, P.L., LaPorte, P., Friedmann, T., Geiduschek, E.P.: Nucl. Acids Res. 9 6439-6456 (1981).

9. Gronenborn, G., Messing, J.: Nature 272, 375-377 (1978).

10. Hamlyn, P.H., Brownlee, G.G., Cheng, C.C., Gait, M.J., Milstein, C.: Cell 15, 1067-1075 (1978).

11. Henikoff, S.: Gene 28, 351-359 (1984).

12. Herrmann, R., Neugebauer, E., Prikl, H., Zentgraf, H., Schaller, H.: Mol. Gen. Genet 177, 231-242 (1980).

13. Hong, G.F.: Biosci. Rep. 1, 243-252 (1981).

14. Hong, G.F.: J. Mol Biol 158, 539-549 (1982).

15. Hu, N.T., Messing, J.: Gene 17, 271-277 (1982).

16. Kaartinen, M., Griffith, G.M., Markham, A.F., Milstein, C.: Nature 204, 320-324 (1983).

17. Kieny, M.P., Lathe, R., Lecoq, J.P.: Gene 26 91-99 (1983).

18. Kitamura, N., Wimmer, D.: Proc. Natl. Acad. Sci. USA 77, 3196-3200 (1980).

19. Maxam, A.M., Gilbert, W.: Proc. Natl. Acad. Sci. USA 74, 560-564 (1977).

20. McGeoch,D.J., Turnbull, N.T.: Nucl. Acids Res. 5, 4007-4024 (1978).

21. Messing, J., Gronenborn, G., Müller-Hill, B., Hofschneider, P.H.: Proc. Natl. Acad. Sci. USA 74 3642-3646 (1977).

22. Messing, J., Crea, R., Seeburg, P.H.: Nucl. Acids Res. 9, 309-321 (1981).

23. Messing, J., Viera, J.: Gene 19, 269-276 (1982).

24. Norrander, J., Kempe, T., Messing, J.: Gene 26 101-106 (1983).

25. O'Hare, K., Levis, R., Rubin, G.M.: Proc. Natl. Acad. Sci. USA 80 6917-6921 (1983).

26. Poncz, M., Solowiejczyk, D., Ballantine, D., Schwartz, E., Surrey, S.: Proc. Natl. Acad. Sci. USA 79 4298-4302 (1982).

27. Queen, C.L., Korn, L.J.: Methods in Enzymology 65, 595-609 (1980).

28. Rocca-Sera, J., Tonnelle, C., Fougereau, M.: Nature 304, 353-355 (1983a).

29. Rocca-Sera, J., Matthes, H.W., Kaartinen, M., Milstein, C., Theze, J., Fougereau, M.: EMBO J. 2 867-872 (1983).

30. Rocca-Sera, J.: Ann. Immunol. (Inst. Pasteur) 135C, 305-317 (1984).

31. Sablitzky, F., Rajewsky, K.: EMBO J. in press.

32. Sanger, F., Nicklen, S., Coulson, A.R.: Proc. Natl. Acad. Sci. USA 74, 5463-5407 (1977).

33. Sanger, F., Coulson, A.R., Barrell, B.G., Smith, A.J.H., Roe, B.: J. Mol. Biol. 143, 161-178 (1980).

34. Sasavage, N.C., Smith, M., Gillam, S., Astell, C., Nilson, J.H., Rottman, F.: Biochemistry 19, 1737-1743 (1980).

35. Schreier, P.J., Cortese, R.: J. Mol. Biol. 129, 169-172 (1979).

36. Staden, R.: Nucl. Acids Res. 4, 4037-4061 (1977).

37. Staden, R.: Nucl. Acids Res 5, 1013-1015 (1977).

38. Staden, R.: Nucl. Acids Res. 6, 2601-2610 (1979).

39. Staden, R.: Nucl. Acids Res 8 817-825 (1980).

40. Tonnelle, C., Rocca-Sera, J., Moulin, M., Moinier, D., Fougereau, M.: J. Exp. Med 158 1415-1427 (1983).

41. Zimmern, D., Kaesberg, P.: Proc. Natl. Acad. Sci. USA 75 4257-4261 (1978).

SECONDARY STRUCTURE DETERMINATION OF MEMBRANE PROTEINS BY RAMAN SPECTROSCOPY

Fritz Jähnig and Horst Vogel[+]

Max-Planck-Institut für Biologie, Correnssstrasse 38, D-7400 Tübingen

Introduction

Without question, X-ray diffraction is the best method to determine the three-dimensional structure of proteins. However, for high resolution of the structure, crystals of good quality are required, which are not easily obtainable in every case. Especially membrane proteins have resisted attempts to grow three-dimensional crystals of sufficient quality and to obtain the necessary heavy-atom derivative through isomorphic replacement (1). Bacteriorhodopsin spontaneously forms a two-dimensional lattice in membranes which permitted the determination of the structure to 5Å resolution by electron diffraction (2). This is just at the limit of resolving secondary structures such as α-helices, but insufficient to resolve individual amino acid residues (3). For two polypeptides, alamethicin and melittin, which strongly bind to membranes, the structure has been determined at 1.5 and 2.0 Å resolution, respectively, by X-ray diffraction from three-dimensional crystals (4,5). However, the crystals were grown in methanol, in the case of alamethicin, and in water, in the case of melittin, hence a priori the structure obtained need not be identical to the structure in membranes.

[+]Present address: Biozentrum der Universität Basel, Klingelbergstrasse 70, CH-4056 Basel

Modern Methods in Protein Chemistry, Volume 2

In this situation, one may turn to other techniques which provide structural information - less than high-resolution X-ray diffraction, but without requiring three-dimensional crystals. The potential techniques fall into two groups: Nuclear magnetic resonance (NMR) spectroscopy and optical spectroscopy including circular dichroism (CD), infra-red absorption (IR), and Raman scattering. The progress in NMR on water-soluble proteins has been reviewed at this conference by Wüthrich (preceding article). The application of NMR to membrane proteins suffers from the extensive immobilization of proteins in membranes, which prevents sufficient motional narrowing to obtain well-resolved spectra (6). This problem can be circumvented by studying proteins in detergent micelles which undergo fast rotational diffusion. Here, the lipid environment of a protein in membranes is replaced by detergent molecules, so that the protein structure is assumed to be largely conserved. But this must be proved in each case.

CD has provided valuable information on water-soluble proteins (7). The most serious problem in the application of CD to membrane proteins arises from light scattering depolarization at the water-dispersed membranes. Ways to eliminate this source of error in the data analysis have been proposed (8) and, actually, secondary structure determinations of membrane proteins by CD have been reported (3). Whereas the α-helix content derived from CD is reliable, the ß-sheet content is subject to large uncertainty (9).

In applying IR to membrane proteins, light scattering is a less serious problem, because membranes scatter infra-red light less than the ultraviolet light used for CD and, furthermore, conventional IR is insensitive to depolarization due to light scattering (10). However, IR is hampered by another problem, the strong absorption of water in the spectral region analyzed for secondary structure. In principle, the water contributions to the spectra can be subtracted, if they are identical to a pure water spectrum, but this must not be so in every case.

The signal detected in Raman scattering is light shifted in frequency relative to the incoming light, hence, light scattering is even less important than for IR. The disadvantage of Raman, on the other hand, is the weakness of the signal compared to CD and IR. Probably, this is the reason why, apart from a conference report (11), Raman investigations of membrane proteins have not been communicated. However, this drawback can partially be overcome by extensive sampling of data. As will be shown in the following for the example of melittin, Raman spectroscopy indeed provides reliable information on secondary structure of membrane-incorporated proteins. In analogy to Raman studies of water-soluble proteins (12), the amide I band in the region 1600-1700 cm^{-1} is evaluated for protein secondary structure. This band arises from vibrations of the repeating $-\overset{\overset{\displaystyle O}{\|}}{C}-\underset{\underset{\displaystyle H}{|}}{N}-$ groups in the backbone of the polypeptide chain and, therefore, is sensitive to secondary structure.

Experimental Method

If a membrane protein can be isolated, purified, and reconstituted into lipid membranes, the preparation of samples for Raman measurements does not pose any further problem. Special care must be taken to avoid any fluorescent impurities. Typically, a protein concentration of 30-300 µg/µl is used, which is fairly high, but only a small volume of about 10 µl is required. An aqueous dispersion of membranes with the reconstituted protein, about 10^{-3}M in lipid, is prepared and concentrated by high-speed centrifugation. The pellet is transferred to glass capillaries, overlayered with buffer, and the capillaries are sealed.

Our spectrometer (13) consists of an Ar^+ ion laser (Spectra Physics model 165) and a double monochromator (Spex model 1402). Photon-counting electronics and monochromator stepper motor are coupled to a microprocessor (Spex Scamp). Measurements are performed at a laser light power of 100 mW at 514,5 nm and slit width 5 cm^{-1}. For each measurement, the spectral region of interest is scanned for several hours, with a step size of 0.2 Å corresponding to 0.65 cm^{-1} and a counting time per step of 0.5 sec. The sample temperature is controlled by a water thermostat connected to a metal sample holder.

Data Analysis

The amide I band of proteins is superimposed by a weak band at 1635 cm^{-1} arising from water. This water contribution was subtracted from the membrane spectra according to a method described by Williams (12). Furthermore, in order to obtain the pure protein spectrum contributions from the lipids must be eliminated. This was performed by recording a Raman spectrum of pure lipid membranes prepared in the same way as the reconstituted membranes except for the omittance of protein. The water contribution in the lipid spectrum was subtracted as in the case of the reconstituted membranes. Lipids show a strong Raman band at 2850 cm^{-1} due to symmetric C-H stretch vibrations in the hydrocarbon chains, which is not superimposed by any protein band. This band was, therefore, used to normalize the spectra of reconstituted and pure lipid membranes. The final pure protein spectra in the amide I region were analyzed according to the method of Williams (12). Each spectrum is described by 15 data points and fitted by a linear combination of the spectra of 15 (the identity of these numbers is purely accidental) reference proteins of known three-dimensional structure. We always use Williams' second method of fitting.

Here, the set of 15 reference proteins is transformed into a set of 15 quasi proteins whose spectra are orthogonal, and an increasing number of quasi proteins is utilized to optimize the fit. In this way, the structure of the protein under study is represented as a superposition of the structures of quasi proteins and finally of the 15 reference proteins. Since their structure is known, the structure of the protein can be calculated. The amide I band is sensitive to the secondary structure of the polypeptide backbone; therefore, the structures of the 15 reference proteins and of the protein are specified as secondary structures, expressed in percentages of the different classes of secondary structure. Williams distinguishes six classes: ordered helix (H_{ord}), when the amino acid residue participates in hydrogen bonds to two sides along the helix; disordered helix (H_{dis}) when less than two hydrogen bonds are formed; antiparallel ß-sheet (S_a), parallel ß-sheet (S_p), turn (T), and undefined (U). The amide I spectra of the pure structure classes have been presented by Williams and Dunker (14). These spectra are discernibly different for each structure class, e.g. in the number and position of maxima.

As a control, we have measured the amide I spectrum of lysozyme, analyzed it according to the above rules omitting lysozyme from the set of reference proteins, and compared the result with the structure given by Williams on the basis of X-ray diffraction. The differences in the percentages of secondary structure classes did not exceed 3%.

Result for Melittin

The Raman spectrum in the range 700-1800 cm^{-1} of melittin in water at a concentration of 300 μg/μl is presented in Fig.1. Assignments of the individual bands are given in Table 1. The region of the amide I band is shown on a larger scale in Fig. 2A. The same region for melittin incorporated in bilayers of ditetradecylglycerophosphocholine at 15°C is presented in Fig.2B.

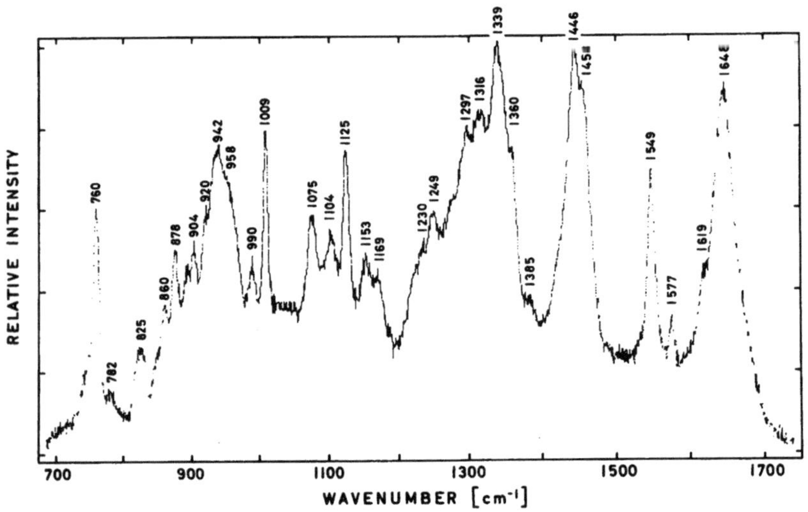

Fig.1. Raman spectrum of melittin in water at a concentration of 300 µg/µl, 15°C.

This spectrum has a higher level of noise than the spectrum in water, because the melittin concentration is 200 times lower. The difference in concentration was partially compensated by prolonged sampling of data for the membrane-incorporated melittin. Apart from noise, the spectra of water-dissolved and membrane-incorporated melittin resemble each other closely. The difference spectrum (Fig.2C), in fact, shows that the amide I bands are identical.⁺ This implies that the secondary structures of water-dissolved and membrane-incorporated melittin are the same.

⁺The peak at 1550 cm^{-1} in the difference spectrum arises from a Trp band indicating that upon incorporation of melittin into a membrane the environment of Trp 19 becomes more hydrophobic.

Table 1. Raman frequencies and assignments of melittin in aqueous solution

$[cm^{-1}]$	assignment	
760	Trp, Ala	
782	Pro	
825		
860	Val, Leu	
878	Trp	
904	Ala	
920	Ile, Glu, Asp	
942	Val, Leu	C^{α}-C stretch (904–990)
958	Val, Leu	
990	Ile	
1009	Trp	
1075		C^{α}-N stretch (1075–1104)
1104	Ala	
1125	Val, Leu, Ileu; C -C stretch	
1153	Val, Leu (CH_3 asym. rock in plane)	
1169	Val, Leu (CH_3 asym. rock out of plane)	
1230	Val	
1249	Trp	amide III (1230–1339)
1297		
1316	CH_2 twist/wag (1297–1339)	
1339		
1360	Trp	
1385		
1446	CH_2 bend (1446–1458)	
1458		
1549		
1577	Trp (1549–1619)	
1619		
1648		amide I

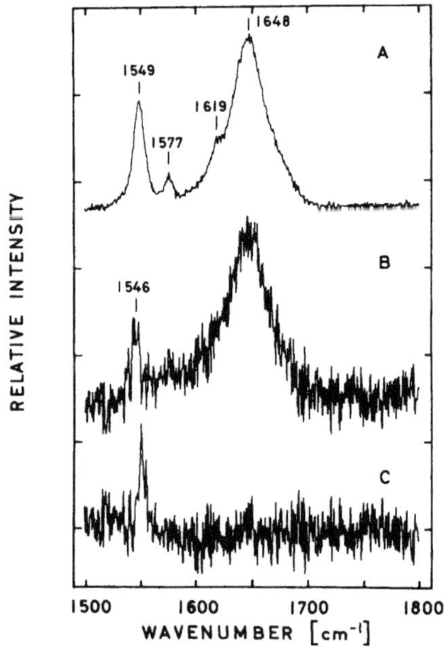

Fig.2. Amide I spectral region of melittin at 15°C (A) in water (300 μg/μl), (B) in lipid membranes of ditetradecylglycerophosphocholine at a molar ratio of lipid/melittin = 100, and (C) the difference spectrum (A) - (B).

This is true whether melittin is in an ordered lipid membrane, at 15°C, or in a fluid membrane, at 35°C (data not shown). The far reaching consequences of this result will be discussed before entering the analysis of the spectra in terms of secondary structure.

Melittin is a polypeptide of 26 amino acid residues whose sequence was determined by Habermann and Jentsch (15) (Fig.4). The first 20 residues are predominantly hydrophobic, the C-terminal 6 residues are hydrophilic with 4 positive charges. In crystals, the structure of the polypeptide chain of melittin consists of two α-helical segments, 1-11 and 12-26, interrupted by a bend at Thr11-Gly12 (5). This gives melittin the form of a bent rod. Apart from the hydrophilic C terminus, the α-helical bent rod is amphipathic in the sense that the concave side is strongly hydrophobic, and the convex side is more hydrophilic including the residues Lys7, Thr10, Thr11, and Ser18. In the crystals which, as mentioned, were grown in water, monomers aggregate to form tetramers with the hydrophobic sides of the monomers facing each other to avoid contact with water, and the monomers being associated antiparallel in pairs to reduce the electrostatic repulsion between the charged C-termini (Fig.3).

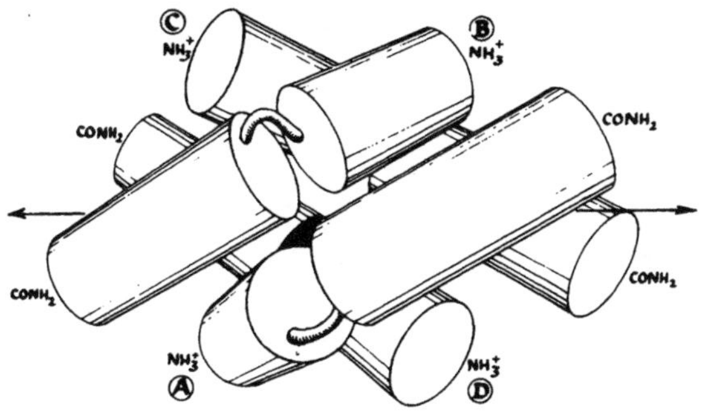

Fig.3. Schematic structure of melittin tetramers in crystals as determined by X-ray diffraction (taken from ref.5 with kind permission of the authors).

Melittin at high concentration in water (or at high ionic strength) also forms tetramers (16). They are assumed to be the same tetramers found in crystals. Indeed, NMR studies of tetrameric melittin in water indicate that the structure is the same as the crystalline structure, except for the hydrophilic C-terminus which in water adopts a random conformation (17). This agrees with the result of CD measurements on tetrameric melittin in water in which an α-helix content of 70% was found, corresponding to 18 amino acid residues (16,18). Hence, the structure of the polypeptide chain of melittin in the tetramer dissolved in water is that schematically portrayed in Fig. 4.

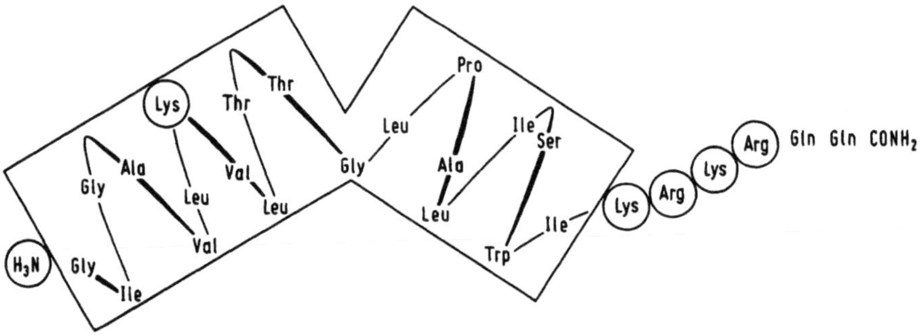

Fig.4. Structure of the melittin polypeptide chain in tetramers in water.

Under the experimental condition of our Raman measurements on melittin in water, melittin was in the tetrameric state. Since the secondary structures of water-dissolved and membrane-incorporated melittin were found to agree, we conclude that the structure of the polypeptide chain of melittin in membranes is also that of Fig.4, an α-helical bent rod with a random C-terminal tail. This knowledge, available due to the exceptional circumstance that the crystal structure of melittin has been determined, will be used as a control for the analysis of

Raman data in terms of secondary structure.

To fit the amide I band of melittin, in water or in membranes, 8 quasi proteins (corresponding to $k_{rank} = 8$ in the notation of Williams) were used. The result for the percentage of secondary structure classes and the corresponding numbers of amino acid residues are given in Table 2. The total α-helix content is 76% corresponding to 20 residues. The remaining 6 residues are distributed among the other structure classes.

Table 2. Percentages and amino acid residues of secondary structure classes of membrane-incorporated melittin.

	H_{ord}	H_{dis}	S_a	S_p	T	U	H_{tot}
%	57.5	18.5	10	-2.5	8.5	8	76
AA	15	5	2	0	2	2	20

Based on the preference of membrane-incorporated protein segments to form α-helices (19,20), the 20 α-helical residues are attributed to the predominantly hydrophobic N-region of melittin and 5 of the non-helical residues to the hydrophilic C-terminus. The sixth non-helical residue is Pro14, since proline is always considered as member of the undefined structure class (12). The distribution of total α-helix content between residues in ordered and disordered helix conformation is 15:5. By definition, each helix contains 4 disordered helical residues, 2 at each end. Five disordered helical residues, therefore, are too many for one helical segment and too few for two. This consideration suggests a bent helix in which one residue at the bend is not fully hydrogen-bonded to its neighbors. Hence, the secondary structure of melittin as determined by Raman spectroscopy is that of a bent α-helix of 21 residues (including Pro14) and 5 non-helical residues. Structural predictions locate the α-helix in the N-terminal hydrophobic segment of melittin, the bend two residues N-terminal to

Pro14, and the non-helical residues in the hydrophilic C terminus. This structure is consonant with the structure deduced above from the crystal structure. Thus, Raman spectroscopy in combination with simple methods of structure prediction (19) provides extensive information on the structure of membrane proteins.

A remark on the reliability of secondary structure determination by Raman spectroscopy is in place. The largest error arises from the arbitrariness of the fit: Equally good fits can be obtained with different combinations of quasi proteins. With increasing number of quasi proteins the result for the structure classes reaches a level of relative constancy, typically from 6 to 8 quasi proteins, and upon further increase of the number of quasi proteins varies irregularly. In the constant region, which we used for evaluation, the percentages vary within $\pm$ 2%, representing a measure for the error in the analysis.

The final question we address is the arrangement of melittin in a membrane and its quarternary structure. An α-helix of 21 amino acid residues, even with a bend, is capable of spanning the hydrophobic core of the membrane bilayer. From IR studies on oriented membranes the average orientation of melittin helices was found to be parallel to the membrane normal (21). If melittin is monomeric in membranes, a transmembrane arrangement implies that the hydrophilic side of the helices face the lipid hydrocarbon chains, which for energetic reasons is highly unfavorable. To avoid this contact melittin may aggregate with the hydrophilic sides facing each other. To investigate this possibility we performed fluorescence energy transfer which indicated that melittin aggregates to a tetramer (22). With the further information from fluorescence quenching that all Trp19 are accessible from the lipid phase and are located on that side of the membrane to which melittin was added (22), one arrives at the model as shown in Fig.5.

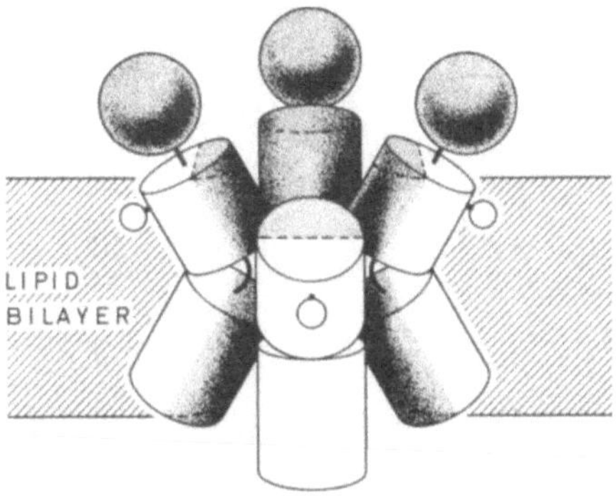

Fig.5. Schematic structure of melittin tetramers in membranes. Hydrophobic surfaces are white, hydrophilic surfaces are dotted. Small spheres projecting from the α-helices represent the side chains of Trp19 residues.

Obviously, this aggregate can act as a pore for permeation of small ions through the membrane and, in fact, melittin is known to increase drastically the conductivity of black lipid membranes (23).

Although the structure of the polypeptide chain of melittin in membranes is largely the same as in the crystal, the aggregation is different. Tetramers in each case, in the crystal the hydrophobic sides of the helices are buried in the interior of the aggregate, whereas in the membrane they form the outer surface facing the lipids. Hence, the aggregate in membranes is the inside-out version of the aggregate in the crystal or water. This is conceivable and may be relevant for other membrane proteins.

References

1. Michel, H.: TIBS 8, 56-59 (1983).
2. Henderson, R., Unwin, P.N.T.: Nature 257, 28-32 (1975).
3. Jap, B.K., Maestre, M.F., Hayward, S.B., Glaeser, R.M.: Biophys. J. 43, 81-89 (1983).
4. Fox, R.O., Richard, F.M.: Nature 300, 325-330 (1982).
5. Terwilliger, T.C., Weissman, L., Eisenberg, D.: Biophys. J. 37, 353-361 (1982).
6. Keniry, M.A., Gutowsky, H.S., Oldfield, E.: Nature 307, 383-386 (1984).
7. Chen, Y.H., Yang, I.T., Chan, K.H.: Biochemistry 13, 3350-3359 (1974).
8. Urry, D.W., Long, M.M.: Meth. Membr. Biol. 1, 105-115 (1974).
9. Hennessey, J.P., Johnson, W.C.: Biochemistry 20, 1085-1094 (1981).
10. Wallach, D.F.H., Verma, S.P., Fookson, J.: Biochim. Biophys. Acta 559, 153-208 (1979).
11. Chang, E.L., Yager, P., Williams, R.W., Dalziel, A.W.: Biophys. J. 41, 65a (1983).
12. Williams, R.W.: J. Mol. Biol. 166, 581-603 (1983).
13. Vogel, H., Jähnig, F.: Chem. Phys. Lipids 29, 83-101 (1981).
14. Williams, R.W., Dunker, A.K.: J. Mol. Biol. 152, 783-813 (1981).
15. Habermann, E., Jentsch, J.: Hoppe Seyler's Z. Physiol. Chemie 348, 37-50 (1967).
16. Talbot, J.C., Dufourcq, J., de Bony, J., Faucon, J.F., Lussan, C.: FEBS Lett. 102, 191-193 (1979).
17. Brown, L.R., Lauterwein, J., Wüthrich, K.: Biochim. Biophys. Acta 622, 219-230 (1980).
18. Lauterwein, J., Bösch, C., Brown, L.R. Wüthrich, K.: Biochim. Biophys. Acta 556, 244-264 (1979).
19. Kyte, J., Doolittle, R.F.: J. Mol. Biol. 157, 105-132 (1982).
20. Jähnig, F.: Proc. Natl. Acad. Sci. USA 80, 3691-3695 (1983).
21. Vogel, H., Jähnig, F., Hoffmann, V., Stümpel, J.: Biochim. Biophys. Acta 733, 201-209 (1983).
22. Vogel, H., Jähnig, F. (submitted).
23. Tosteson, M.T., Tosteson, D.C.: Biophys. J. 36, 109-116 (1981).

STANDARD METHODS FOR THE CHEMICAL SYNTHESIS OF PEPTIDES

R. Geiger
Hoechst Aktiengesellschaft, 6230 Frankfurt (Main) 80, FRG

Introduction (1 - 4)

Peptides, the low-molecular relatives of proteins, are involved in numerous physiological processes. Synthesis of naturally occurring peptides and of partial structures and analogues thereof with modified biological activities is for this reason of great importance in obtaining meaningful answers to physiological questions.

Starting materials are usually naturally occurring L-amino acids and their D-enantiomers. With the progress in the design of biologically active peptides as hormonal agonists and antagonists or substrates for enzymes, non-natural amino acids which are only accessible through synthesis are also used as constituents. They have to be resolved into their L- and D-forms either at the stage of a suitable amino acid derivative or later by separation of diastereomeric peptides.

Peptides formally result from the condensation of amino acids with elimination of water (1 - 4):

$$\mathrm{H_2N\text{-}\overset{R}{\overset{|}{C}}H\text{-}CO\text{-}COOH + H_2N\text{-}\overset{R'}{\overset{|}{C}}H\text{-}COOH \longrightarrow H_2N\text{-}\overset{R}{\overset{|}{C}}H\text{-}CO\text{-}NH\text{-}\overset{R'}{\overset{|}{C}}H\text{-}COOH}$$

In this paper the abbreviations generally accepted in peptide synthesis are used (see l.c. 1 - 4).

Modern Methods in Protein Chemistry, Volume 2

If the reaction is to yield a single product, the functional groups which are not to react must be blocked by protective groups. Side-chains R may also contain reactive groups which have to be blocked, too.

We therefore need two types of protective groups: one which intermittently protects the α-amino and -carboxyl groups and liberates them selectively, and another which blocks further functional groups throughout the synthesis. It must be possible to remove these protective groups at the end of the synthesis under conditions that do not affect the sensitive final product.

The peptides are either synthesized in solution or bound to an insoluble resin (Merrifield's solid-phase technique). Because of the repetitive steps in peptide synthesis, the solid phase synthesis (SPS) has successfully been automatized. Peptide synthesizers have meanwhile found acceptance in many laboratories.

The strategy for the synthesis of peptides in solution follows two different paths, either in steps, amino acid by amino acid, or in segments of increasing size (Fig. 1). In SPS, the step-by-step procedure is preferred.

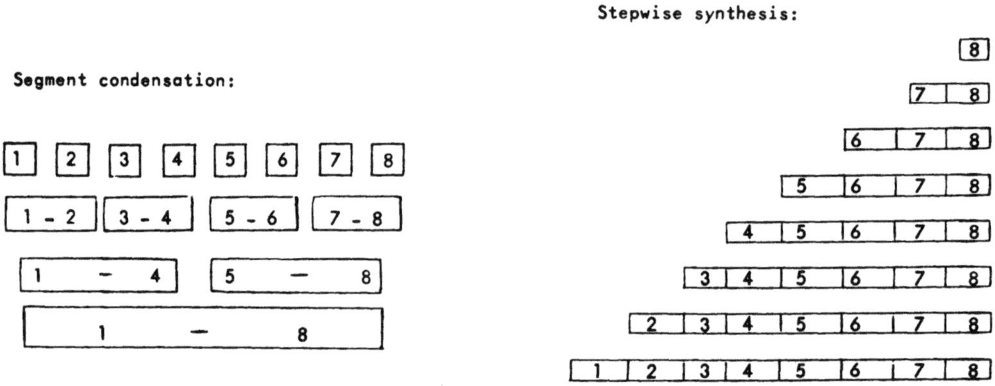

Fig. 1 Segment and Stepwise Synthesis of Peptides

Protective Groups (5 - 9)

Though many protective groups have been recommended, only few have found general application. Since most peptides are reasonably stable in acidic media, combinations of protective groups with graded acid labilities are preferred.

Two important types are shown in Table 1: the benzyl type and the tert-alkyl or aralkyl type. O-benzyl residues can be removed by catalytic hydrogenation, by protonic solvolysis in HBr/acetic acid, hydrogen fluoride or strong sulfonic acids, and be sodium in liquid ammonia.

Table 1: Frequently used N- and O-Protecting Groups

Amine protecting groups	Abbrev.	preferred cleavage
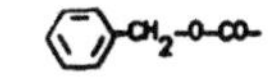	Z	H_2/Pd, strong acids (HBr, HF, sulfonic acids) Na/NH_3
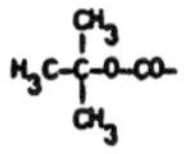	Boc	CF_3COOH, diluted HCl in organic solvents
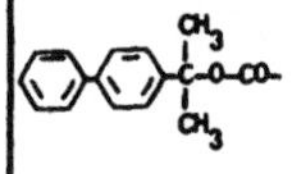	Bpoc	weak acids, diluted CF_3COOH in CH_2Cl_2
Carboxyl and hydroxyl protecting groups		
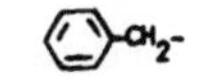	Bzl	H_2/Pd, strong acids Na/NH_3
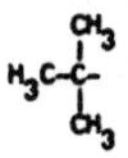	Bu^t	CF_3COOH, diluted HCl in organic solvents

O-Tert-butyl residues are used in a similar manner. They are already cleaved by weaker acids, e.g. trifluoroacetic acid, and are not susceptible to catalytic hydrogenation and Na/NH_3. Tert-butyl esters also resist alkaline saponification which cleaves the frequently used methyl or ethyl esters and benzyl esters as well.

Benzyl and tert-butyl residues not only protect carboxyl groups but also hydroxyl groups of serine, threonine and tyrosine. For amine protection, they are linked to nitrogen as urethanes thus showing susceptibility toward cleaving reagents similar to that of corresponding ethers and esters. Whereas substitution of the benzene ring with halogen and nitro groups renders benzyl groups more stable to acids, the substitution of a methyl group of tert-butyl by aryl leads to increased sensitivity to acids. Diluted trifluoracetic acid in methylene chloride cleaves Bpoc 3,000 times faster than Boc. This difference provides a sufficient selectivity (5).

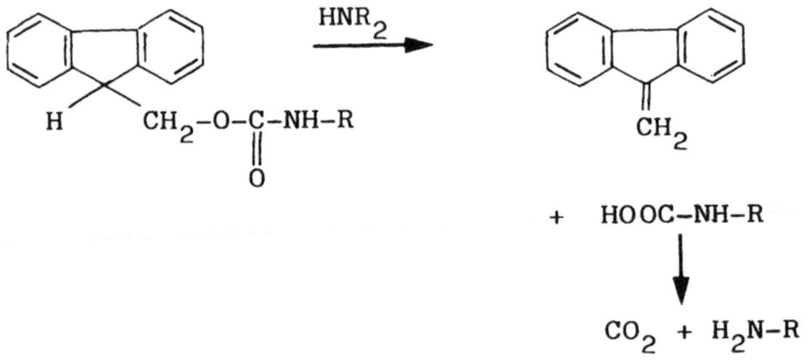

Fig. 2: The Fmoc-Group and its Removal by Secondary Amines

More recently, the set of N-protecting groups has been complemented by the 9-fluorenylmethyl-oxycarbonyl residue (Fmoc). It is split off by ß-elimination (Fig. 2). The cleavage conditions are mild enough to avoid side-reactions.

The described residues can be combined as shown in Table 2, in both strategies of peptide synthesis: the segment condensation, mostly in solution, and the step-by-step approach, both in solution and in solid phase synthesis. In particular, Fmoc for temporary N^{α}-protection and tert-butyl residues for permanent protection is promising because of the comparatively mild deblocking conditions for tert-butyl residues.

Table 2: Combination of Temporary and Permanent Protecting Groups

N^{α}- Protection	Side-chain Protection	Deblocking conditions for N^{α}	Deblocking conditions for Side-chain
Z	Boc, Bu^t	H_2(Pd)	CF_3COOH
Bpoc	Boc, Bu^t	diluted acid	CF_3COOH
Boc	Z, Bzl	CF_3COOH	strong acids H_2(Pd) Na/NH_3
Fmoc	Boc, Bu^t	base	acid

The protection of functional groups in the side-chains deserves particular attention.

In the synthesis of arginine peptides in solution, we can forgo protecting groups since protonation of the strongly basic guanido group is sufficient if the condensation step is carried out under slightly acidic conditions. During solid phase synthesis, in which an approximately threefold excess of the acylating amino acid derivative is applied, the guanido group is mostly blocked. If benzyl or chlorobenzyl residues are employed for permanent protection, tosyl and 4-methoxy-benzenesulfonyl are convenient and effective residues. Tosyl is split off by hydrogen fluoride, 4-methoxybenzenesulfonyl by HF and strong sulfonic acids (5).

With tert-butyl residues for side-chain protection and Fmoc for N^{α}-protection, the Mtr group has been recommended (Fig. 3). It is split off by trifluoracetic acid under slightly stronger conditions than those for tert-butyl residues (10).

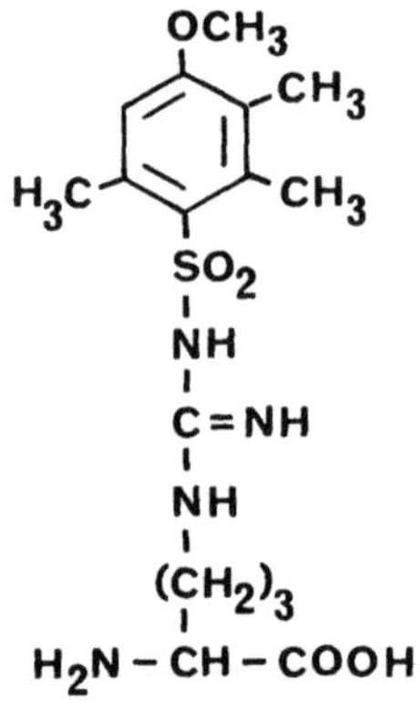

Fig. 3: N^G-4-Methoxy-2,3,6-trimethylbenzenesulfonyl-L-arginine (H-Arg(Mtr)-OH)

Protection of the sulfhydryl group of cysteine is crucial. A selection of S-protecting groups is shown in Table 3.

For strong deblocking conditions the highly stable benzyl or even better 4-methoxybenzyl residues are appropriate. If milder cleavage conditions can be applied, trityl or acetamidomethyl (Acm) are used and mostly cleaved with iodine under simultaneous formation of disulfides. S-tert-butyl forms a sufficiently stable asymmetric disulfide which, however, should not be exposed to stronger bases (7).

Table 3: A Selection of Sulfhydryl Protecting Groups

Group	Abbrev.	Cleavage
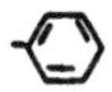	Bzl	Na/NH_3
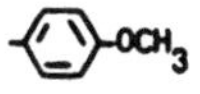	Mob	Na/NH_3, HF, CF_3SO_3H
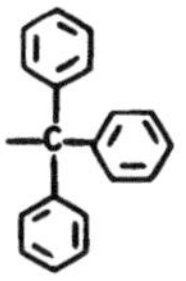	Trt	$H^{\oplus}/Hg^{2\oplus}$ I_2 (→ disulfide)
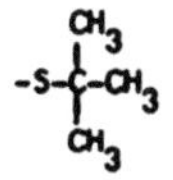	SBu^t	$P(C_4H_9)_3$
$-CH_2-NH-CO-CH_3$	Acm	$Hg^{2\oplus}$(pH 4) I_2 (→ disulfide)

Histidine remains problematic due to the pronounced tendency to racemization if its carboxyl group is activated to form the peptide bond. This tendency is reduced if electronegative substituents such as 2.4-dinitrophenyl (Dnp) are used for N^{im}-protection. The Dnp group, however, is susceptible to nucleophiles already under the conditions of peptide synthesis. Benzyl or 4-methoxybenzyl, on the other hand, is stable during the synthesis but does not safely suppress racemization nor is it easily cleaved. Most other traditional N^{im}-protecting groups are not sufficiently stable or have other disadvantages. Synthesis in solution via segments is therefore frequently carried out without N^{im}-protection. Acylation as a minor side-reaction does not severely affect the procedures since acyl imidazoles are already cleaved by methanol as nucleophile (5).

Figure 4: Particular Caboxyl Protecting Groups

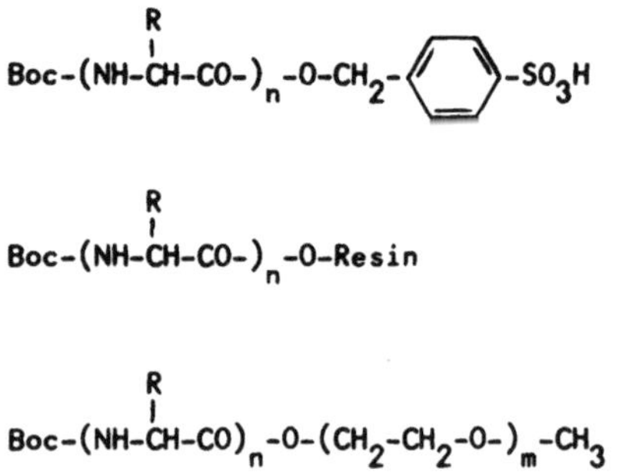

Solid Phase

Liquid Phase

Medium-sized protected peptides are often insoluble in all suitable solvents, and this can prematurely terminate a synthesis. Solubilizing protecting groups (11) help to prevent such a problem (Figure 4).

Polyethylene glycol as carboxyl protecting group transmits its excellent solubility and crystallization property to the growing peptide chain and makes possible the elaboration of a repetitive "liquid phase method" (12). Solid phase synthesis (13) circumvents solubility problems since every reaction occurs in the swollen resin. The reagent penetrates the solid phase by diffusion.

Formation of the peptide bond

A central problem of peptide synthesis is the condensation of the protected amino acids or peptide segments without racemization. The formation of a peptide linkage between a carboxyl group and an amino group with elimination of water requires energy. This energy is supplied through the chemical activation of the carboxyl group. In this high-energy state, optically active amino acids readily undergo racemization (16).

This constant hazard in peptide chemistry is caused by a 5(4H)-oxazolone-mediated mechanism involving an acylamino acid or the ultimate amino acid of a peptide acid (Scheme 1).

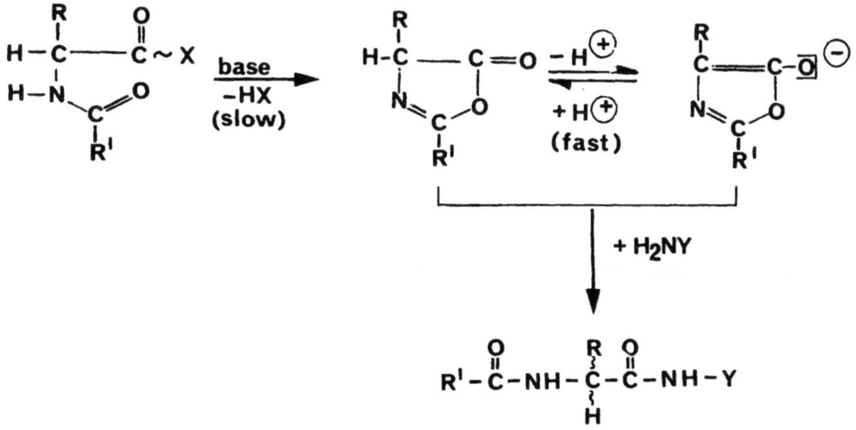

Scheme 1: Racemization During Peptide Synthesis

Therefore most of the simple reagents for the activation of carboxyl groups lead to extensive epimerization at the α-carbon.

Only if R' in scheme 1 is -OR", i.e. a urethane N^{α}-protecting group has been introduced, are oxazolones still formed but the tendency of epimerization is extremely low. Consequently, N^{α}-urethane-protected amino acid derivatives are in practice racemization-safe, a fact which favours the step-by-step procedure of peptide synthesis.

Numerous studies on racemization depending on different methods of peptide coupling have revealed that only a few techniques are suitable for routine synthesis. The most prominent methods are:

1. Azide coupling (17,18)
2. Mixed carbonic anhydride procedure (19)
3. Active esters (20)
4. Condensation with dicyclohexyl-carbodiimide (DCC) and additives (21).

The azide coupling (17,18) has turned out to be one of the safest methods of avoiding racemization. The formation of the hazardous HN_3 renders it unsuitable for technical production but it can nevertheless be effectively applied to laboratory synthesis. Azides are particularly attractive as activated peptide derivatives because they originate from commonly used carboxyl protecting groups (methyl, ethyl, and benzyl) via hydrazides (Scheme 2).

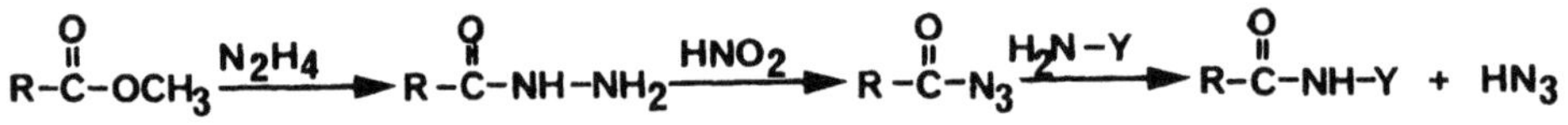

Scheme 2: Peptide synthesis by azide coupling. R represents a protected peptide chain.

Hydrazides as intermediates can often be purified easily by crystallization which is an additional advantage of this method. A disadvantage is the possible formation of many side-products, although, if they do appear, they are usually in minor amounts and can be removed by effective purification procedures such as HPLC.

The most frequently used conditions for azide formation have been elaborated by Rudinger and Honzl (22). They showed that optimal formation is obtained in anhydrous organic solvent at high acidity with organic nitrite at low temperature.

A solution of the protected peptide hydrazide (10 mmol) in about 30 ml dimethylformamide is cooled to -20°C, and 3N HCl in dioxan (30 mmol) is added, followed by 1.48 ml tert-butyl nitrite. The mixture is left at -10°C for 10 min. A solution of the amino component (7 mmol) in about 120 ml dimethylformamide is added dropwise at -15°C, then ethyldiisopropylamine (30 mmol) is added and the mixture kept at 0°C for 24 h. During the first hours, 0.2 ml of the base are added every hour. The product is precipitated by pouring into ice-cold 1 % acetic acid.

$$\text{R-}\overset{\text{O}}{\overset{\|}{\text{C}}}\text{-OH} + \text{Cl-}\overset{\text{O}}{\overset{\|}{\text{C}}}\text{-O-Alkyl} \xrightarrow{\text{base (B)}} \text{R-}\overset{\text{O}}{\overset{\|}{\text{C}}}\text{-O-}\overset{\text{O}}{\overset{\|}{\text{C}}}\text{-O-Alkyl} + \text{BCl} \xrightarrow{\text{H}_2\text{N-Y}}$$

$$\text{R-}\overset{\text{O}}{\overset{\|}{\text{C}}}\text{-NH-Y} + \text{CO}_2 + \text{Alkyl-OH}$$

Scheme 3: Mixed Anhydride Method with Alkyl Chloroformate as Activating Agent

The mixed carbonic anhydride method (Scheme 3) was first a popular reaction for step-by-step elongation of peptides. Extensive racemization which had been observed previously could be overcome using well-defined conditions, thus extending the scope of the mixed carbonic anhydride method to segment condensation as well (3,19).

The N-methylmorpholinium salt of a protected amino acid or peptide acid is reacted with isobutylchlorformate at -15°C in dry tetrahydrofurane, acetonitrile, or dimethylacetamide. After 1 - 2 min the reaction is mostly complete (amino acids with bulky side-chains such as isoleucine or valine may require some more time). The amine is added in said solvents or dimethylformamide. The main reaction occurs within 15-30 min at -15°C and is completed when the mixture is allowed to warm to room temperature. The solvent is removed under vacuum and the residue dissolved in ethyl acetate, washed with $NaHCO_3$, 1N HCl, and water, and dried over anhydrous Na_2SO_4. Evaporation of the solvent yields the desired peptide in a reasonably pure state.

Since the formation and reaction of the mixed carbonic anhydride proceeds rapidly, this procedure has been developed into an effective repetitive procedure (23). Excess mixed anhydride accelerates the reaction. It is rapidly destroyed after each reaction cycle with aqueous $KHCO_3$ at pH 8 and 0°C. After cleavage of the urethane N^{α}-protecting group, the next amino acid is coupled without extensive purification. The most impressive achievement with this procedure has been the synthesis of biologically fully active 27-peptide secretin (24).

The formation of mixed anhydrides has to be closely observed. N-Ethyloxycarbonyl-2-ethyloxy-1,2-dihydroquinoline (EEDQ) or the more recently prepared N-isobutyloxycarbonyl-2-isobutyloxy-1,2-dihydroquinoline allows the mixed anhydride method to be used in a more convenient way, i.e. by simply adding the reagent to a mixture of the carboxyl and the amino component at ambient temperature (25,26).

Despite all refinement of the mixed carbonic anhydride method, it is hampered by the problem that the amine can attack two carbonyl-C-atoms

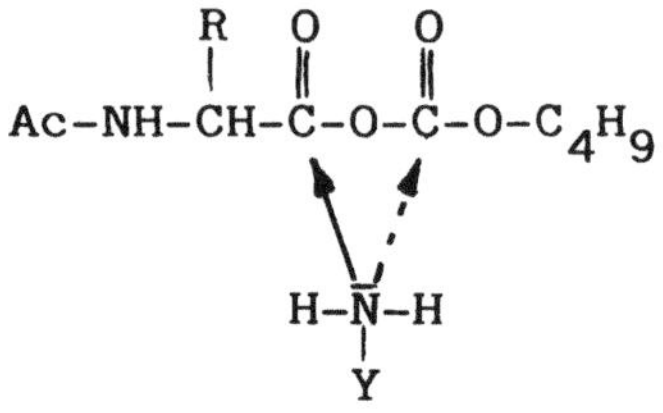

Normally the amino acid moiety is clearly favoured, however, in the case of sterically hindered carboxyl components, this side-reaction becomes apparent and can even be the main reaction with amino-terminal proline.

In the synthesis of secretin, the crude product was contaminated with about 40 % truncated peptides, presumably urethane side products.

These problems may account for the limited use of this elegant and economical method and for the preference given to the convenient and relatively safe carbodiimide method with addition of N-hydroxy compounds.

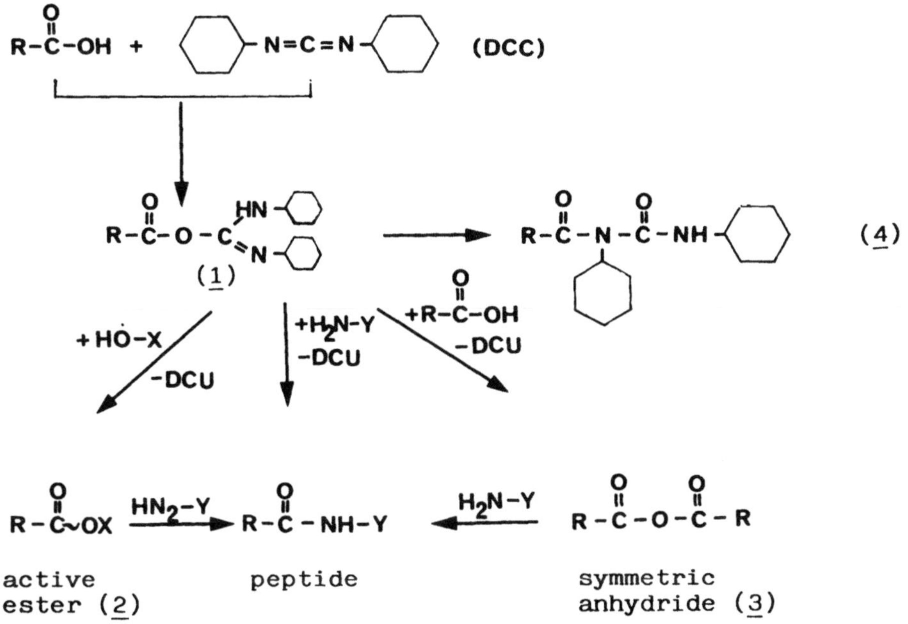

Scheme 4: Peptide formation with DCC (DCU = 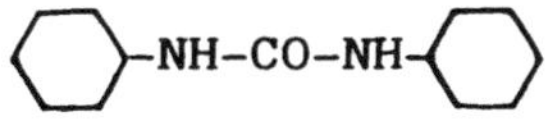)

<u>Dicyclohexylcarbodiimide (DCC)</u> as a coupling reagent is highly reactive and its ease of application has made it the most frequently used reagent in peptide synthesis. It is the reagent of choice for the preparation of active esters (2) and symmetric anhydrides (3) which both play an important role in the formation of peptide bonds. DCC reacts with an N-protected amino acid to form an O-acyl-isourea (1). This highly reactive intermediate can undergo aminolysis directly or via formation of a symmetric anhydride in situ.

There are, however, several disadvantages which limit the use of DCC. A rearrangement of O-acyl-isourea leads to non-reactive N-acylurea (4) which contaminates the desired peptides, as does the formed sparingly-soluble dicyclohexylurea (DCU).

If the carboxyl group of N-protected glutamine or asparagine is activated, nitriles and cyclic imides are formed by dehydration.

The extensive racemization which has been found with this method also limits its application. In practice, it is only used for the synthesis of active esters and of symmetric anhydrides and in solid phase synthesis where no racemization of the N^{α}-urethane protected amino acids occurs and N-acylureas are washed out together with excess amino acid derivatives and reagents.

Active esters (20) of N-urethane protected amino acids with e.g. 4-nitrophenol (HONp), pentafluorophenol (HOPfp) or N-hydroxysuccinimide (HONSu) are prepared with DCC in tetrahydrofurane or ethyl acetate without racemization as crystalline solids and can be purified by recrystallization. They can be stored over a long period in a refrigerator.

With peptide acids, 1,1-2 equivalents of N-hydroxsuccinimide form stable active esters mostly without racemization (see below).

The elongation of the peptide chain by excess active esters is a smooth and elegant method which can also be applied to peptides containing free carboxyl groups. The formation of active esters from O-acyl-isoureas in situ avoids N-acyl-urea formation.

HONSu additionally suppresses racemization (27,28). By this feature, the additive-assisted coupling with DCC became the most frequently applied method in routine peptide synthesis in solution. A side-reaction of HONSu and DCC, however, sometimes leads to ß-alanine-containing peptides as impurities. The search for a safer reagent resulted in 1-hydroxybenzotriazole (HOBt) as a most effective reagent (29). It suppresses not only racemization but also nitrile and imide formation from asparagine and glutamine as carboxyterminal amino acids.

HOBt also strongly catalyzes the aminolysis of 4-nitrophenyl esters in solvents such as dimethylformamide (30).

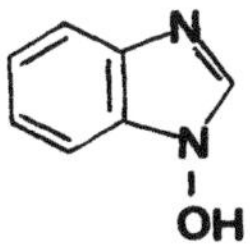

HOBt

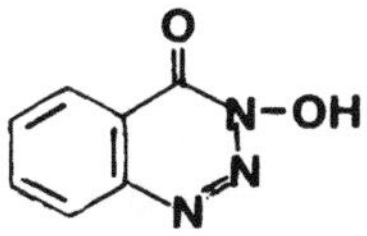

HOObt

In some cases racemization even occurs with HOBt as an additive. Fortunately, a related reagent, 3-hydroxy-4-oxo-3,4-dihydro-1,2,3-benzotriazine (HOObt), was revealed to be superior to all N-hydroxy compounds in suppressing racemization (31). A side-reaction analogous to that with HONSu can occur leading to 2-azido-benzoic-peptides. It does not diminish its value, however, since it occurs, if at all, only to a minute extent and the formed truncated peptides are not reactive. Therefore, HOObt is recommended in all cases in which racemization has to be expected.

The coupling procedure is convenient: equimolar amounts of the carboxyl and amino component and HOBt (or HOObt) are mixed in a suitable solvent, mostly dimethylformamide, and one equivalent DCC, dissolved in the same solvent, is added at

0°C. After 30 min stirring, the reaction is allowed to proceed for 3 - 4 hrs at room temparature. Dicyclohexylurea is removed by filtration, the solvent ist evaporated in vacuo, and the residue is worked up in an appropriate manner.

Side-reactions

Side-reactions in peptide synthesis are numerous (3,32). They begin during the very preparation of N^{α}-protected amino acids under Schotten-Baumann conditions. An N^{α}-protected amino acid can react with excess acylating reagent forming an asymmetrical anhydride which reacts with the amino acid to yield a dipeptide:

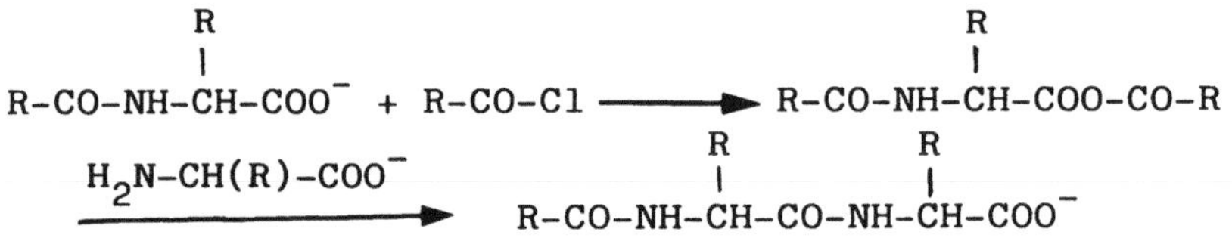

This side-reaction is suppressed if the reagent is added slowly at constant alkaline pH in the cold under vigorous stirring.

The purification of alkali-stable N-protected amino acids is best achieved as dicyclohexylamine salt.

Fmoc-chloride always produces considerable amounts of Fmoc-dipeptides, which can be avoided if Fmoc-ONSu is used.

Racemization as a side-reaction has already been discussed. It is circumvented with N-urethane-protected amino acids in the step-by-step procedure or with peptide segments containing glycine and proline, respectively, as C-terminal amino acids.

In all other cases, racemization has to be expected to a different degree. No method exists which is absolutely racemization-safe, however, DCC/HOObt already mentioned is the most reliable coupling reagent in routine syntheses to achieve a high degree of chiral homogeneity. EEDQ and coupling via azides are similarly safe.

Cyclization reactions occur with ease in several cases. ß-Benzyl- or ß-alkyl-aspartyl peptides yield aminosuccinimide derivatives upon saponification.

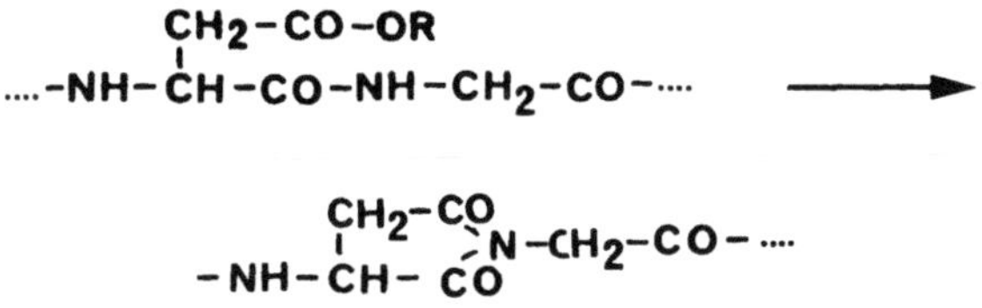

The reaction proceeds fast with glycine as subsequent amino acid and slow with sterically hindered amino acids such as valine. The same reaction also occurs after protonation, i.e. during deblocking of protected peptides by strong acids.

Dipeptide esters with free α-amino groups form dioxopiperazines.

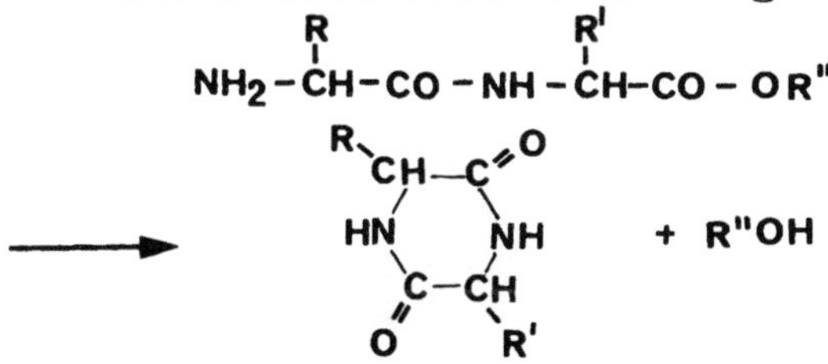

This reaction is suppressed if the amino group is acylated in a fast reaction. In SPS, the reaction is favoured by the relatively slow penetration of acylating reagents.

It can therefore be advantageous to couple a dipeptide with the amino acid attached to the resin, thus avoiding dioxopiperazine formation. Chiral purity has to be ascertained in this case.

Activated N-protected Gly-Pro cyclizes under the influence of bases:

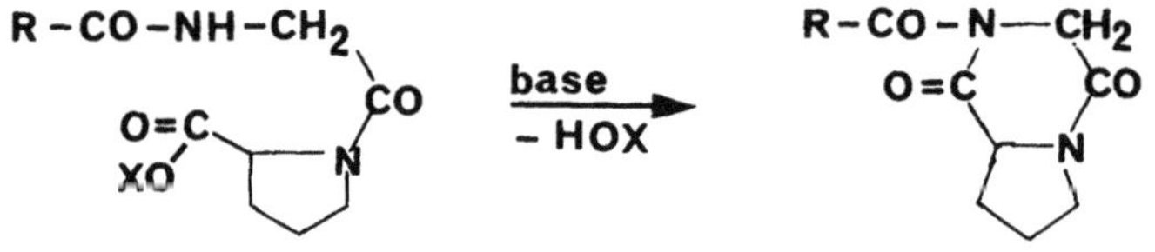

Hydantoins are formed if a benzyloxycarbonyl amino acid glycine ester is treated with bases, e.g. with ammonia to form an amide:

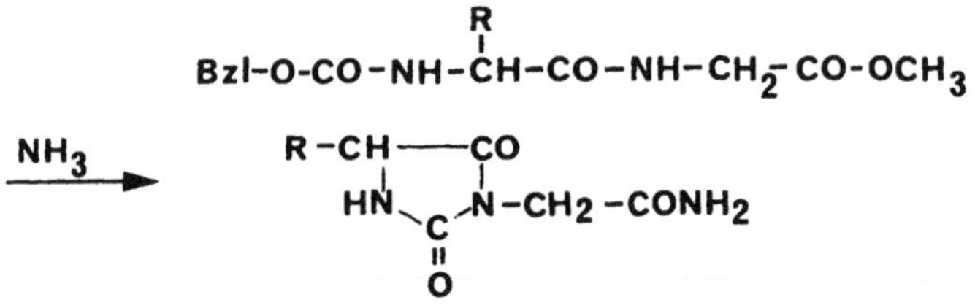

Peptides with free N-terminal Gln or the frequently used Glu(OBzl) tend to cyclize to pyroglutamyl peptides but again, this side-reaction is avoided by fast coupling.

Many side-reactions can occur during deblocking of the protected end product. Strong acids cause side-reactions of tryptophan. Acidolytic cleavage of benzyl and tert-butyl residues produces cations which lead to alkylation products at amino acid side-chains:

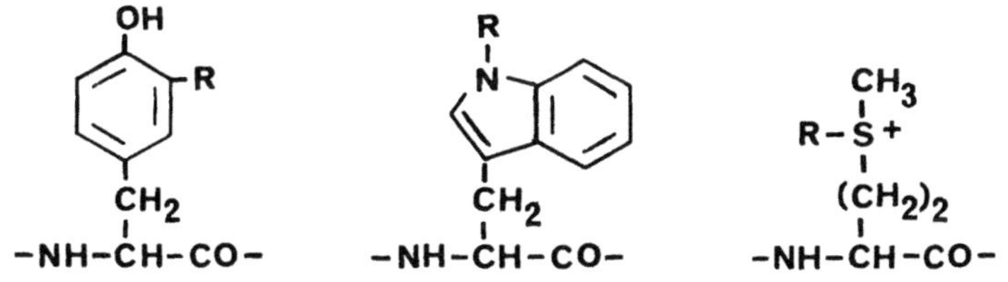

In tryptophan, R migrates to C-atoms 2,5, and 7 of the indole ring, thus generating additional side-products. Suppression succeeds by adding scavengers. The most effective ones are thioethers (thioanisole, methylethyl sulfide) and dithioglycol.

A two-step procedure has been proposed for the final cleavage of benzyl groups by HF (33). Most benzyl residues are split off in the presence of a high amount (75 %) of scavengers. The reaction rate is reduced by the additive so that in a second step few remaining groups are to be eliminated with 90% HF.

Methionine can also be protected from benzyl or tert-butyl cations by transient oxidation to sulfoxide (34).

Treatment of peptides with strong acids also results in chain fragmentation via N ⟶ O shift which is not quantitatively reversible under slightly alkaline conditions.

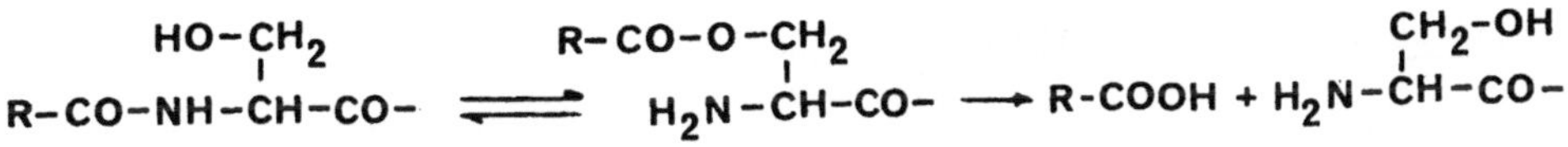

Peptide bonds of aspartic acid and particularly the Asp-Pro bond are susceptible to aqueous acid, the latter already at room temperature.

Another chain fragmentation is observed with sodium in liquid ammonia which cleaves the X-Pro bond to a certain extent. This reaction is suppressed by addition of sodium amide.

During peptide synthesis from only partially protected segments, some over-reaction occurs. Excess acylating agents react with hydroxyl groups, preferably with the hydroxy group of tyrosine and with the imidazole ring. Carbodiimides react with imidazole to form substituted guanidines. The aforementioned reactions with histidine are reversible.

Not only mixed anhydrides are a source of faulty coupling. Boc-Pro-ONSu has been described as reacting with proline or sarcosine in two different ways:

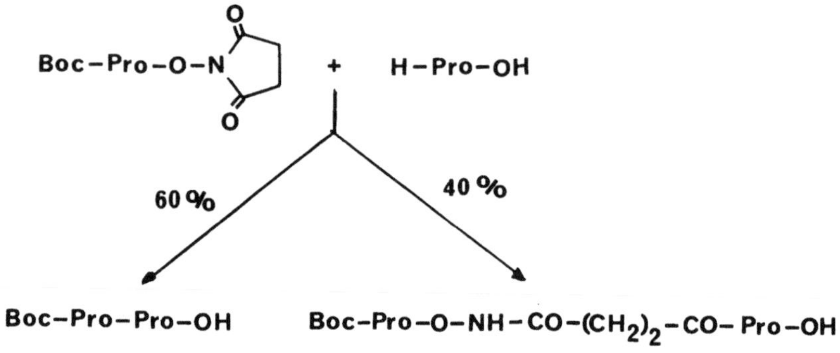

As we have seen, side-reactions can occur at every stage of peptide synthesis. They have been described in more detail by Bodanzsky (3) and Kisfaludy (32).

Analysis and purification

Progress of synthesis in solution and the stages of purification are followed by thin-layer chromatography (tlc). The simultaneous use of different solvents, the recognition of free amino groups by ninhydrin and subsequent staining of N-protected peptides with chlorine-tolidine or related reagents (35) already provide good information. In SPS, a ninhydrin reaction with a small sample of resin allows the degree of coupling to be detected (36, 37).

Tlc also helps to establish chiral homogeneity of segment condensation products. For this purpose, coupling is additionally carried out with a segment containing a C-terminal amino acid residue with a chirality opposite to that of the original component. Both reaction products are compared in tlc and, if necessary, modified without loss (e.g. deblocked) until a solvent is found in which the diasteromers separate. To quantify the results, HPLC is then applied (38).

This procedure is strongly recommended for segment condensation since even reactions which are considered to be racemization-safe are often not able to suppress it sufficiently. Process-control of elaborated steps is usually performed by HPLC which is also the method of choice for preparative purification (39). Columns with capacities between milligrams and several hundred grams are now commercially available.

Totally or partially protected peptides are effectively purified on unmodified silica gel whereas free peptides are preferably separated on reverse-phase material (39).

Established pathways to synthetic peptides

Methods in peptide chemistry are far from perfect. The numerous protecting groups, coupling procedures, and deblocking agents proposed are evidence of this.

This extension of methods, however, has been reversed and the number of procedures used in practice comprises only a few established principles. The most important ones have been described in this article.

Fundamental problems in peptide chemistry remained unchanged over the years: Selective removal of protecting groups, side-reactions including racemization, steric hindrance and insolubility of intermediates during synthesis in solution. They are inherently connected with the properties of amino acids and peptides. It is therefore not surprising that we have to discuss them as long as we work with peptides. This discussion of the problems was not intended to dishearten the beginner but to protect him from trivial pitfalls.

The first question in synthesizing a peptide is which strategy should be applied, classical synthesis in solution or SPS. The answer depends on the aim of the synthesis.

Short peptides are for the most part easier to prepare in solution than on a solid phase. Moreover the beginner should start with this method to get a better feel for these compounds. Longer chains are now fairly well accessible by SPS. The method has improved remarkably during the 20 years since Merrifield's ingenious invention. Careful control of the completeness of coupling reactions (and, if necessary, capping with acetic acid anhydride), accumulated experience in the choice of protecting groups, improved conditions for deblocking, and in particular improved analysis and purification by HPLC, rendered this method an elegant procedure for routine synthesis. Automatic synthesizers relieve the chemist of the burden of carrying out the same laborious operations repeatedly. Although expensive, they now provide a perfected self-controlling procedure.

Cheaper semi-automatic machines work reliably round a reaction cycle. Only the addition of reagents and the control of completeness of the reactions have to be carried out by the operator.

Analysis of the raw material and the purified peptide has been described in the preceding chapter. Combination of two or more different purification steps is always advisable.
If the route of the synthesis has been followed carefully,

evidence of the correct structure and reasonable purity of a compound may be obtained from relatively simple analytical procedures.

References

1. Wünsch, E.. 1974. In: Houben-Weyl, Methoden der Organischen Chemie (E. Müller ed.). Thieme, Stuttgart, Vol. 15.
2. Jakubke, H.-D., H. Jeschkeit. 1982. Aminosäuren, Peptide, Proteine. Verlag Chemie, Weinheim.
3. Bodanszky, M.. 1984. Principles of Peptide Synthesis. Springer-Verlag, Berlin - New York.
4. Gross, E., J. Meienhofer eds. 1975. The Peptides.Analysis, Synthesis, Biology. Academic Press, New York, London. Vol. 1 (continued).
5. Geiger, R., W. König. 1981. In: l.c. (4), Vol. 3, p. 1.
6. Roeske, R.W.. 1981. In: l.c. (4), Vol. 3, p. 102.
7. Hiskey, R.G.. 1981. In: l.c. (4), Vol. 3, p. 137.
8. Stewart, J.M.. 1981. In: l.c. (4), Vol. 3, p. 170.
9. Fauchère, J.-L., R. Schwyzer. 1981. In: l.c. (4), Vol. 3, p. 203.
10. Fujino, M., M. Wakimasu, C. Kitada. 1981. Chem. Pharm. Bull. 29, 2825.
11. Hubbuch, H., R. Bindewald, J. Föhles, V.K. Naithani, H. Zahn. 1980. Angew. Chem. Int. Ed. Engl. 19, 394.
12. Mutter, M., E. Bayer. 1980. In: l.c. (4), Vol. 2, p. 286.
13. Barany, G., R.B. Merrifield. 1980. In: l.c. (4), Vol. 2, p. 3.
14. Jones, J.H.. 1979. In: l.c. (4), Vol. 1, p. 66.
15. Klausner, Y.S., M. Bodanszky. 1972. Synthesis, p. 453.
16. Kemp, D.S.. 1979. In: l.c. (4), Vol. 1, p. 317.
17. Meienhofer, J.. 1979. In: l.c. (4), Vol. 1, p. 198.

18. Klausner, Y.S., M. Bodanszky. 1974. Synthesis, p. 549.

19. Meienhofer, J.. 1979. In: l.c. (4), Vol. 1, p. 264.

20. Bodanszky, M.. 1979. In: l.c. (4), Vol. 1, p. 106.

21. Rich, D.H., J. Singh. 1979. In: l.c. (4), Vol. 1, p. 242.

22. Honzl, J., J. Rudinger. 1961. Coll. Czech. Chem. Commun. 26, 2333.

23. Tilak, M.A.. 1970. Tetrahedron Letters, p. 849.

24. van Zon, A., H.C. Beyerman. 1976. Helv. Chim. Acta 59, 1112.

25. Belleau, D., G. Malek. 1968. J.Am. Chem. Soc. 90, 1651.

26. Kiso, Y., Y. Kai, H. Yajima. 1973. Chem. Pharm. Bull. 21, 2507.

27. Wünsch, E., F. Drees. 1966. Chem. Ber. 99, 110.

28. Weygand, F., D. Hoffmann, E. Wünsch. 1966. Z. Naturforsch, Teil B 21, 426.

29. König, W., R. Geiger. 1970. Chem. Ber. 103, 788.

30. König, W., R. Geiger. 1973. Chem. Ber. 106, 3626.

31. König, W., R. Geiger. 1970. Chem. Ber. 103, 2034.

32. Kisfaludy, L.. 1979. In: Peptides 1978 (I.Z. Siemion, G. Kupryszewski ed.). Wrocław University Press, p. 25.

33. Tam, J.P., W.F. Heath, R.B. Merrifield. 1983. J.Am. Chem. Soc. 105, 6442.

34. Izeboud, E., H.C. Beyerman. 1978. Recl. Trav. Chim. Pays-Bas 97, 1.

35. von Arx, E., M. Faupel, M. Brugger. 1976. J. Chromatography 120, 224.

36. Kaiser, E., R.-L. Colescott, C.D. Bossinger, P.I. Cook. 1970. Anal. Biochem. 34, 595.

37. Sarin, V.K., S.B.H. Kent, J.P. Tam, R.B. Merriefield, 1981. Anal. Biochem. 117, 147.

38. Wissmann, H., W. König, V. Teetz, R. Geiger. 1981. In: Peptides 1980 (E. Brunfeldt ed.) Scriptor, Copenhagen, p. 174.

39. Rivier, J., R. McClintock, R. Galyean, H. Anderson. 1984. J. Chromatography 228, 303.

40. Frank, H., G.J. Nicholson, E. Bayer. 1977. J. Chromatog. Sci. 15, 174.

41. Buko, A.M., L.R. Phillips, B.A. Fraser. 1983. Biomedical Mass Spectrometry 10, 324.

42. Amster, I.J., M.A. Baldwin, M.T. Cheng, C.J. Proctor, F.W. Mc Lafferty. 1983. J. Am. Chem. Soc. 105, 1654.

Author Index

Subject Index

Bei Fragen zur Produktsicherheit wenden Sie sich bitte an:
If you have any questions regarding product safety,
please contact:

Walter de Gruyter GmbH
Genthiner Straße 13
10785 Berlin
productsafety@degruyterbrill.com